MECHANICAL IDENTIFICATION OF COMPOSITES

Proceedings of the European Mechanics Colloquium 269, 'Experimental Identification of the Mechanical Characteristics of Composite Materials and Structures' held in Saint-Étienne, France, 3–6 December, 1990.

MECHANICAL IDENTIFICATION OF COMPOSITES

Edited by

A. VAUTRIN

Department of Mechanical and Materials Engineering, École Nationale Supérieure des Mines, Saint-Étienne, France

and

H. SOL

Department of Structural Analysis, Vrije Universiteit Brussel, Brussels, Belgium

ELSEVIER APPLIED SCIENCE
LONDON and NEW YORK

ELSEVIER SCIENCE PUBLISHERS LTD
Crown House, Linton Road, Barking, Essex IG11 8JU, England

Sole Distributor in the USA and Canada
ELSEVIER SCIENCE PUBLISHING CO., INC.
655 Avenue of the Americas, New York, NY 10010, USA

WITH 43 TABLES AND 270 ILLUSTRATIONS

© 1991 ELSEVIER SCIENCE PUBLISHERS LTD

British Library Cataloguing in Publication Data

European Mechanics Colloquium 269, 'Experimental
Identification of the Mechanical Characteristics
of Composite Materials and Structures (1990:
Saint-Etienne, France)
Mechanical identification of composites.
I. Title   II. Vautrin, A.   III. Sol, H.
620.118

ISBN 1-85166-694-X

Library of Congress CIP data applied for

INTRODUCTION

The widespread idea that composite materials are materials which can be tailored to achieve any desired structural and functional performances is wrong, since the present commonly used mechanical characterization and designing procedures do not incorporate the two prominent features of these materials, their *anisotropy* and *heterogeneity*, from the starting point of the analysis. *Anisotropy* and heterogeneity involve particular effects, at different scales of observation, that cannot be regarded as simple extensions of any phenomena occurring in homogeneous or isotropic media. So the challenge is real and further industrial applications of composites are closely related to new ways of approaching the *experimental identification* of such materials or structures in the rational framework of the mechanics of heterogeneous and anisotropic media.

The European Mechanics Committee sponsored a Colloquium on Experimental Identification of the Mechanical Characteristics of Composite Materials and Structures in Saint-Étienne, France, on 3–6 December 1990. Its main purpose was to present, criticize and survey new developments in analysis techniques, data treatments, mechanical modelling and identification approaches regarding the anisotropic behaviour of composite structures.

The fourty nine papers, including the general lectures, touched four main aspects:

- quasi-static and dynamic mechanical tests suited to anisotropic media;
- identification methodologies based on new data treatments;
- displacement and strain fields measurement based on optical methods;
- damage and cracking modelling of anisotropic and heterogeneous media.

The papers have been arranged according to six classical scientific topics for convenience sake. As a matter of fact, most of the paper titles do not refer clearly to any methodological approach and therefore any Proceedings arrangement based chiefly on analysis techniques would have been less relevant and less convenient for the reader. The six thematic sections are the following:

I.   Identification
II. Static and Dynamic Characterization
III. Impact Behaviour and Damage Characterization
IV. Micromechanics and Interfaces
V.   Fracture and Fatigue Damage Mechanics
VI. Damage Modelling and Non-Destructive Testing.

Two basic experimental approaches can be distinguished to assess the mechanical behaviour and performances of composites: specimen testing which produces *uniform deformation fields* to identify with reasonable accuracy the material behaviour and to determine specific properties, such as shear properties; and structure testing which did not necessarily produce uniform deformation but enabled *practical identification of the complete set of structural parameters* to be determined in the framework of a well-established anisotropic model.

Recent advances in numerical methods, strain analysis and data acquisition and treatment on personal computers enable efficient data analyses and multi-parameter optimization procedures to be performed. The direct consequence is that novel experimental procedures can be set up to identify the complex mechanical composite behaviour through *structure testing*. Nevertheless, since the relevance of any rheological model has to be firmly assessed beforehand, it is thought that testing material specimens under highly controlled conditions should not be completely avoided. This latter approach is obviously required to check the various assumptions on boundary conditions and specimen size influences for instance.

A large number of papers deals with *specific tests* and methodologies to determine at best particular characteristics, such as shear or bending plate characteristics, and/or to achieve overall anisotropic behaviour modelling, based on wave propagation, steady state dynamic vibration or quasi-static experiments. Dynamic testing and modal analysis look quite suited to optimizing complete sets of anisotropic characteristics, and relevant identifications of orthotropic thin plates or shells are presented in the framework of linear elasticity or viscoelasticity.

Identification procedures mainly deal with *heterogeneous displacement* or *strain fields* and data treatment technique involve material parameters adjustment by finite element calculation routines and iteration processes in many cases. The least squares method with weights and the familiar Euclidian distance are quite often used.

Optical techniques, such as moiré, holographic interferometry and speckle, can be used to provide displacement or deformation field mapping. Furthermore they do not affect the system response and give a prominent insight into the overall behaviour of the structure, which is relevant to both mechanical modelling and damage assessment. Several examples emphasized the high interest of optical techniques for dynamic and quasi-static testing and obviously proved that such techniques can now be practised without advanced training in optics, even if prior skill in optical applications is clearly needed. The dissemination of optical techniques is greatly facilitated by the enhancement of personal computer capabilities and image processing which lead to routine fringe pattern treatment and first analysis.

Personal computer facilities can be used with profit to continuously monitor the running of a given experimental procedure. Several papers point out that real time data preanalysis can be introduced. The constantly increasing capabilities of personal computers provide fast data acquisition, treatment and analysis, involving image processing and finite element calculations. Parameter adjustments requiring numerical simulations and iterations can now be performed on low-priced compatible systems.

Large discrepancies in recently published papers obviously show that present knowledge in mechanics of composites has still to be highly improved. In particular, it should be necessary:

- to set up new testing approaches based on *heterogeneous cinematical fields*; these tests seem to be suited to identifying the *complete anisotropic behaviour*, taking into account the structural coupling effects due to the anisotropy and two-dimensional loadings; advances in the range of large displacements and large deformations of structures should be required as well, since the present works are chiefly restricted to small displacements and deformations;
- to found *relevant rheological models* in agreement with the specific mechanical responses and microstructures of composites: what is the true nature of the nonlinear mechanical responses? What is plasticity or viscoplasticity for polymer matrix composites? Can mechanical modelling ignore the internal state of stress, strain and defects distributions? Can damage occurring in composite laminates be modelled in the same way as metals? How to characterize fibre/matrice interface?
- to identify *transverse mechanical properties* of plates; much work has been done in the framework of the classical laminated plate theory; however, transverse properties of laminates are probably the most restrictive structural parameters in practice and transverse shear properties of laminated and sandwich plates are to be assessed;
- to introduce *optical extensometry* in mechanics of composite structures and to encourage any joint work with specialists of optical technique in order to develop specific tools as novel optical methods, in the field of fracture mechanics or NDT for instance, or user-friendly software for fringe pattern analysis on personal computers.

This volume provides the reader with a current assessment of methods of mechanical analysis, experimental procedures and techniques and overall conclusions dealing with the mechanical properties of composites, including fracture and damage mechanics. It will be valuable to engineers and scientists in industry, as well as to teachers and students at universities, to keep them up to date in this rapidly developing and essential field.

ALAIN VAUTRIN
Saint-Étienne

CONTENTS

xii

ACKNOWLEDGMENTS

The European Mechanics Colloquium 'Experimental Identification of the Mechanical Characteristics of Composite Materials and Structures' was sponsored by the European Mechanics Committee and was held in the École des Mines de Saint-Étienne from 3 to 6 December 1990.

Financial support was provided by national and local French organizations:

- the French Ministry of Research and Technology
  Ministère de la Recherche et de la Technologie (MRT),
- the Research Agency of the French Ministry of Defence
  Direction des Recherches, Études et Techniques (DRET),
- the French University Mechanics Society
  Association Universitaire de Mécanique (AUM),
- the French Composite Materials Society
  Association pour les Matériaux Composites (AMAC),
- the Loire County Council
  Conseil Général de la Loire,
- the Saint-Étienne City Council
  Conseil Municipal de Saint-Étienne,
- the Saint-Étienne & Montbrison Chamber of Commerce and Industry
  Chambre de Commerce et d'Industrie de Saint-Étienne et de Montbrison,
  avec le concours de son 'Point Accueil Matériaux',

The selection of the scientific programme was possible thanks to the International Scientific Committee, composed of H. Sol and A. Vautrin, Chairmen of the Colloquium, A. H. Cardon, W. P. De Wilde, R. Dechaene, J. M. Rigo and D. Van Gemert for Belgium and Y. Chevalier, A. Gérard, A. Lagarde and G. Verchery for France.

The French Local Organizing Committee was chaired by M. Grédiac and composed of S. Aivazzadeh, Y. Surrel and myself. I am greatly obliged to my colleagues for their constant and efficient involvement in the Colloquium organization and wish to express my thanks to all of them.

The assistance of the École des Mines de Saint-Étienne in holding the conference was appreciated and the participation of the members of the Mechanics and Materials Engineering Department, especially B. Ferent, J. François-Brazier, T. Lachi, D. Marty, B. Mohamadou, C. Toukourou, and L. Santangelo, secretary of the Department, was extremely helpful.

ALAIN VAUTRIN
Saint-Étienne

ACKNOWLEDGMENTS

The European Mechanics Colloquium Experimental Identification of the Mechanical Characteristics of Composite Materials and Structures, was sponsored by the European Mechanics Committee and was held in the "Locré de Mitterie Saint-Étienne from 1 to 3 December 1986.

Financial support was provided by national and local French organisations:

- the French Ministry of Research and Technology,
  Ministère de la Recherche et de la Technologie (MRT)
- the Research Agency of the French Ministry of Defence,
  Direction des recherches, Études, Techniques (DRET)
- the French University Mechanics Society,
  Association Universitaire de Mécanique et Ma...
- the French Composite Materials Society,
  Association pour les Matériaux Composites (AMAC)
- the Loire Country Council,
  Conseil Général de la Loire,
- the Saint-Étienne City Council,
  Conseil Municipal de Saint-Étienne
- the Saint-Étienne & Montbrison Chamber of Commerce and Industry,
  Chambre de Commerce et d'Industrie de Saint-Étienne de Montbrison
  avec le concours de ... Jean-René Auguel Mutuelles.

The selection of the scientific programme was possible thanks to the International Scientific Committees composed of H. Lipson and A. Vautrin, Chairmen of the Colloquium, V. H. Cardon, A. P. De Wilde, E. Deschamps, M. Rigaud, D. Van Hemelrijck for Belgium and Y. G. Bershov, A. Gérard, A. Lagarde and G. Verchéry for France.

The French Local Organizing Committee was chaired by M. Grédiac and composed of F. Azzazadeh, L. Sarraf and myself. I am grateful to all of my colleagues for their constant and efficient involvement in the Colloquium organisation and wish to express my thanks to all of them.

The assistance of the École des Mines de Saint-Étienne in holding this conference was important and the participation of the members of the Mechanics and Materials Engineering Department, especially, B. Bérenbach, J. Francois-Brazier, T. Lachat, D. Néully, B. Moh...enot, C.J. Athanase and L. Sarasecto, secretary of their eminent, was extremely helpful.

ALAIN VAUTRIN
Saint-Étienne

Identification of the rigidities of composite systems by mixed numerical/experimental methods.

W.P. De Wilde

Civ. Eng., Ph.D., Full Professor,
Hd of Structural Analysis - Faculty of Engineering
Brussels Free University (V.U.B)
Pleinlaan, 2
1050 Brussels (Belgium)

Abstract

The author presents techniques of structural properties identification which were developed during the past ten years at Brussels University (VUB) by the Composite System and Adhesion Research Group. The paper is concerned with the identification of stiffness in structural elements and elaborates quite extensively on a method developed by the author and H. Sol.

Introduction

The determination of *material parameters* (e.g. those characterizing stiffness) of structural components has very often been done using simple test specimens; however, when using composite materials, these experiments very often yield poor results, especially for the transverse or shear properties. This is due to the fact that complex stress fields are created through the heterogenity of the system. Moreover, many composite systems also need (sometimes scheduled) non-destructive inspection, which necessitates measurements on the structure itself and interpretation of the results.

It has thus been quite logical that attempts to establish these properties would been using *"experimental information"* taken from the structural element and link it with *"numerical results"* output by a computer model. Several research groups have been investigating this possibility over the last years and some of them have achieved promising results.

The author does not claim to give a complete review of these attempts and results, but wants to concentrate on work done in the field of *stiffness estimation of simple structural elements*, as well as on more complex elements and the interpretation of the experimental data. A similar analysis is presented by **De Visscher et al.** [DE VISSCHER, J., 1990] with emphasis on the damping properties.

Identification of stiffness properties by strain measurements: the early attempts

Assume that *strain measurements* can be made on an experimental specimen (which of course can as well be a simple structural element): it is obvious that these strains will be depending on both the material and geometrical properties of the specimen.

An idea to establish the correct values of the *stiffness properties* of the specimen can be to compare the output of a numerical (eg. finite element, boundary element, Rayleigh-Ritz,...) model with these strain measurements. Confrontation of both experimental and numerical values yields corrected values of the stiffness:

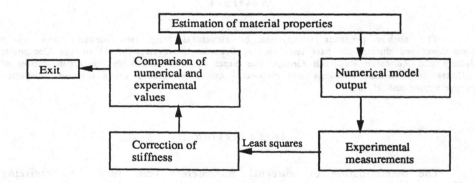

Mathematically, the procedure can be formulated as follows, supposing one tries to establish the stiffness properties of an *anisotropic thin layer* (see eg. HERMANS, P. et al, 1984):

$$\left\{ \begin{array}{c} \sigma_x \\ \sigma_y \\ \tau_{xy} \end{array} \right\} = \left[\begin{array}{ccc} C_{11} & C_{12} & C_{16} \\ & C_{22} & C_{26} \\ & & C_{66} \end{array} \right] \left\{ \begin{array}{c} \varepsilon_x \\ \varepsilon_y \\ \gamma_{xy} \end{array} \right\} \quad (1)$$

Consider now a point in which an experimental measurement has been done:

$$\begin{Bmatrix} \sigma_x \\ \sigma_y \\ \tau_{xy} \end{Bmatrix} = \begin{bmatrix} \varepsilon_x & \varepsilon_y & \gamma_{xy} & 0 & 0 & 0 \\ 0 & \varepsilon_x & 0 & \varepsilon_y & \gamma_{xy} & 0 \\ 0 & 0 & \varepsilon_x & 0 & \varepsilon_y & \gamma_{xy} \end{bmatrix} \begin{Bmatrix} C_{11} \\ \cdots \\ C_{66} \end{Bmatrix} \qquad (2)$$

If one expands the previous relation for M experimental points, it can be written :

$$\{ \sigma \} = [E] \{C\} \qquad (3)$$

in which $\{ \sigma \}$ has dimensions (3M,1), $[E]$ is (3M,6) and $\{ C \}$ is (6,1). A least squares technique can now be applied in order to find $\{ C \}$, by pre-multiplication of both members by $^t[E]$:

$$\{ C \} = [({}^t[E][E])^{-1}] . {}^t[E] \ \{ \sigma \} = [A]\{ \sigma \} \qquad (4)$$

One thus has to choose both a numerical and experimental method, as the latter yields the matrix $[E]$ and thus also $[A]$.

This technique was - as far as the author could check - first proposed by Kavanagh [KAVANAGH, K.T., 1971; KAVANAGH, K.T., 1972] who combined *finite element* techniques and strain measurements with *strain gauges*. The method - only illustrated for isotropic materials - showed shortcomings inherent to both (i.e. numerical and experimental) used methods: both finite element techniques and strain gauge measurements tend to average the calculation and measurement over a finite domain of the structural element. The result of it is a lack of precision in the technique, depending on the complexity of the structural element to be investigated, the refinement of the finite element grid (and performance of the selected element !) and the physical dimension of the strain gauge (cf. for instance problems of cross-sensitivity with very small gauges).

In 1982, Hermans, De Wilde and Hiel [HERMANS, P. et al, 1982] tried to combine *boundary element techniques* and *strain gauge measurements* on anisotropic materials, hereby using a fundamental solution developed by Muskheliskvili [MUSKHELISKVILI, V.I., 1953] and implemented by Hermans in a boundary element progam. The use of the boundary element method was inspired by the fact that it allows to compute stresses in internal *points*, and not *domains* like it is the case with finite elements. This should thus have given better results: although these were obtained, they were not yet very convincing. Additionally the use of the boundary element method was tricky and the experimentation was taking quite a long time.

This method also yielded the *in-plane stiffness*, which allowed to find the values of material parameters by some back-calculations but in any case yielded values of *compression* or *extensional* stiffness. As one knows, the behaviour of composites in *bending* can be quite different than that forecasted with the in-plane values.

Finally, it should be noted that the numerical methods - both finite element and boundary element methods - yield solutions in terms of the *displacements*; the strains and thus stresses are found by taking the *derivatives* of the unknown field, which is also an additional source of error (or slower convergence tot the solution). On the other hand, one is obliged to compare experimental values of strains, as a relation linking the "numerical" stresses to "experimental" displacements does not involve *explicitly* the material properties (like it is the case in e.g. equation (2)).

Techniques based on the modal analysis of orthotropic plates

A mathematical model of an elastic system can be characterized - through discretization - by a finite number of parameters. A discipline called *system analysis* allows to calculate the response of this discretized system to a given input signal: in this sense different methods like *finite element techniques, boundary elements, and also analytical or semi-analytical methods* - e.g. obtained by Rayleigh-Ritz techniques - may be considered as system analysis techniques.

Following figure gives a scheme of this approach:

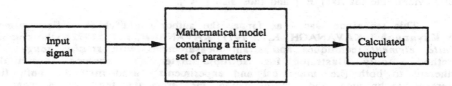

For our purpose, the parameters to be considered are related to the *stiffness* of the structure: it could be so-called elastic constants, extensional or in-plane stiffness, bending and/or torsional rigidities.

Another possible approach is *system identification*, in which the structure is considered as a *"black box"*: in this case the dynamical behaviour is established with experimentally measured values of both the input and the output. A commonly used technique is the modal analysis.

Formally the technique described in the previous section combines a technique of system analysis and another of system identification: the *parameter estimation technique* then proceeds in following steps, like suggested by Sol [SOL, H., 1986] :

i)   system model identification: the selection of a system which through its structural behaviour, described by adequate partial differential equations (or - of course - variational principles), boundary conditions, initial conditions, will allow to extract the required mechanical information;

ii)   error function: the selection of an error function which must be minimized: it evaluates the distance between numerically obtained results (on the basis of the previous - or initial - estimation) and the experimentally obtained measurements;

iii) <u>system parameter identification</u>: an algorithm is established to identify the parameters by minimizing the error.

After the rather disappointing attempts explained in the previous section, **De Wilde, Narmon, Roovers and Sol** [DE WILDE, W.P. et al., 1984] looked in another direction: during the preceding years there had been indications - during a quite extensive research project on optimization of tennis rackets - that modal analysis could yield useful information on the stiffness of a structure. The above mentioned paper certainly set the foundations for the identification method, although the real breakthrough to a really practical method was set later by **Sol** [SOL, H., 1986] in his doctoral thesis: he showed that an additional ingredient was mandatory, i.e. *sensitivity analysis*. Finally, a technique was developed which allows for the complete stiffness identification of orthotropic plates; to-day the technique has been implemented in a measurement system, which is commercially available; attempts are now in the extension of the ideas in the measurement of damping properties (see e.g. paper by **De Visscher, J.** et al.) and/or more complex structural elements.

In the subsequent sections this technique is detailed.

Eigenvalue problem for an undamped mechanical system

For a linear, undamped and discretized system - e.g. through finite element techniques, or Rayleigh-Ritz techniques - the natural vibrations - and their frequencies - can be found by solving the associated *discrete eigenvalue problem*:

$$\left([K] - \lambda [M]\right)\{q\} = \{0\} \tag{5}$$

in which :

$[K]$: stiffness matrix of the system
$[M]$: mass matrix of the system
λ : eigenvalue (square of the eigenfrequency)
$\{q\}$: eigenmode of the structure

The developments on this well-known technique can be found in any standard textbook on discretization techniques, e.g. by **Zienkiewicz** [ZIENKIEWICZ, O.C., 1977]. The properties of the eigenvalues and eigenmodes (real and positive frequencies, orthogonality of the modes,...) are also well known. One will consider that the modes are orthonormalized with respect to the mass matrix, in which case:

$$\begin{align}
[Q][M][Q] &= [I] \\
[Q][K][Q] &= [\Lambda]
\end{align} \tag{6}$$

in which [I] is the identity matrix and [Λ] a diagonal matrix containing the eigenvalues; [Q] is a matrix having the eigenmodes in its columns.

As one can see, the eigenvalues carry information about the stiffness of the structure. This will be used further.

Sensitivity of the eigenmagnitudes, Bayesian estimation

If one modifies any parameter p_i of the considered mechanical system, the eigenmagnitudes will be modified:

$$\lambda_i = \lambda_{i0} + \sum_{j=1}^{k} \frac{\partial \lambda_i}{\partial p_j} \Delta p_j + \ldots \tag{7}$$

which has been obtained using a Taylor expansion in the vicinity of the previously established value of the parameter (indicated by a subscript "$_0$"), and limited to the linear expansion terms.

The same expansion could be done for the eigenmodes, but as one will see further, it is not mandatory to take these modes into the analysis (they are anyway affected by quite important experimental measurement errors).

The equation (7) can be written in matrix notations:

$$\{ \Delta \lambda \} = [S] . \{ \Delta p \} \tag{8}$$

in which [S] is a (rectangular) *sensitivity matrix*, linking the variation of parameters to variations of eigenvalues.

Combining numerical and experimental techniques, one has now following situation:

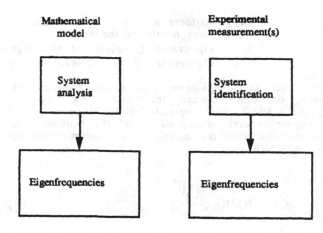

A parameter estimation method will adapt the parameters in the numerical model such that the newly calculated eigenfrequencies match the experimental values. The most general procedure for that purpose is *Bayesian parameter estimation*, which also takes into account relative confidence in the estimation of the model parameters: in this sense, both parameters and response are considered as stochastic variables with a certain probability to take a correct value.

In the Bayesian estimation technique, the problem is solved by minimizing an error function combining the discrepancies between the model prediction and the test data. The error is weighted taking into consideration the relative confidences in the different values, i.e. test data and model parameters:

$$E = {}^{t}\{\Delta\lambda\}[C_\lambda]\{\Delta\lambda\} + {}^{t}\{\Delta p\}[C_p]\{\Delta p\} \tag{9}$$

In this case, one finds an equation of the system by taking the partial derivatives of (9), with respect to the parameters p_j, and annihilating it:

$$\frac{\partial E}{\partial p_j} = \frac{\partial \Delta\lambda_i}{\partial p_j}(C_\lambda)_{ik}\Delta\lambda_k + \Delta\lambda_i(C_\lambda)_{ik}\frac{\partial \Delta\lambda_k}{\partial p_j} +$$
$$+ \delta_{ij}(C_p)_{ik}\Delta p_k + \Delta p_m(C_p)_{mk}\delta_{kj} = 0 \tag{10}$$

(a repeated index is implicitly supposed to be summed over all its possible values)

Finally, one finds:

$$\{\Delta p\} = \left(2[C_p] + {}^{t}[S][C_\lambda][S]\right)^{-1}\left({}^{t}[S][C_p](\Delta\lambda)\right) \tag{11}$$

Stiffness identification of orthotropic plates

Orthotropic thin plates, subjected to small deflections, may be modelled with *classical laminated plate theory* (CLT), which is based on Love-Kirchhoff assumptions. Essentially, these assumptions - the normal to the middle plate remaining straight, normal and unstrained - neglect the shear effects in the plate, which cause deformation of the normal.

Starting from the equilibrium equations of an orthotropic plate (in which the stiffness coefficients C_{16} and C_{26} vanish) and applying the Ritz technique (by starting from a weighted residual formulation and integrating by parts twice), one finds [DE WILDE et al., 1984, SOL, H., 1986; LEKHNITSKII, S.G., 1968] following generalized eigenvalue problem:

$$\int_V \left[\begin{array}{c} D_{11}\dfrac{\partial^2 W}{\partial x^2}\dfrac{\partial^2 V}{\partial x^2} + D_{22}\dfrac{\partial^2 W}{\partial y^2}\dfrac{\partial^2 V}{\partial y^2} \\[2ex] + D_{12}\left(\dfrac{\partial^2 W}{\partial y^2}\dfrac{\partial^2 V}{\partial x^2} + \dfrac{\partial^2 W}{\partial x^2}\dfrac{\partial^2 V}{\partial y^2}\right) + 4D_{66}\dfrac{\partial^2 W}{\partial x \partial y}\dfrac{\partial^2 V}{\partial x \partial y} \end{array} \right] dxdy = \lambda \, \rho \, h \int_V W.V.dxdy$$

$$(12)$$

in which W is one of the (time-independent) eigenmodes, associated with the eigenvalue λ; V is a weighting function, ρ the mass per unit volume of the plate and h its thickness.

For 'experimental errors to be *minimal*, the boundary conditions to take into consideration are F-F-F-F (4 free boundaries); the drawback is that the associated eigenvalue problem does not yield a closed form solution in this case. Solutions of this problem are then sought using a Rayleigh-Ritz method, by writing:

$$W(x,y) = \Sigma \; A_{pq} . X_p(x) . Y_q(y) \qquad (13)$$

Most of the authors used orthogonal sets of beam functions or Fourier series. Sol [SOL, H., 1986] proposed to use a model which is a superelement, an isoparametric element with 7*7 nodes, regularly spaced over the plate in its two directions, with shape functions given by:

$$N_i(x,y) = L^7{}_I(\xi) . L^7{}_J(\eta) \qquad (14)$$

in which $L^7{}_I(\xi)$, $L^7{}_J(\eta)$ are Lagrange polynomials of the 7-th order. A computer program can then be developed and run on a PC (the system of linear equations has 49 unknowns and yields the first modes and eigenvalues easily and with the necessary accuracy). This model has been choosen as a compromise between computing time (in order to have it run on a PC) and precision of results, which was checked with refined finite element models and results from Leissa [LEISSA, A.W., 1973]. Comparisons between calculated frequencies and measured frequencies, made on aluminium and steel plates, showed discrepancies which were small, as can bee seen on the results for e.g. steel in the table (see next page)

Comparison with other authors, e.g. Deobald [DEOBALD, L.R., 1986] showed also excellent agreement. The essential reasons are to be found in the F-F-F-F experimental set-up which allows easy to perform experiments, gives boundary conditions which are easy to model numerically, is not too much influenced by transverse shear deformations and is very sensitive to small variations of Poisson's ratio.

| Frequency | Steel | | |
	Experiment	Finite el.	Lagr. model
1	214	212	213
2	292	289	290
3	404	404	405
4	524	520	524
5	580	573	578

For orthotropic plates, and using eq. (14) substituted in eq. (12), one ends up with a *discrete eigenvalue problem* :

$$(K_{ij} - \lambda M_{ij}) \{ W_j \} = \{ 0 \} \qquad (15)$$

in which [DE WILDE et al., 1984, SOL, H., 1986], a and b being the dimensions of the plate in the 1- and 2-direction, the D_{ij} are the plate rigidities :

$$K_{ij} = \frac{4b}{a^3} D_{11} A_{ij} + \dots + \frac{16}{ab} D_{66} E_{ij} \qquad (16)$$

with:

$$A_{ij} = \int_S \frac{\partial^2 N_i}{\partial \xi^2} \frac{\partial^2 N_j}{\partial \xi^2} d\xi d\eta \qquad (17)$$

...

$$E_{ij} = \int_S \frac{\partial^2 N_i}{\partial \xi \partial \eta} \frac{\partial^2 N_j}{\partial \xi \partial \eta} d\xi d\eta \qquad (18)$$

and the mass matrix M_{ij}, which can be written :

$$M_{ij} = \rho h \int_S N_i N_j \frac{ab}{4} d\xi d\eta \qquad (19)$$

This now allows to set up the *identification problem* :

$$K_{ij} = \left[\frac{4b}{a^3} A_{ij} + \ldots + \frac{16}{ab} E_{ij}\right] \cdot \left\{\begin{array}{c} D_{11} \\ \ldots \\ D_{66} \end{array}\right\} \tag{20}$$

The iteration scheme takes the form shown in next page. Any norm, e.g. a max-norm can control the process.

The key issue is - of course - the conditioning of the sensitivity matrix, as its pseudo- inverse is required; numerically speaking it should thus be well conditioned.

An important result was proven by Sol [SOL, H., 1986] and yields a completely automated procedure: the sensitivity for D_{12} is maximal for a ratio between a and b equal to :

$$a/b = (D_{11}/D_{22})^{1/4} \tag{21}$$

which yields following procedure - now completely implemented in a PC:

i) cut two beams in the 1- and 2-direction and establish their resonant frequencies; this yields a first estimate of D_{11} and D_{22}, together with the optimal ratio a/b of a test plate (= "Poisson"-plate);

ii) measure the two first resonant frequencies - after eliminating the zero-frequencies due to rigid body motions - to establish the other unknowns by initializing the iteration procedure.

iii) it has been shown by Sol [SOL, H., 1986] that ν_{12} - for orthotropic materials - can be found through following procedure :

$$\nu_{12} = \frac{\left|\sqrt{K_3^2 + |4K_2 K_1|} - |K_3|\right|}{|2K_2|}$$

$$\text{with } K_1 = \left(\lambda_2 - \lambda_1\right)/\lambda_1$$

$$\log K_2 = 0.42 - 1.55\, x + 0.2128\, x^2$$

$$K_3 = 1.17 - 0.987\, x + 0.2457\, x^2 \tag{22}$$

in these relations:

$$x = \log\left(D_{11}/D_{22}\right) \tag{23}$$

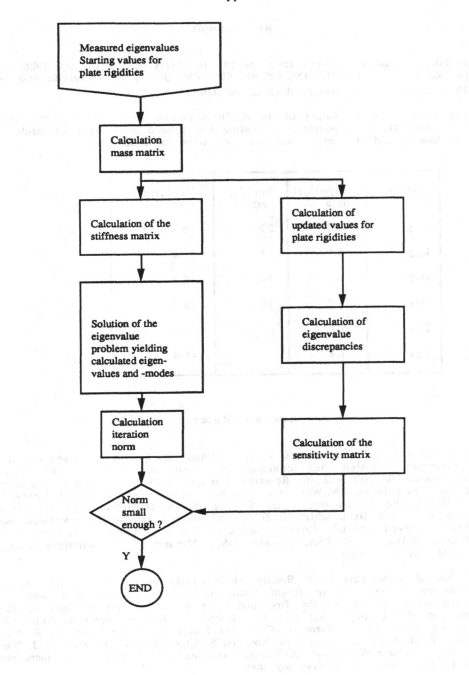

An example

Following example is taken from the Ph. D. thesis of **Sol** [Sol, 1986]: a laminated plate [0°, -45°, 45°, 90°,90°, 45°, -45°, 0°] $_s$ with a specific mass of 1578 kg/m^3 and an average thickness of 0.00217 m.

The table gives the values of the rigidities obtained with the suppliers data, with classical test experiments (bending and torsion tests made on machined test beams) and the results obtained with Bayesian estimation.

Rigidity	Supplier data	Statical tests	Vibration tests
D_{11}	62.8	55	56
D_{22}	45.9	43	41.2
D_{12}	16.2	14.3	14.4
D_{66}	17.7	15	15.3
D_{16}	-1.4	-	-1.4
D_{26}	-1.4	-	-1.4

Acknowledgments

It is not possible to write such a paper without being in a privileged research environment, which has continuously been provided by the Vrije Universiteit Brussel and its Research Council (OZR), the Belgian National Science Foundation (NFWO), the National Institute for Industrial Research (IWONL), the Administration for Science Policy (DPWB). They sponsored research in this field, which at first sight seemed somewhat conjectural, but finally proved fecund. Private companies, among which Snauwaert, FN, Donnay, Solvay, Look, PSA, Vynckier, DSM, Mecanova, CRIF Plastique (Liège) should also be thanked.

A lot of co-workers have directly or indirectly contributed to this paper. Although I am sure to forget some of them, I nevertheless want to acknowledge H. Sol in the first place for his major contributions to the contents of this paper, but also A. Cardon, C. Hiel, P. Hamelin (Univ. Cl. Bernard, Lyon, F), G. Verchery (Ecole des Mines de Saint-Etienne, F), R. Van Geen, P. Hermans, B. Narmon, M. Roovers, S. Claeyssens, J. De Visscher, J. Van Tomme, the late R. Snoeys (KULeuven) and his collaborators. All of them are to be mentioned and deserve my thanks.

References

Berman, A., Nagy, E.J.
Improvement of a large analytical model using test data
AIAA Journal, vol. 21, n° 8, Aug. 1983

Caesar, B.
Update and Identification of dynamic mathematical models
Proc. 4th Int.Mod.Analysis Conf., LA, Ca, 1986, pp. 394-401
Union College, Shenectady, N.Y. 12308

De Visscher, J., Sol, H.
Identification of orthotropic complex material properties using a mixed
numerical/experimental technique
Euromech 229, Saint-Etienne, Dec. 1990.

De Wilde, W.P., Narmon, B., Sol, H., Roovers, M.
Determination of material constants of an anisotropic lamina by free
vibration analysis
Proc. 2nd Int.Mod.Analysis Conf., Orlando, Fa, 1984, pp. 44-49.
Union College, Shenectady, N.Y. 12308

De Wilde, W.P., Sol, H.
Identification of anisotropic plate rigidities using experimental free
vibration analysis.
Internal Report J1985/03, Structural Analysis, VUB, 1985.

De Wilde, W.P., Sol, H., Van Overmeire, M.
Coupling of Lagrange interpolation, modal analysis and sensitivity analysis
in the determination of anisotropic plate rigidities
Proc. 4th Int.Mod.Analysis Conf., LA, Ca, 1986, pp. 1058-1063
Union College, Shenectady, N.Y. 12308

Deobald, L., Gibson, R.
Determination of elastic constants of orthotropic plates by a modal
analysis/Rayleigh Ritz technique
Proc. 4th Int.Mod.Analysis Conf., LA, Ca, 1986, pp. 682-690
Union College, Shenectady, N.Y. 12308

Fox, R.L., Kapoor, M.P.
Rate of changes of eigenvalues and eigenvectors
AIAA Journal, vol. 6, n° 12, Dec. 1968, pp. 2426-2429

Gysin, H.
Attempts to localize finite element modelling errors
13th Int. Sem. on Modal Analysis (in memory of Prof. R. Snoeys), Sep. 1988,
Leuven, Belgium.
Catholic University Leuven, Dept. of Mech. Eng.

Hendriks, M.A.N., Oomens, C.W.J., Kok, J.J.
A numerical experimental approach for the mechanical characterization of
composites
Manuscript.

Hermans,P., De Wilde, W.P., Hiel, CC.
Boundary integral equations applied in the characterization of elastic materials
Computational methods and experimental measurements,
Springer Verlag, 1982   Keramidas, G.A., Brebbia, C.A. (ed.)

Ibanez, P.
Review of analytical and experimental techniques for improving structural dynamic models
Welding Research Council Bulletin, n° 249, 1979.

Kavanagh, K.T.
Finite element applications in the characterization of elastic solids
IJSS, 1971, vol.7, pp. 11-23

Kavanagh, K.T.
Extension of classical experimental techniques for characterizing composite material behaviour
Experimental Mechanics, 1972, pp. 50-56

Larsson, P.O.
Determination of Young's and shear moduli from flexural vibrations of beams
15th Int. Sem. on Modal Analysis, Sep. 1990, Leuven, Belgium.
Catholic University Leuven, Dept. of Mech. Eng.

Leissa, A.W.
The free vibration of rectangular plates
Journal of Sound and Vibration, 31 (3), 1973, pp. 257-293.

Lekhnitskii, S.G.
Anisotropic plates
Gordon&Breach, 1968.

Link, M.
Identification of physical system matrices using incomplete vibration test data
Proc. 4th Int.Mod.Analysis Conf., LA, Ca, 1986, pp. 386-393
Union College, Shenectady, N.Y. 12308

Muskheliskvili, V.I.
Some basic problems on the mathematical theory of elasticity
Noordhoff, 1953

Snoeys, R., De Wilde, W.P., Verdonck, E., Sol, H.
Design optimization of dynamical structures by the combined use of modal analysis and the finite element method
Proc. 2nd IMAC, Orlando, Fla, 1984

Sol, H.
Identification of anisotropic plate rigidities using free vibration data
Ph.D. thesis, October 1986
Structural Analysis, V.U.B.

Van Tomme, J., Sol, H., De Wilde, W.P., De Visscher, J., Bosselaers, R.
Evaluation of damping measurements in materials and structures
13th Int. Sem. on Modal Analysis (in memory of Prof. R. Snoeys),
Sep. 1988, Leuven, Belgium.
Catholic University Leuven, Dept. of Mech. Eng.

Vanhonacker, P.
The use of modal parameters of mechanical systems in sensitivity analysis,...
Ph.D. thesis, Catholic University of Leuven (B), 1980.

Wang, J., Sas, P.
A method for identifying unknown physical parameters by mechanical systems
13th Int. Sem. on Modal Analysis (in memory of Prof. R. Snoeys),
Sep. 1988, Leuven, Belgium.
Catholic University Leuven, Dept. of Mech. Eng.

Wolf, J.A., Carne, T.G.
Identification of the elastic constants of composites using modal analysis
Meeting of SESA, San Francisco, Ca, paper A-48, May 1979.

MEASUREMENT OF COMPLEX MODULI OF COMPOSITE MATERIALS AND DISCUSSION OF SOME RESULTS

HUGO SOL, JOELLE DE VISSCHER
Vrije Universiteit Brussel (V.U.B.)
Departement Analyse van Strukturen
Fakulteit Toegepaste Wetenschappen
Pleinlaan 2, B-1050 Brussel, BELGIUM

JOHNNY VANTOMME
Koninklijke Militaire School
Departement Burgerlijke Bouwkunde
Renaissancelaan 30, B-1040 Brussel, BELGIUM

ABSTRACT

This paper describes a new method for the characterization of the engineering constants of linear visco-elastic orthotropic materials.
The method is an application of the principle of mixed numerical/experimental techniques: measured responses are compared to responses calculated from a numerical model and the parameters of the numerical model are then updated until the two types of responses match.
For the technique discussed in this paper, the parameters of the numerical model are the complex engineering constants and the measured and calculated responses are vibrational data. Emphasis is put on the characterization of the damping behaviour, the discussion of all possible error sources and some interesting results.

INTRODUCTION

The method discussed in this paper was especially developed to meet the requirements of engineers working and designing with composite materials.It is a very fast and non-destructive technique that combines the characterization of both the elastic behaviour and the damping behaviour of the material. It provides global and representative results since it is based on the measurement of global quantities on relatively large test specimen.

The reason why the technique should also be able to characterise material damping is that, in composite structures, very little energy is dissipated within the mechanical joints, as for example in metal structures. That is because the number of joints in composite structures is smaller and because they are generally of the adhesive type. So, the inherent damping of the material itself plays a very important role and a simple and fast method to determine this material damping in an experimental way is needed.

The method can be applied to all materials that satisfy the following three conditions:
- linear visco-elastic material behaviour;
- orthotropic material properties;
- available under the form of thin plates (Love-Kirchhoffhypothesis).

When these three conditions are satisfied, the mechanical behaviour of the material can be characterized by four complex engineering constants:

E_1^*	$= E_1(1+i.tg\delta(E_1))$	: Complex Young's modulus (1-direction)
E_2^*	$= E_2(1+i.tg\delta(E_2))$	: Complex Young's modulus (2-direction)
v_{12}^*	$= v_{12}(1+i.tg\delta(v_{12}))$	: Complex Poisson's ratio
G_{12}^*	$= G_{12}(1+i.tg\delta(G_{12}))$	: Complex inplane shear modulus

The real parts of these engineering constants are called the elastic moduli since they characterize the elastic behaviour, while the imaginary parts are called the loss moduli and they characterize the damping of the material.

The technique for the determination of the purely elastic material properties was developed first and is now a very reliable and widely used technique [1]. The technique for the determination of the visco-elastic material properties can be seen as an expansion, since it makes use of the same numerical model and the same experimental set-up.

THEORETICAL BACKGROUND

The equilibrium equation of a single degree of freedom system with viscous damping (see figure 1):

$$m\ddot{x} + c\dot{x} + kx = 0 \qquad (1)$$

yields the complex stiffness and the loss tangent:

$$k^* = k+i.\omega c = k(1+i.tg\delta(k)) \qquad (2)$$

where: $tg\delta(k) = \omega c/k$

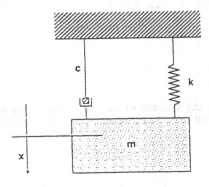

Figure 1. Single degree of freedom system with viscous damping

The free damped vibrations of this single degree of freedom system can be expressed by the following function:

$$w(t) = A\, e^{-\xi\omega t} \sin(\omega t - \phi)$$

(3)

Four parameters determine this function:

A: Modal displacement or amplitude

ϕ: Phase angle

ω: Circular resonant frequency

ξ: Modal damping ratio

It can be shown that the modal damping ratio can be expressed in terms of the loss tangent:

$$\xi = tg\delta(k)/2$$

(4)

or in terms of the maximum potential energy P in one oscillation cycle and the amount of energy D dissipated during that cycle:

$$\xi = D/4\pi P$$

(5)

A thin plate vibrating in one of its eigenmodes can also be regarded as a single degree of freedom system, where the transverse displacement w is now also a function of the plate coordinates (x,y):

$$w(x,y,t) = A(x,y)\, e^{-\xi\omega t} \sin(\omega t - \phi)$$

(6)

A(x,y) represents the modeshape and the only degree of freedom is the amplitude of the modeshape.

So, for an arbitrary structure, vibrating in only one of its eigenmodes, equations (4) and (5) remain valid.

Considering a thin plate with free boundary conditions, vibrating in a vacuum chamber, the only source of energy dissipation of the system is the internal friction of the material.

The energy dissipated within the material per unit volume and per cycle of oscillation is given by:

$$d = \oint \sigma_1^* \, d\varepsilon_1^* + \oint \sigma_2^* \, d\varepsilon_2^* + \oint \tau_{12}^* \, d\gamma_{12}^*$$

(7)

where: $\quad \{\sigma^*\} = \{\sigma\} \sin(\omega t - \phi)$

$\quad\quad\quad \{\varepsilon^*\} = \{\varepsilon\} \sin(\omega t - \psi)$

denote the complex stresses and strains.

Integrating the dissipated energy over the volume of the plate and taking into account the assumptions of thin plate theory, orthotropy and linear visco-elasticity, one obtains:

$$D= \int_A \left[\oint D_{11}{}^* \chi_1{}^* d\chi_1{}^* + \oint D_{22}{}^* \chi_2{}^* d\chi_2{}^* + \oint D_{12}{}^* \chi_2{}^* d\chi_1{}^* + \oint D_{21}{}^* \chi_1{}^* d\chi_2{}^* + \oint D_{66}{}^* \chi_6{}^* d\chi_6{}^* \right] dA$$

(8)

where A is the plate surface,

$$D_{11}{}^* = E_1{}^* t^3/12(1-v_{12}{}^* v_{21}{}^*)$$
$$D_{22}{}^* = E_2{}^* t^3/12(1-v_{12}{}^* v_{21}{}^*)$$
$$D_{12}{}^* = v_{12}{}^* E_2{}^* t^3/12(1-v_{12}{}^* v_{21}{}^*)$$
$$D_{66}{}^* = G_{12}{}^* t^3/12$$

(9)

are the complex plate rigidities and

$$\chi_1{}^* = \frac{\partial^2 w^*}{\partial x^2}$$

$$\chi_2{}^* = \frac{\partial^2 w^*}{\partial y^2}$$

$$\chi_6{}^* = 2 \frac{\partial^2 w^*}{\partial x \partial y}$$

(10)

are the plate curvatures.

The dissipated energy can be seen as the sum of four contributions, associated with $D_{11}{}^*$, $D_{22}{}^*$, $D_{12}{}^*$ and $D_{66}{}^*$ respectively.

$$D = D(11) + D(22) + D(12) + D(66)$$

(11)

Expressing the maximum potential energy of the plate in terms of plate rigidities and curvatures yields:

$$P = \frac{1}{2} \int_A \left[D_{11} \chi_1{}^2 + D_{22} \chi_2{}^2 + 2D_{12} \chi_1 \chi_2 + D_{66} \chi_6{}^2 \right] dA$$

$$= P(11) + P(22) + P(12) + P(66)$$

(12)

where only the real parts of the plate rigidities appear, since only the in-phase components of stresses and strains contribute to the potential energy .

Assuming $D_{12}=0$, the plate shows three fundamental vibration patterns (see figure 2).

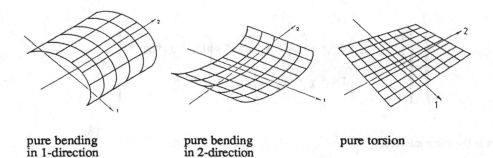

pure bending
in 1-direction

pure bending
in 2-direction

pure torsion

Figure 2. Fundamental vibration patterns $(D_{12}=0)$

For the pure bending in the 1-direction, one can write:

$$\xi = D/4\pi P = D(11)/4\pi P(11) = tg\delta(D_{11})/2 \qquad (13a)$$

since D11* is the only rigidity affecting the mode.
For the two other modeshapes, one can write:

$$tg\delta(D_{22}) = D(22)/2\pi P(22) \qquad (13b)$$
$$tg\delta(D_{66}) = D(66)/2\pi P(66) \qquad (13c)$$

When $D_{12}\neq 0$, the pure bending modeshapes are also affected by the interaction between stresses and strains in orthogonal directions and the term

$$tg\delta(D_{12}) = D(12)/2\pi P(12) \qquad (13d)$$

has to be added.

Making use of expressions (13), the modal damping ratio of an arbitrary modeshape can be written as:

$$2\xi = \frac{P(11)}{P}tg\delta(D_{11}) + \frac{P(22)}{P}tg\delta(D_{22}) + \frac{P(12)}{P}tg\delta(D_{12}) + \frac{P(66)}{P}tg\delta(D_{66})$$

$$(14)$$

The conclusion is that the modal damping ratio of an arbitrary modeshape of an orthotropic plate consists of contributions from each of the plate rigidities.
The contribution factor of a plate rigidity is the ratio of the potential energy associated to that rigidity to the total potential energy of the modeshape.

DETERMINATION OF THE ELASTIC MODULI

The determination of the elastic part of the 4 engineering constants of orthotropic plates is based on the comparison of experimentally measured resonant frequencies and numerically calculated resonant frequencies of a test plate with free boundary conditions.
The elastic plate rigidities of the plate, which are the unknowns of the numerical model, are iteratively updated untill the two sets of resonant frequencies match.
The numerical model that is used is a Rayleigh-Ritz model with 49 degrees of freedom, which prooved to be both simple and extremely accurate.

The iterative updating procedure is illustrated by the flowchart in figure 3.
The mass and stiffness matrix of the plate are calculated from initial values of the plate rigidities and the solution of the eigenvalue problem:

$$([K]-\lambda[M])\{\Phi\} = \{0\} \tag{15}$$

yields the calculated resonant frequencies.
From the discrepancy between calculated and measured resonant frequencies and the sensitivity matrix (expressing the sensitivity of resonant frequencies to plate rigidities), better values of the plate rigidities can be calculated.
The whole procedure is then repeated untill the two sets of resonant frequencies match and the correct plate rigidities are known.

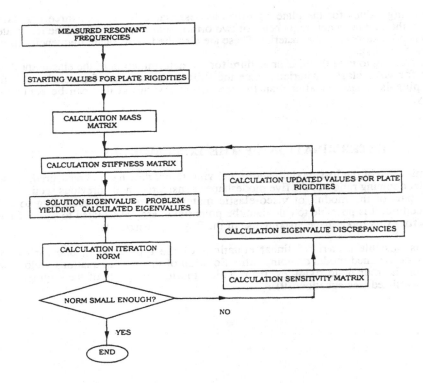

Figure 3. Flowchart for the iterative determination of the plate rigidities

It has been shown in [1] that the sensitivity matrix is best conditioned when the measurement is performed on a plate with an aspect ratio:

$$\frac{a}{b} = \sqrt[4]{\frac{E_1}{E_2}}$$

(16)

Such a plate, which is called a "Poisson plate" always yields the three modeshapes, presented in figure 4, as first modeshapes.

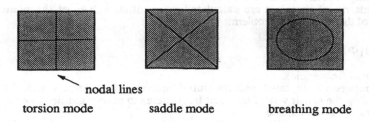

nodal lines

torsion mode   saddle mode   breathing mode

Figure 4. Three first modeshapes of a Poisson plate

The starting values for the plate rigidities are estimated from these three resonant frequencies and the first resonant frequencies of two orthogonal beams cut from the test plate in the directions of symmetry of the material. These are also the five resonant frequencies used in the updating procedure.

It is interesting to note that this procedure for the determination of the elastic moduli remains valid for visco-elastic materials, since the shift in the resonant frequencies due to material damping is usually smaller than the frequency resolution that can be achieved experimentally.

DETERMINATION OF THE LOSS MODULI

For the determination of the loss moduli, additional vibrational data have to measured, these being the modal damping ratio's of the five modeshapes considered in the previous section.
With the real part of the moduli of visco-elastic materials determined according to the described procedure, it is possible to calculate the potential energy associated to each of the plate rigidities for every modeshape, making use of the Rayleigh-Ritzmodel.

So, it is possible to write 5 linear equations of the type of (14), one for every experimentally determined modal damping ratio. Solution of this set of equations yields the loss tangents of the four plate rigidities, from which the loss tangents of the engineering constants are calculated (see equations (9)).

EXPERIMENTAL SET-UP

Figure 5 shows the complete experimental set-up, used for the determination of the complex engineering constants.

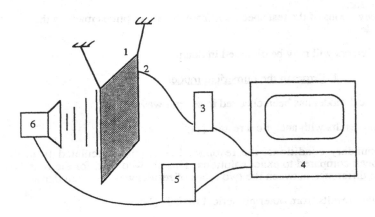

Figure 5. Experimental set-up

The testspecimen -two beams and one plate- are suspended by two very thin threads, fixed at the nodal lines of the modeshape to be measured. This set-up simulates the free boundary conditions in the best possible manner. An accelerometer is used to measure the response of the specimen, which is sent through a charge amplifier and sampled by a Personal Computer. For this purpose, the P.C. is equipped with two measurement cards: one low-pass filtercard and one A/D card.

In order to measure the resonant frequencies, the test specimen is subjected to an impact excitation (hammer excitation). The acceleration signal of the test specimen, which is sampled and stored in the memory of the P.C., thus contains all the resonant frequencies of the specimen that are within the frequency range of the impact. So, taking the Fast Fourier Transform of this time signal yields the lower resonant frequencies of the specimen.

The most accurate way to measure the modal damping ratio's is to curve-fit the measured damped free vibrations of the test specimen, vibrating in the associated mode [3]. In order to generate this type of vibration, the test specimen has to be subjected to an excitation signal containing only the resonant frequency of the considered modeshape. A second requirement is that the excitation device is non-contacting, since a contacting device would add damping to the system.
The excitation technique that is used therefore is acoustic excitation. A sine signal at the resonant frequency of the associated mode is generated by the P.C., sent through an audio-amplifier and going to a loudspeaker. The acoustic waves force the test specimen to vibrate in the desired modeshape. When the loudspeaker is stopped, the P.C. starts sampling the acceleration signal. Curve-fitting the measured signal with expression (6) finally yields the modal damping ratio of the mode.

ERROR DISCUSSION

In general, errors in mixed numerical/experimental techniques can be classified in three categories:

1. Errors in the numerical model;
2. Experimental errors;
3. Errors due to deviations of the test specimen from the assumptions made in the numerical model.

These three categories of errors will now be dicussed in detail.

1. Errors in the numerical model

The validity of the numerical model has been checked in several ways:

- By comparing numerical results with analytical results [1];

For several types of boundary conditions, the resonant frequencies calculated from the numerical model have been compared to exact solutions. This showed that, for the four first resonant frequencies, the difference between the two types of results was less than 0.5%.

- By comparing results with results from other numerical models [1];

The results of the Rayleigh-Ritzmodel have been compared with finite element results on plates with four free boundaries. The finite element model consisted of (6x6) 9-node Mindlin plate elements in which transverse shear deformation was suppressed. This comparison showed once again differences of less than 0.5% for the four first resonant frequencies, except for cases where the fourth modeshape became too complicated and a finer finite element mesh was required.

- By comparing results with experimental results [2];

For two types of test plates, an aluminium and a steel plate, with well-known material properties, the resonant frequencies were both calculated making use of the numerical model and measured. The differences for the three first resonant frequencies were less than 3%.

2. Experimental errors

- Errors in the measurement chain;

Under this header falls the distortion of the accelerometer signal due to for example nonlinear amplification. The effect of this type of error on the measured resonant frequencies is neglectable since the resonant frequencies are determined by the frequency content of the signal, which is fairly insensitive to nonlinearities in the amplification.
But for the modal damping ratio's, the linearity of the measurement chain has to be checked since modal damping ratio's are determined by the vibrational decay of the amplitude of the signal.
The second type of error in this category is the digitalisation error due to the limited sampling rate. This error can be neglected if the sampling frequency and the number of samples are sufficiently high.

- Errors due to badly simulated boundary conditions;

The consideration of this error was one of the reasons for choosing free boundary conditions. Free boundary conditions are the only type of boundary conditions that can be achieved in practice in an acceptable manner, while simply supported and clamped edges are purely mathematical idealizations.

When the free boundary conditions are simulated by suspending the test specimen by two threads fixed at the nodal lines so that they do not disturb the vibration, this type of error can be neglected. This set-up is possible since the modeshapes are exactly known, which is another advantage of using Poisson plates.

- Errors due to added mass and damping from measurement devices.

The accelerometer which is fixed on the test specimen modifies both mass and damping of the system. The added mass decreases the resonant frequencies while the added damping increases the modal damping ratio's.

The effect of the mass of the accelerometer is accounted for by adding a concentrated mass in the numerical model too.

But it is impossible to account for the added damping. Indeed, it depends not only on a simple quantity like the accelerometer mass but on several factors like the length and weight of the accelerometer wire, the amount of wax used for the fixation of the accelerometer etc...The added damping can be minimised by fixing the accelerometer wire to the plate at the height of the nodal lines, so that only a small part of the wire is vibrating and dissipating energy during the measurement.

The only way to eliminate this additional source of damping is to use a non-contacting measurement device. Experiments with a microphone have shown that this type of sensor is insufficiently sensitive for the transverse plate displacements of most composite material plates. Optical measurement techniques however proove to be very accurate, though they are expensive. In the short future, the accelerometer in the present set-up will be replaced by an optical sensor based on the Doppler effect, which measures velocities, so that much more reliable damping measurements are expected.

3. Errors due to deviations of the test specimen from the assumptions made in the numerical model

- The Love-Kirchhoff hypothesis;

In the numerical model, it is assumed that the test plate is sufficiently thin, so that transverse shear deformation can be neglected. This assumption is acceptable if the thickness to length and width ratio of the test plate is smaller than a certain value and when only the lower modes are considered.

This maximum allowable ratio depends on the type of material. Finite element simulations using Mindlin plate elements, in which transverse shear was first suppressed and then taken into account, allowed to see for what ratio the error due to the suppression becomes higher than 1%. In this way, a list of allowable ratio's for a few representative materials was made. For aluminium for example, a ratio of 1/50 is sufficient but for composite materials (especially the uni-directionally reinforced) the ratio is smaller. A safe value is about 1/100.

- The flat plate assumption;

Very often, test plates are not completely flat and the most common type of curvature is purely cilindrical. Numerical tests have shown that the first bending mode of beams and the torsion mode of plates are practically insensitive to cilindrical curvature. But the influence on the saddle and breathing mode is relatively large, depending on two parameters: Poissons ratio v_{12} and the ratio of D_{11}/D_{22}. Since the value of v_{12} can not be anticipated, it is only possible to make a worst case estimation of the error on the resonant frequencies, that is, in the case that v_{12} equals zero. A detailed discussion of this error can be found in [5].

- The assumption of homogeneous material;

This assumption is of course never strictly valid for composite material plates. But when the order of magnitude of the scale of the inhomogeneities is smaller than the distance between the nodal lines, an averaging effect is achieved and the measured resonant frequencies and modal damping ratio's are not affected.

- The assumption of uniform thickness;

As for the inhomogeneous material properties, variability in thickness is averaged when the order of magnitude of the scale of the thickness variation is smaller than the distance between nodal lines.

The conclusion of this error discussion is that, for the determination of the elastic moduli, the accuracy and reliability of the results is mainly dependent on the quality of the testplate. It has to be checked if the plate is sufficiently thin, flat, homogeneous, of uniform thickness and if it is cut under the right angle and at the right size. For every type of error, it was possible to estimate the error on the resonant frequencies, so that corrections can be carried out.

The determination of the loss moduli however is sensitive to errors on the elastic moduli and to experimental errors.

RESULTS

The following example shows the effect of glassfiber reinforcement on the visco-elastic properties of polyester plates.

Four rectangular reinforced polyester test plates were produced, each with a different fiber volume fraction. The reinforcement was bi-directional glasscloth.

The following table shows the volume fractions of the 4 manufactured test plates.

Plate number	1	2	3	4
number of layers glasscloth	0	4	5	6
volume fraction glass (%)	0	24.1	29.2	35.7
volume fraction resin (%)	100	70.0	65.8	60.4
volume fraction air (%)	0	5.9	5.0	3.9
specific mass (kg/m³)	1237.3	1452.7	1531.6	1631.5

The amplitudes and phase angles of the 4 orthotropic moduli are represented as a function of fiber volume fraction in figure 6.

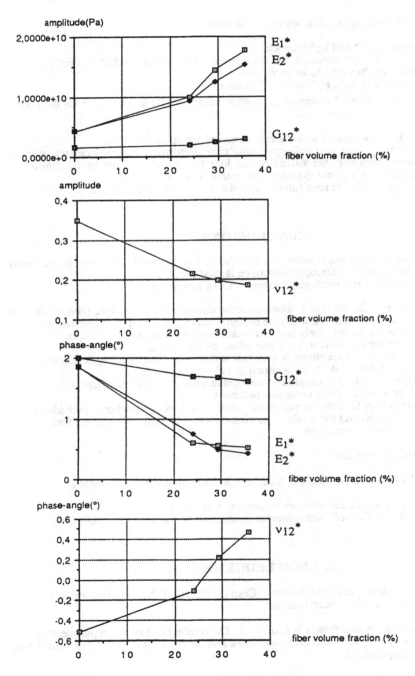

Figure 6. Amplitude and phase angle of the complex engineering constants as a function of fiber volume fraction

From figure 6, the following conclusions can be drawn:

*   the amplitude of E_1^* and E_2^* increase;
*   the amplitude of G_{12}^* increases, but the slope is smaller than for E_1^* and E_2^*;
*   the amplitude of Poisson's ratio decreases;
*   the phase-angles of E_1^*, E_2^* and G_{12}^* decrease;
*   the phase-angles of v_{12}^* increases to positive values, while it is negative for the pure resin.

The presented results are obtained with the experimental set-up of figure 5.
Since the measurements of the modal dampingratio's are performed with an accelerometer and in air at atmospheric pressure, the values of the loss moduli should not be interpreted as absolute values but they only allow to make a comparison between materials.
Better results are expected in the near future when the set-up will be improved.

CONCLUSIONS

The method described in this paper offers the possibility to completely characterise the visco-elastic material properties of orthotropic composite laminates.
Compared with traditional test methods, it shows a lot of advantages:

-   The method provides averaged values over the domain of the test specimen, so that the results are still representative when the material is not completely homogeneous.
-   The test specimen are relatively large, so that boundary effects (for example, delamination near boundaries) have less effect on the results.
-   The preparation of test specimen is very fast and simple (no strain-gages needed) and so is the fixation of the test specimen (no clamping needed).
-   The experimental set-up is simple and the measurement procedure is highly automated by a menu-driven computer program.
-   The method is very fast: the preparartion, set-up and measurement only take about 15 minutes (or less) and the result is the simultaneous determination of the four complex engineering constants.

    Future developments are:

-   The use of an optical transducer in order to eliminate all additional damping sources;
-   The building of a vacuum chamber around the test specimen;
-   The study of the effect of environmental conditions on the material properties.

ACKNOWLEDGEMENTS

The authors wish to acknowledge the Research Council (OZR) of the Vrije Universiteit Brussel for its support in this research project.

A person who deserves special thanks is Prof. W.P. De Wilde, head of the department of Structural Analysis, for his deep interest in our work and for the creation of a very stimulating environment at the department.

REFERENCES

[1] Sol H., <u>Determination of the anisotropic plate rigidities by free vibration analysis</u>, PhD dissertation, Free University of Brussels,1986

[2] Sol H., <u>Experimental verification of plate vibration models</u>, Internal report 1986/06, Dept. of Structural Analysis, Free University of Brussels

[3] De Visscher J., Bosselaers R., <u>Experimentele bepaling van de demping van komposietmaterialen</u>, Thesis, Free University of Brussels, 1988

[4] Sol H., De Visscher J., <u>Identification of the orthotropic complex engineering constants by free vibratuion analysis</u>, Internal report, Free University of Brussels, 1989

[5] De Visscher J., <u>Influence of curvature on the resonant frequencies and modeshapes of Poisson plates</u>, Internal report, Free University of Brussels, 1990

MOIRE INTERFEROMETRY FOR COMPOSITES

Daniel Post
Virginia Polytechnic Institute and State University
Engineering Science and Mechanics Department
Blacksburg, Virginia 24061–0219, U.S.A.

ABSTRACT

With its high sensitivity and high spatial resolution, moire interferometry is capable of analyzing strains in laminated composites on a ply–by–ply basis. It is an optical technique that produces contour maps of in–plane U and V displacement fields. The paper describes the experimental technique and presents diverse applications. Examples are taken from studies of free–edge effects, where compressive loading causes severe transverse and shear strains; shear tests, that show the importance of whole–field analysis of the strain distributions; metal–matrix specimens, which document concentrations of strains in the matrix and show anomalous movements between groups of fibers; thermal strains, which result from steady–state changes of temperature.

INTRODUCTION

Thank you for the very kind invitation to present a paper on moire interferometry. It is a relatively new method of experimental mechanics, but it has already been applied by several laboratories throughout the world. In my laboratory at VPI&SU my students and I have applied moire interferometry to numerous studies of composite materials. With this method, the data are extracted in the form of whole–field maps of IN–PLANE displacements. Moire interferometry is characterized by subwavelength displacement sensitivity, high spatial resolution, high signal–to–noise ratio and very large displacement range. The complete state of strain on the specimen surface — the normal strains and the shear strains — is determined easily from the displacement field.

This paper describes the basic nature of the experimental method and it reviews diverse applications to composite materials. A comprehensive treatment of the method appears in Ref. [1], and practical suggestions for laboratory practice are offered in Ref. [2].

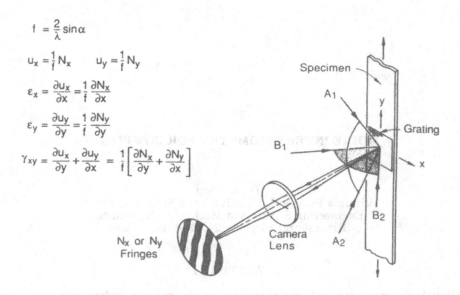

$$f = \frac{2}{\lambda}\sin\alpha$$

$$u_x = \frac{1}{f}N_x \qquad u_y = \frac{1}{f}N_y$$

$$\varepsilon_x = \frac{\partial u_x}{\partial x} = \frac{1}{f}\frac{\partial N_x}{\partial x}$$

$$\varepsilon_y = \frac{\partial u_y}{\partial y} = \frac{1}{f}\frac{\partial N_y}{\partial y}$$

$$\gamma_{xy} = \frac{\partial u_x}{\partial y} + \frac{\partial u_y}{\partial x} = \frac{1}{f}\left[\frac{\partial N_x}{\partial y} + \frac{\partial N_y}{\partial x}\right]$$

Fig. 1. Optical system and small–strain equations for moire interferometry. Interaction of the specimen grating with coherent beams A_1 and A_2 produces the N_y or V pattern; interaction with B_1 and B_2 produces the N_x or U pattern.

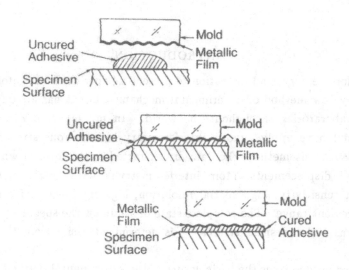

Fig. 2. Steps to replicate a crossed–line grating on the specimen.

THE EXPERIMENTAL METHOD

The principle of moire interferometry is depicted in Fig. 1. A crossed—line diffraction grating of frequency f/2 is replicated on the specimen and it deforms together with the loaded specimen. Two coherent beams, indicated by their central rays A_1 and A_2, produce closely spaced horizontal bands of constructive and destructive interference. These function as a reference grating, and the array of bands (or "lines") is called a virtual reference grating. This virtual reference grating, of frequency f lines per mm, interacts with the horizontal set of lines of the specimen grating to form a moire fringe pattern, namely the contour map of fringe order N_y depicting the V displacement field. Similarly, the virtual reference grating created by beams B_1 and B_2 combines with the vertical set of lines on the specimen to form the contours of N_x that depict the U displacement field. The fringe patterns are viewed at normal incidence by a camera focused on the specimen grating.

The relevant equations are listed in Fig. 1. Here, f is the frequency of the reference grating; λ is the wavelength of the coherent laser light; and ϵ and γ are normal and shear strains, respectively. For all the illustrations shown here, f was 2400 lines/mm and the corresponding displacement sensitivity was 0.417 μm per fringe order.

This brief description emphasizes the analogous nature of moire interferometry and geometrical moire [3]. A more rigorous treatment addresses the diffraction of incident beams by the specimen grating and optical interference of the diffracted light [1].

A technique for applying the crossed—line diffraction grating to the specimen is illustrated in Fig. 2. It is a replication process in which a mold is used to form a corresponding system of hills and valleys in a thin adhesive layer (e.g., epoxy) on the specimen [1]. The mold is a phase—type crossed—line grating of frequency f/2. Before the replication, an ultra—thin film of reflective metal, usually aluminum, is applied to the mold by vacuum deposition. The mold is pried off after the adhesive hardens, but the reflective film remains with the adhesive. The result is a reflective crossed—line grating of 10—25 μm thickness on the specimen.

COMPRESSION OF THICK COMPOSITES, FREE—EDGE EFFECTS

The specimen illustrated in Fig. 3 was cut from a thick—walled cylinder of graphite/epoxy IM6/2258. It had a quasi—isotropic stacking sequence $[90_2/0_2/+45_2/-45_2]_n$ with ply thickness of 0.19 mm. The specimen dimensions were 25 x 13 x 13 mm. A crossed—line grating was applied to one face to determine details of the deformation field and strain gages were applied to the other free faces [4].

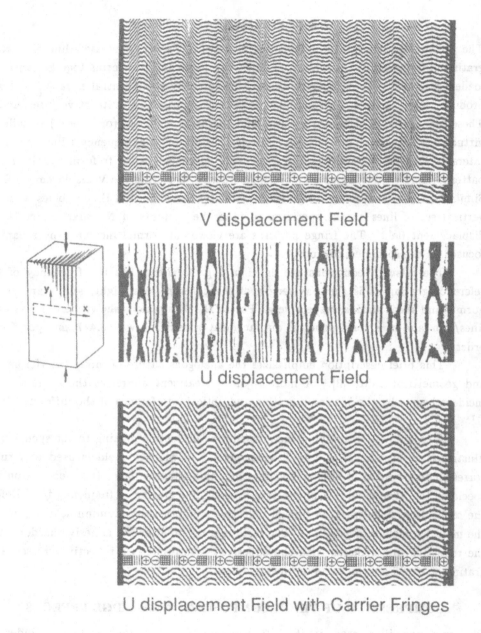

V displacement Field

U displacement Field

U displacement Field with Carrier Fringes

Fig. 3   Moire patterns for a thick $[90_2/0_2/45_2/-45_2]_n$ graphite/epoxy laminate in compression. The symbols represent fiber directions, where ±45° plies are indicated by plus and minus signs. Sensitivity is 0.417 μm per fringe order. Deformations are revealed on a ply–by–ply basis.

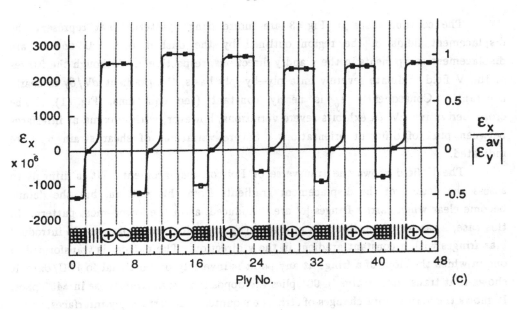

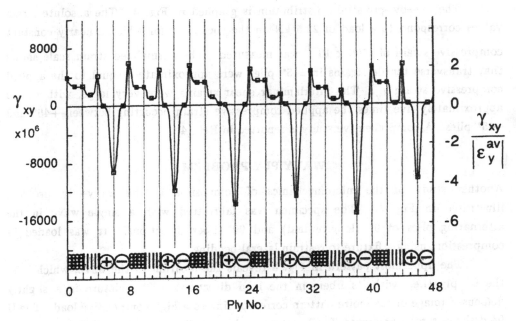

Fig. 4. Normal and shear strain distributions across the specimen width. Compressive strain ϵ_y is 2700×10^{-6}.

The contour maps in Fig. 3 are moire fringe patterns that represent the displacement fields in the region outlined by the dashed box. U and V are displacement components in the x and y directions, respectively. Although the fringes of the V field undulate severely on a ply–by–ply basis, the gradient $\partial V/\partial y$ is nearly constant. Consequently ϵ_y is nearly constant (see equations, Fig. 1). The cross–derivative $\partial V/\partial x$ exhibits severe variations, however, with maxima at interfaces between plies of different orientations. Severe variations of shear strain γ_{xy} are indicated.

The U field shows the transverse or Poisson displacements. It is difficult to assess the details of the displacement gradients from this pattern, but the details become clear when carrier fringes [5] are introduced, as in the lower–most pattern. In this case, a carrier of rotation was produced (by adjusting the apparatus) to introduce bias fringes with a constant gradient in the y direction. The pattern is transformed to one in which the slope of a fringe at any point is inversely proportional to $|\partial U/\partial x|$. It shows that transverse strains in 90° plies are opposite in sign from those in ±45° plies. It shows the very abrupt changes of strains encountered at certain ply interfaces.

The ply–by–ply strain distribution is graphed in Fig. 4. The absolute strain values correspond to a load of 22.5 kN on the specimen, for which a nearly constant compressive strain of 2700 x 10^{-6} was measured. The normalized strain scale shows that transverse tensile strains in ±45° plies were approximately equal to the applied compressive strain ϵ_y. The interlaminar shear strains were very large, with values approximately five times the applied compressive strain occurring between +45° and –45° plies. A comprehensive study is reported in Ref. [4].

WAVY PLY PROBLEM

Another study of the micromechanics of composites is the wavy–ply problem illustrated in Fig. 5. The specimen was fabricated with a single wave in the alternating plies of 0° (longitudinal) and 90° fiber directions. It was loaded in compression, using a fixture to restrain lateral bending.

The uppermost pattern in Fig. 5 reveals the location of the wave, which is in the 0° ply (i.e., with its fibers in the load direction). The picture is a slightly defocused image of the moire pattern corresponding to a high compressive load. The U field shows strong gradients $\partial U/\partial y$ extending beyond the wave region. The pattern is made less dense with carrier fringes of extension [5], but the gradient $\partial U/\partial y$ remains unchanged. The V field is relatively sparse, containing smaller gradients $\partial V/\partial x$ which have only minor influence on γ_{xy}. Thus, large $\partial U/\partial y$ signifies large shear strain.

WAVY-PLY PROBLEM

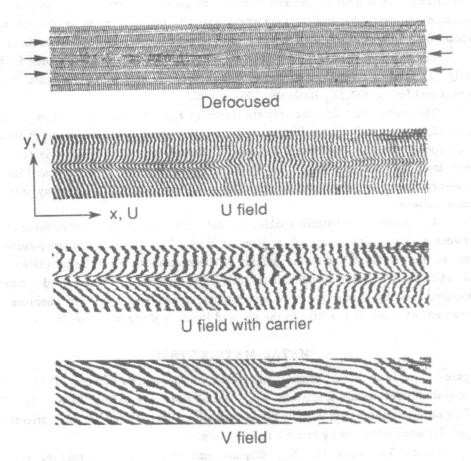

Defocused

U field

U field with carrier

V field

Fig. 5. Deformation of a graphite/polysulfone cross—ply compression specimen with a wavy ply. The wave causes high interlaminar shear strains far beyond the wave region.

The main interest lies in the shear strains. Ideally, the 0/90 laminate should show zero γ_{xy}. It is clear that the wave produces strong shears. It is especially curious that the strongest shear strains occur outside the wave region and extend to a substantial distance beyond the wave. This is part of an ongoing investigation to determine the extent of the shear zone and the effect of other wave geometries.

SHEAR TESTS

Two critical issues arise in the measurement of shear properties of composites: (1) specimen geometry to provide pure shear and (2) nonuniform shear distributions. Whole–field high–sensitivity measurements are essential to evaluate these parameters.

The compact shear specimen geometry illustrated in Fig. 6a was chosen for this study. It exhibited very low normal strains in the test zone between notches (nearly pure shear) for the $[90_2/0]_9$ laminate of Fig. 6b [6].

The whole–field deformations are shown in Figs. 6c and d for a more complex material, where the fibers are intertwined in a three–dimensional weave and imbedded in an epoxy matrix. The general features of the deformation are shown in (c) for a lower load level and the details in (d) for a higher load level. It is clear that the cross–derivatives and the shear strains vary along the specimen in harmony with the weave pattern.

The whole–field pattern enables the determination of the average shear strain between notches. The average shear stress is known from the load and cross–sectional area, so the representative shear modulus or shear compliance can be calculated. On the other hand, when such nonuniform fields are considered, the accepted modes of measurements by strain gages or extensometers are prone to error because the measurement is highly sensitive to the size and location of the instrumentation.

METAL–MATRIX TESTS

Figure 7 reviews some aspects of recent metal–matrix studies [7,8]. The boron/aluminum tensile specimens, with central slots, are defined in Fig. 7a. The deformations of outer plies with 0° fiber orientation (Spec. I) and 45° fiber orientation (Spec. II) were measured by moire interferometry.

Figure 7b shows the $N_{x'}$ displacement field, i.e., the contour map of displacements in the fiber direction (the x′ direction) for Spec. II. Patterns of displacement components in the x, y, x′ and y′ directions can be obtained readily using moire interferometry [9]. The zig–zag nature of the fringes is not optical noise, but real information. It depicts the larger shear strains in the matrix and smaller shear strains in the fibers. A most interesting micromechanical feature is highlighted by cross–hatching between two neighboring fringes in the field. It shows that large displacements occur in blocks of several fibers, instead of smoothly varying distributions of displacement across all fibers.

Figure 7c is a plot of the shear strains along the 45° line L3. The shear strains were concentrated in the aluminum matrix material. The spikes indicate large plastic shear strains between randomly spaced blocks of fibers. These zones of anomalous

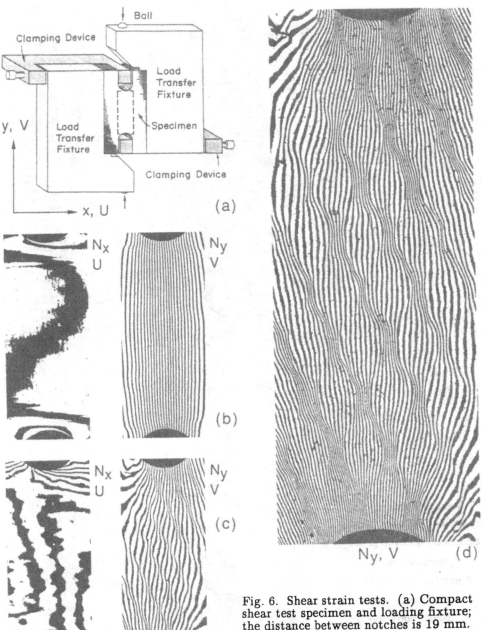

Fig. 6. Shear strain tests. (a) Compact shear test specimen and loading fixture; the distance between notches is 19 mm. (b) Graphite/epoxy $[90_2/0]_9$ specimen. (c) Glass/epoxy specimen with a 3—dimensional woven fiber construction at a lower load and (d) at a higher load level.

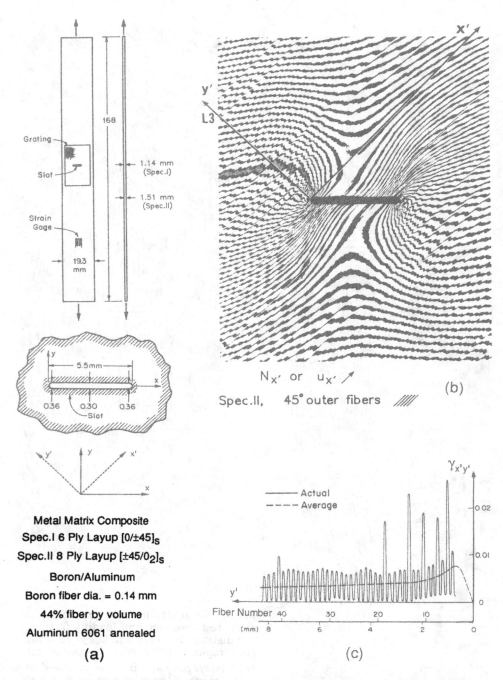

Metal Matrix Composite
Spec.I 6 Ply Layup [0/±45]$_S$
Spec.II 8 Ply Layup [±45/0$_2$]$_S$
Boron/Aluminum
Boron fiber dia. = 0.14 mm
44% fiber by volume
Aluminum 6061 annealed

(a)

$N_{x'}$ or $u_{x'}$ ↗
Spec.II,   45° outer fibers

(b)

(c)

Fig. 7 Micromechanical deformation of boron/aluminum specimen. (a) Specimen geometry and material. (b) Contours of constant displacement component in fiber direction x'. (c) Shear strains along line L3.

plastic slip did not extend along the entire length of the interface between neighboring fibers; instead, the extreme strains gradually diminished, and sometimes reappeared at interfaces between nearby fibers. In addition to the interfiber deformations, severe interlaminar shears were deduced from the results.

THERMAL STRAINS

Thermal strains that result from a change of temperature of a multidirectional laminate can be determined by moire interferometry [10,11]. A simple technique is to apply the specimen grating at an elevated temperature and observe its deformation after the specimen is cooled to room temperature. The elevated temperature operation is illustrated in Fig. 8a, wherein all the steps of specimen grating applications are done inside an oven. The crossed—line diffraction grating mold is carried on a zero—expansion substrate so that the grating frequency of the mold is the same at room temperature and elevated temperature. In this experiment, the substrate was Corning Ultra—Low Expansion Glass (ULE).

The specimen was a thick graphite/epoxy cross—ply laminate $[90_2/0]_n$, approximately 13 mm on each side and 0.2 mm ply thickness. It is illustrated in Fig. 8b. The specimen was heated to 120°C and the grating was applied at that temperature. The specimen grating was undeformed at that temperature, but when it was cooled to 20°C room temperature it recorded the deformation associated with the 100°C change. The in—plane U and V fields were revealed by a four—beam moire interferometer (Fig. 1), and the out—of—plane W field was revealed by a Twyman—Green interferometer. They are shown in Figs. 8c, d, e for the portion of the specimen outlined by the dashed box. A free—edge effect, or a surface—skin effect is evident. The scallop—like fringes in the V field document high values of $\partial V / \partial x$ and corresponding strong interlaminar shear strains γ_{xy} near the free edge. The W field exhibits ridges (90° plies) and valleys (0° plies) in the specimen surface which must be accompanied by strong shear strains γ_{yz} in a narrow interlaminar zone below the free surface.

CONCLUSIONS

This review emphasizes the diversity of moire interferometry. It can be used to study the behavior of composites on the macromechanics level and also on the ply level of micromechanics. It is applicable to many cases of mechanical and thermal loading. Its whole—field character provides a great advantage for complex states of deformation. Normal strains and shear strains are extracted easily. The high contrast of fringes and the high spatial resolution allow tiny details to be studied. The method is compatible

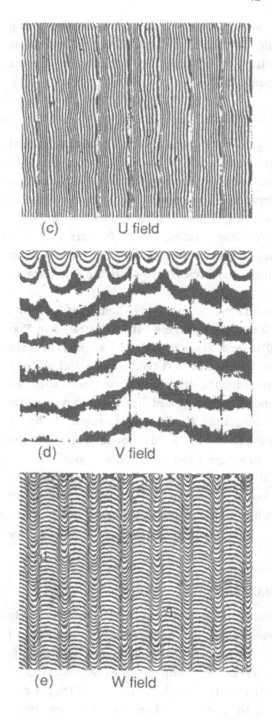

(c)   U field

(d)   V field

(e)   W field

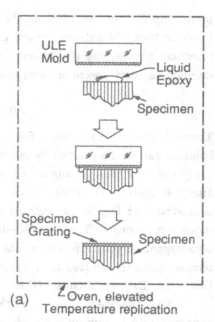

ULE
Mold

Liquid
Epoxy

Specimen

Specimen
Grating

Specimen

(a)   Oven, elevated
Temperature replication

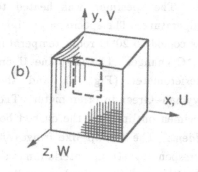

y, V

(b)

x, U

z, W

Fig. 8. Thermal strains induced by
cooling $[90_2/0]_n$ laminate by 100°C.
(a) Specimen grating is applied at
elevated temperature. (b) Specimen;
fringe patterns represent region in
dashed box. (c,d) Moire patterns.
(e) W field, documenting ridges (90°
plies) and valleys.

with very large displacements and ＿y high strain gradients. Moire interferometry offers an excellent capability for the study of the mechanics of composites.

ACKNOWLEDGEMENTS

Special thanks are extended to students, colleagues and sponsors who contributed to the development of moire interferometry in recent years. The encouragement and support by the Office of Naval Research under grants N00014—86—K0799 and N00014—90—J1688 is warmly acknowledged.

REFERENCES

[1] D. Post, "Moire Interferometry," Chap. 7, *Handbook of Experimental Mechanics*, A. S. Kobayashi, Editor, Prentice—Hall, Englewood Cliffs, NJ, pp. 314—387 (1987).

[2] D. Post, "Moire Interferometry for Composites," Chap. IV, *SEM Manual on Experimental Methods for Mechanical Testing of Composites*, M. E. Tuttle and R. L. Pendleton, Eds., Soc. Exp'l. Mech., Bethel, CT (1989).

[3] A. Livnat and D. Post, "The Governing Equations for Moire Interferometry and Their Identity to Equations of Geometrical Moire," *Experimental Mechanics*, 25(4), pp. 360—366 (Dec. 1985).

[4] Y. Guo and D. Post, "Thick Composites in Compression: an Experimental Study of Micromechanical Behavior and Smeared Engineering Properties," J. Composite Materials (submitted for publication).

[5] Y. Guo, D. Post and R. Czarnek, "The Magic of Carrier Fringes in Moire Interferometry," *Experimental Mechanics*, 29(2), pp. 169—173 (June 1989).

[6] P. Ifju and D. Post, "A Compact Double—Notched Specimen for In—Plane Shear Testing," *Proc. 1989 Spring Conf. of the Society for Experimental Mechanics*, pp. 337—342 (May 1989).

[7] D. Post, R. Czarnek, D. Joh, J. Jo and Y. Guo, "Deformation of a Metal—Matrix Tensile Coupon with a Central Slot: An Experimental Study," *Jl. of Composites Technology and Research (ASTM)*, Vol. 9, No. 1, pp. 3—9 (Spring 1987).

[8] D. Post, Y. Guo, and R. Czarnek, "Deformation Analysis of Boron/Aluminum Specimens by Moire Interferometry," *Metal Matrix Composites: Testing, Analysis and Failure Modes*, ASTM STP 1032, W. S. Johnson, Editor, American Society for Testing and Materials, Philadelphia, 1989.

[9] R. Czarnek and D. Post, "Moire Interferometry with ± deg Gratings," *Experimental Mechanics*, 24(1), 68—74 (March 1984).

[10] D. Post and J. D. Wood, "Determination of Thermal Strains by Moire Interferometry," *Experimental Mechanics*, 29(3), pp. 318—322 (Sept. 1989).

[11] D. Post, J. Morton, J. D. Wood, M. Y. Tsai, V. J. Parks and F. P. Gerstle, Jr., "Thermal Stresses in a Bimaterial Joint," ASME *J. Applied Mechanics* (submitted for publication).

A DAMAGE APPROACH FOR COMPOSITE STRUCTURES :
THEORY AND IDENTIFICATION

Pierre LADEVEZE

Laboratoire de Mécanique et Technologie
ENS Cachan/CNRS/Université Paris 6
61 avenue du Président Wilson 94235 CACHAN CEDEX - FRANCE

ABSTRACT

Damage generally refers to the more or less gradual development of microdefects. A general modelling of these phenomena is proposed for composites structures. It is a kind of behavior model which, when included in a structural analysis code, allows to simulate the damage state at any point of the structure until the fracture of the structure.

The purpose of the present study is to try to focus on the main features of this general Damage Mechanics approach for composites. Identification problems are also discussed.

INTRODUCTION

Damage could well be the main mechanical phenomenon in composite structures. For such materials, damage is of a highly complex nature. There are not a single but several damage mechanisms. The mechanisms are highly anisotropic and present a strong unilateral feature depending on whether the microdefects are closed or open. To precise the aim of a damage mechanical approach, let us consider the tensile test :

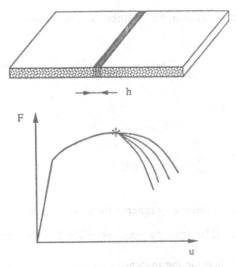

figure 1 : The tensile test

The fracture phenomenon happens after two phases. In a first step, the micro-voids and micro-cracks growth is nearly uniform locally : it is the initiation stage. From the critical point (or from a point just beside it) the strain and also the damages become more and more localized ; a macro-crack appears and grows until becoming unstable. The purpose of Damage Mechanics and in particular, of our approach devoted to Composite Structures, is to get a modelling of these phenomena. It is a kind of constitutive relation which, when included in a structural analysis code, allows to simulate and predict the damage state at any point of the structure until the complete fracture.

The objective of the paper is to try focus on the general tools and concepts and then to show the theory itself. We will also insist on the experimental aspects particularly on the identification of the material curves and constants. Two different materials are detailed, the three-dimensionnal composites (ex : 3D Carbon - Carbon) and the laminate composites (ex : T 300 - 914). The last part deals with the numerical simulation of such a model, i.e. the computation of damages. The challenge is to build up a true rupture theory which includes, as a simplified approach, the classical linear fracture mechanics. This study has been supported for 10 years by works done in joint collaboration with Aerospatiale - Les Mureaux and more recently with CNES and BERTIN.

A DAMAGE THEORY : TOOLS AND CONCEPTS

The scale

Before deriving damage models, it is essential to specify the chosen scale. For composite materials, between the macro and micro scales, there exists an intermediate and preferential scale where damage phenomena can be described in a rather simple way. It is named the meso scale. Let us consider first the example of 3D composites as Carbon/Carbon or Ceramic/Ceramic.

These materials (figure 2) can be defined at the meso-scale by the following constituents :

- fibre-yarns

- matrix-blocks

- interfaces

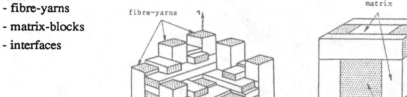

figure 2 : Geometry scheme for a 3D material

At the level we called meso-level, laminate composites as T300 - 914 may be described by :

- a homogeneous single layer in the thickness

- an interface i.e.a mechanical surface connecting two adjacent layers which depends on the relative direction of their fibers

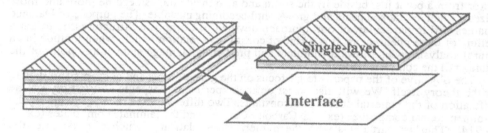

Figure 3 : Laminate modelling

Damage concept

The ideas we are following have been introduced by Kachanov and Rabotnov [14], [28] for metallic materials :

- First, we stay in the Continuum Mechanics framework by adding a new internal variable to describe the damage state, to the classical internal variables as plastic strain, hardening variable,....

- We take as a damage indicator the relative variation of Young's modulus.

This simple scheme which is the classical theory of isotropic damage has been more particulary developed in [24] [5] [8] [25] [27] for metallic materials and in [26] for concrete. For composites, it is far from being sufficient : the damage state is of a highly complex nature. Then, we have to consider the problem again and several proposals can be found in [1] and [32].

The central problem is for us to build up reasonable "kinematics" of damage, i.e. kinematics with a minimum of damage variables. The general theory [16] is the tool we have proposed [18] [19].

It has been shown that the damage state at a given scale is completely defined through two damage functions :

$$d(\vec{n}) = \frac{E_o(\vec{n}) - E(\vec{n})}{E_o(\vec{n})} \qquad \delta(\vec{n}) = \frac{\gamma_o(\vec{n}) - \gamma(\vec{n})}{\gamma_o(\vec{n})}$$

where E, γ denote Young's modulus and the volumic modulus of the damaged material. E_o, γ_o are initial values and $\vec{n}$ an unitary vector which gives the space direction. Futhermore, d and δ are represented in the three-dimensionnal space by two surfaces, S_d, S_δ called damage surfaces :

$$\vec{OM} = \vec{n}\, d(\vec{n}) \qquad \vec{OM} = \vec{n}\, \delta(\vec{n})$$

To derive damage kinematics is then equivalent to approximate the damage surfaces. If , for example, the damage surfaces may be approximated by means of two spheres, one obtains, for the damage state, a two variable description which constitutes the real isotropic damage theory. More generally, the d and δ functions may be expanded into Fourier series. Thus, depending on the choice of the truncation, we obtain different possible damage kinematics.

<u>Modelling of the Damage Evolution. Coupling Damage-Plasticity (or viscoplasticity)</u>

Let us consider a damage kinematics defined by the set of scalar damage variables :

$$d \; ; \delta$$

The free energy $\rho\psi$ is a function of :

$$\rho\psi \, (\varepsilon_e, d, \delta, X)$$

where X denotes the hardening variables or any other variable. The conjugate quantities to d, δ, i.e. the quantities which govern the damage evolution and then the rupture are :

$$Y_d = -\rho \, \frac{\partial \psi}{\partial d} \, \sigma, X : cst$$

$$Y_\delta = -\rho \, \frac{\partial \psi}{\partial \delta} \, \sigma, X : cst$$

where $\tilde{\sigma}$ denotes the chosen effective stress. For many cases, we have also :

$$Y_d = -\rho \, \frac{\partial E_D}{\partial d} \, \sigma : cst \qquad\qquad (E_D : \text{strain energy})$$

$$Y_\delta = -\rho \, \frac{\partial E_D}{\partial \delta} \, \sigma : cst$$

The micro-defects lead to sliding with friction and then to anelastic strains. A way to describe these phenomena is to use plasticity or viscoplasticity mechanical modelling. The idea which seems to work quite well is to build the modelling upon quantities which are called "effective" :

- effective stress $\tilde{\sigma}$

- effective anelastic strain rate $\tilde{\dot{\varepsilon}}_p$

which verify : $Tr \, [\, \sigma \, \dot{\varepsilon}_p] = Tr \, [\tilde{\sigma} \, \tilde{\dot{\varepsilon}}_p]$

A particular choice is :

$$\tilde{\sigma} = K_0 \, K^{-1} \, \sigma \qquad ; \qquad \tilde{\dot{\varepsilon}}_p = K^{-1} \, K_0 \, \tilde{\dot{\varepsilon}} \qquad (K \, : \, \text{Hooke's}$$
tensor)

Remark : The micro-voids and micro-cracks may open or shut depending on the case. Various possible modellings are given in [16]. The idea is to express this unilateral character in terms of energy.

Homogenization - Local stress and strain distributions

It is clear that in a composite many levels may be distinguished. It is then interesting to transfer information from one scale to another. For example the proposed single layer model results for one part from a homogenization at a smaller scale, the scale of the fiber diameter.

Homogenization techniques have become classical, that is why we shall not emphasize them.

EXAMPLE 1 : 3D - COMPOSITES

Damage kinematics - Macro-modelling

We introduce only one scalar damage variable such that :

$$G_{12} = (1 - \underline{d}) \, G_{12}^0$$

$$G_{23} = (1 - \underline{d}) \, G_{23}^0$$
$$G_{31} = (1 - \underline{d}) \, G_{31}^0$$

The elastic energy is then :

$$E_D = \frac{1}{2} \frac{\sigma_{11}^2}{E_1^0} + \frac{\sigma_{22}^2}{E_2^0} + \frac{\sigma_{33}^2}{E_3^0} - (\frac{v_{12}^0}{E_1^0} + \frac{v_{21}^0}{E_1^0}) \sigma_{11} \sigma_{22} - (\frac{v_{23}^0}{E_2^0} + \frac{v_{32}^0}{E_3^0}) \sigma_{22} \sigma_{33}$$
$$- (\frac{v_{31}^0}{E_3^0} + \frac{v_{12}^0}{E_1^0}) \sigma_{33} \sigma_{11} + \frac{1}{(1-d)} + \left[\frac{\sigma_{12}^2}{G_{12}^0} + \frac{\sigma_{23}^2}{G_{23}^0} + \frac{\sigma_{31}^2}{G_{31}^0} \right]$$

This expression agrees with experimental results ($\delta(\vec{n}) = 0$). It can be also derived by homogenization techniques from a damage model at the meso scale [9].

Damage evolution for static loadings - Macro -modelling

The conjugate quantity to $\underline{d}$ is

$$Y = \frac{\partial E_D}{\partial \underline{d}} \Big|_{\sigma: cst} = \frac{1}{(1-\underline{d})E_\tau}$$

where E_τ is the shear energy. The quantity Y is smilar to the energy release rate introduced in Fracture Mechanics ; it governs the damage evolution and then the rupture. For 3D Carbon-Carbon we propose the following model :

$$\underline{d} = h [\underline{Y} + g(\underline{Y}, Z)] \text{ if } \underline{d} < 1 \text{ otherwise } \underline{d} = 1$$

$$Z = < \sigma_{11} + \sigma_{22} + \sigma_{33} >_+$$

$$\underline{Y_t} = \sup_{\tau \leq t} Y_\tau$$

h, g are functions depending on the material.

Damage and inelastic strains - Macro-modelling

The effective stress and strain are :

$$\tilde{\sigma}_{ii} = \sigma_{ii} \qquad\qquad \tilde{\sigma}_{ij} = \frac{\sigma_{ij}}{1 - \underline{d}} \qquad\qquad \text{for } i \neq j$$

$$\tilde{\dot{\varepsilon}}_{iip} = \dot{\varepsilon}_{iip} \qquad\qquad \tilde{\dot{\varepsilon}}_{ijp} = \dot{\varepsilon}_{ijp} (1 - \underline{d}) \qquad\qquad \text{for } i \neq j$$

One uses a plasticity model with isotropic hardening. The elasticity domain is :

$$(\tilde{\sigma}_{12}^2/G_{12}^0 + \tilde{\sigma}_{23}^2/G_{23}^0 + \tilde{\sigma}_{31}^2/G_{31}^0)^{1/2} - R(p) \leq 0$$

where $p \rightarrow R(p)$ is the hardening function.
To be complete, we have to add rupture criteria in the fiber directions : $-\varepsilon_c \leq \varepsilon_{ii} \leq \varepsilon_t$

Identification and check up of the Macro-modelling

The model depends on the following material curves and constants :
- damage evolution : two functions h, g
- inelastic strain : one hardening function R
- fracture criteria in the fiber direction - $\varepsilon_c^* < \varepsilon_{ii} < \varepsilon_T^*$

The functions h, g, R are identified from one tensile test and one compression test. The direction may be 45° in an orthropic plane. These have been carried out by Poss [31] and Remond [30].

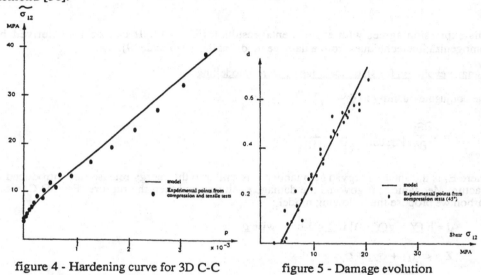

figure 4 - Hardening curve for 3D C-C   figure 5 - Damage evolution

The model has been checked out on various tests [9].

Meso-modelling

A damage macro-modelling has several limits :
- the failure prediction in the edge zone is quite poor (ex : fiber pull out)
- the model is not suitable for materials with variable characteristics of the reference volume
- only the initiation stage can be predicted. Classical continuum mechanics models are not sufficient to predict the complete fracture phenomenon.

Then refined models are necessary. A first modelling on the meso-scale, unfortunately a continuum mechanics model, has been derived in [9]. Recently , the concept of meso-modelling has been proposed [Ladeveze 21]. The material is not only described at the meso scale ; the effective meso damage state is locally taken as uniform within each constituent (blocks and interfaces). This model is not a continuum mechanics model. It allows the complete simulation of the fracture phenomenon.

The identification of such a modelling involves several serious difficulties because usual experimental macro-information are not sufficient. A general idea is to use macro-information for a set of materials which differ from known characteristics of the reference volume. To get the first modelling at the mesoscale that we derived in [9], bending and torsion tests were used on fibre yarn specimens which were taken out of the material by machining. Non-standard compression tests were performed on specimens for which one fibre-yarn direction was eliminated by machining. Figure 6 shows one of these tests.

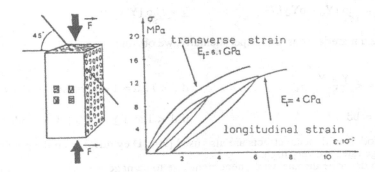

Figure 6 : Compression specimen - shear test

EXAMPLE 2 : MESOMODELLING OF LAMINATE COMPOSITES

Modelling of the single layer

It is to be noticed that we limit ourselves to single layers with only one reinforced direction (fiber direction : 1). The single layer is also analyzed at a smaller level, the level of its constituents : fibers, matrix, interfaces. Some more or less qualitative information are thus transfered at the single layer level by means of a homogenization process. A simulation with FE method has been recently carried out by Anquez [4].

The transverse rigidity in compression being supposed equal to E_2^0, we obtain the following energy for the damaged material :

$$E_D = \frac{1}{2} \left(\frac{\sigma_{11}^2}{E_1^0} - \left(\frac{v_{12}^2}{E_1^0} + \frac{v_{21}^2}{E_2^0} \right) \sigma_{11} \sigma_{22} + \frac{<-\sigma_{22}>_+^2}{E_2^0} + \frac{<\sigma_{22}>_+^2}{(1-d')E_2^0} + \frac{\sigma_{12}^2}{(1-d)G_{12}^0} \right)$$

where d and d' are two scalar damage variables which are constant within the thickness. The conjugate variables associated with dissipation are :

$$Y_d = \frac{[E_D]m}{\partial d} \Big|_{\tilde{\sigma}:cst} = \frac{1}{2} \frac{[\sigma_{12}^2]m}{G_{12}^0(1-d)^2}$$

$$Y_{d'} = \frac{[E_D]m}{\partial d}\bigg|_{\tilde{\sigma}:cst} = \frac{1}{2}\frac{[<\sigma_{22_+}^2]m}{G_{12}^0(1-d')^2}$$

where [ ]m denotes the mean value through the thickness. From experimental results there follow the governing quantities of damage evolution :

$$\underline{Y} = \sup_{\tau \leq t}[Y_d + bY_{d'}]^{1/2} \qquad ; \qquad \underline{Y} = \sup_{\tau \leq t}[Y'_d]^{1/2}$$

where b is a material constant. Experimentally, we obtain :

$$d = <\frac{\underline{Y} - Y_0}{Y_c}>_+ \qquad\qquad \text{if } d < 1 \text{ ; } d = 1 \text{ otherwise}$$

$$d' = bd \qquad\qquad \text{if } d' < 1 \text{ and } \underline{Y'} < Y'_c \text{ ; } d' = 1 \text{ otherwise}$$

Models with delay effects are also used [20]. They differ from the previous one only if the damage rates are very high.

To describe the anelastic phenomena due to damage we use a plasticity model. Details can be found in [20] [22].

Remarks : Near the edges it is necessary to take out-plane stresses into account. For the sake of simplicity, Young's modulus E_3 and the shear modulus G_{13} and G_{23} are taken constant and thus damage effects of out-plane stresses are assumed to affect interface behavior only.

Identification of the single layer's model (static loadings)

The material parameters are :

- for the damage modelling
  . progressive mode
      1 function (linear) : 2 constants
      1 coupling constant $\sigma_{22} <-> \sigma_{12}$

  . brittle mode
      1 critical value

- for the inelastic strains modelling
      1 hardening function
      1 coupling constant $\sigma_{22} <-> \sigma_{12}$

We also have the tensile and compression limit values for the strain in the fiber direction, which does not involve any difficulties. The main test we propose is $[\pm 45°]_{2s}$ tensile test. It allows the complete identification of the model excepted the coupling constants. For that ,a test such as $[\pm 65°]_{2s}$ is suitable. Results for T300-914 and IM6 - 914 are given in [20] [22]. Figure 7 shows the damage critical values Y_c, Y'_c respectively associated to the progressive and brittle modes for several materials.

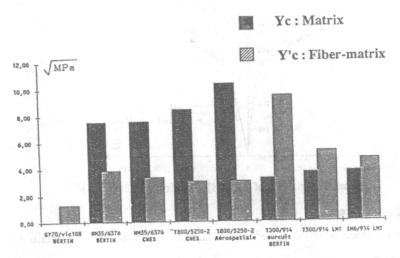

Figure 7 : Critical damage values related to the matrix and the fiber-matrix degradations
Le Dantec [Bertin 1990]

The model has been checked out on numerous experimental tests. Figures 8 and 9 give two comparisons.

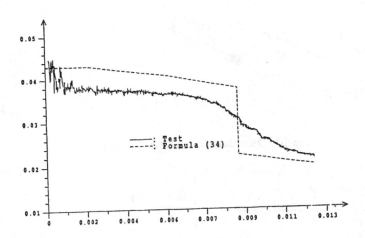

Figure 8 : Poisson's ration versus strain for [090]$_{2s}$ laminate
(IM6 914)

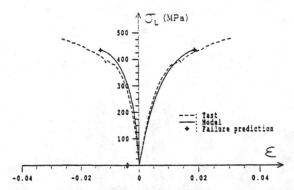

Figure 9 : Analysis of [-12,78]$_{2s}$ laminate (T300-914)

Modelling of the interface

For the interface which is a mechanical surface entity, a similar modelling is used [20] [18].

FRACTURE COMPUTATION

As an example of the method possibilities, let us consider the laminate structure defined by figure 10 where a tension loading is applied. The edge effects are the only ones to be computed taking the interface damage modelling into account.The results are given figure 11 et 12.

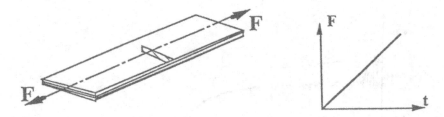

Figure 10 : Delamination analysis of a tension test

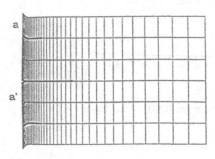

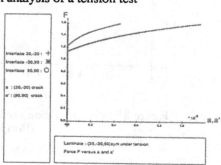

Figure 11 : Delamination pattern

Figure 12 : Delamination analysis

Then, there appears a crack first in the central interface and after a certain tension load, two new cracks in the interfaces (+ 30, - 30). All these cracks grow together until rupture. These first results have been obtained with a certain range of material parameters [10]. Comparisons with experimental results are currently under way. Other problems have been simulated in [2] [3] . It is to be noted that with our meso-modelling the usual mesh sensitivity difficulties are not present.

CONCLUSION

The proposed damage mechanics approach seems to be a powerful tool for the prediction of complex structure deteriorations. The use of a such damage mesomodelling avoids the main computational difficulties and involves a finer modelling of physical phenomena. Nevertheless, this kind of modelling needs a multiscale computational approach. To be complete, the sensitivity to imperfections i.e. to large defects has to be studied. Other further researches are of course necessary to achieve a true Fracture Theory for Composites.

REFERENCES

[1]   ALLEN D.H., HARRIS C.E., GROVES S.E., "A Thermomechanical constitutive theory for elastic composites with distributed damage", Int. J. of Solids and Structures, Vol. 23-9, 1987, p 1301 - 1338.

[2]   ALLIX O., LADEVEZE P., GILLETTA D., OHAYON R., "A Damage Prediction Method for Composite Structures, Int. Journal for Num. Meth. in Engineering, Vol 27, 1989, p 271-283.

[3]   ALLIX O., "Délaminage par la mécanique de l'Endommagement, "Calcul des structures" et Intelligence Artificielle", Vol. 1, Pluralis, Ed FOUET J.M., LADEVEZE P., OHAYON R., 1989.

[4]   ANQUEZ L., "Modélisatin du comportement d'un composite carbone-epoxy", to appear in "Calcul des Structures et Intelligences Artificielles", Pluralis Paris, Ed FOUET J.M., LADEVEZE P., OHAYON R., 1991

[5]   BAZANT Z.P., PIJAUDIER-CABOT G., Non local damage : Continuum Model and Localization Instability Report n°87-2/ Northwestern University, Evanston, also in J. of Appl. Mech., ASM, 1987

[6]   BELYTSCHKO T., LASRY D., Localization limiters and numerical strategies for strain-softening materials in Cracking and Damage, Edited by MAZARS J. and BAZANT J.P., Elsevier, 1988, p 349-362.

[7]   CLUZEL C., POSS M., REMOND Y., "Prévision du comportement Mécanique d'un Composite 3D en fonction de la géométrie de sa maille élémentaire", Colloque Matériaux Composites pour Applications à Hautes Températures, Bordeaux, 29-30 Mars 1990.

[8] CORDEBOIS J.P., SIDOROFF F., "Endommagement anisotrope en élasticité et plasticité", Journal de Mécanique Théorique et Appliquée, numéro spécial, 1982, p 45-60.

[9] DUMONT J.P., LADEVEZE P., POSS M. and REMOND Y., "Damage mechanics for 3D composites", Composite Structure 8, 1987, p 119-141.

[10] DAUDEVILLE L., Thèse (à paraître), 1991

[11] GILLETTA D., GIRARD H., LADEVEZE P., "Composites 2D à fibres à haute résistance : modélisation mécanique de la couche élémentaire", JNC 5, Pluralis, Paris, 1986, p 685-697.

[12] GILORMINI P., LICHT C., SUQUET P., "Growth of voids in a ductile matrix : a review", Arch. Mech., 40, 1988, p. 43-80.

[13] HASHIN Z., "Failure criteria for unidirectional fiber composites", J. of Appl. Mechanics, Vol. 47., 1980.

[14] KACHANOV L.M., "Time of the rupture process under creep conditions", Izv Akad Nauk S.S.R. otd Tech Nauk, 8, 1958, pp 26-31.

[15] KIM R.Y., SONI S.R., "Experimental and analytical studies on the onset of delamination in laminated composites, Journal of Composite Materials, 18, 1984, p 70-80.

[16] LADEVEZE P., "Sur une théorie de l'endommagement anisotrope", Rapport Interne n°34, Laboratoire de Mécanique et Technologie, Cachan, 1983.

[17] LADEVEZE P., Sur la Mécanique de l'Endommagement des Composites, JN5, Pluralis, Paris, 1986, p 667-683

[18] ALLIX O., LADEVEZE P., LE DANTEC E., VITTECOQ E., "Damage Mechanics for Composites Laminates under/Complex Loading", in Yielding, Damage and Dailure of Anisotropic Solids Ed BOEHLER J.P., MEP London, 1990, p 551-571.

[19] LADEVEZE P., "About a Damage Mechanics Approach", Proceeding Int. Conf. on Mechanics and Mechanisms of Damage in Composites and Multimaterials - MECAMAT - St Etienne, 1989.

[20] LADEVEZE P., ALLIX O., DAUDEVILLE L., "Mesomodelling of Damage for laminate composites", Proceeding IUTAM Symposium on Inelastic Deformation of Composites Materials, Troy, June 1990.

[21] LADEVEZE P., "A Damage Computatinal Method for Composite Structures", Proceeding of the Dutch "National Mechanica Congress" 2-4 April 1990 Rolduc Kerkrade, Dijkmans J.F. and Nieuwstadt F.T.M.(eds), Integration of Theory and Applications in Applied Mechanics, 13-24. 1990 Kluwer Academic Publishers. Printed in Netherlands.

[22] LADEVEZE P., LE DANTEC E., "Damage modelling of the elementary ply for laminated composites", to appear in Composites Science and Technology, 1991.

[23] LE DANTEC E., "Contribution à la Modélisation du Comportement Mécanique des Composites Stratifiés", Thèse de l'Université Paris 6 - Cachan, 1989.

[24]   LEMAITRE J., "How to use Damage Mechanics",Nuclear Engineering and Design, 80, 1984, p 233-245

[25]   LECKIE F.A., ONAT E.T., "Tensorial nature of damage measuring internal variable", Proceedings IUTAM Symposium Physical Non-Linearities in Structural Analysis, Springer-Verlag, 1980.

[26]   MAZARS J., "Application de la mécanique de l'endommagement au comportement non-linéaire et à la rupture du béton de structure", Thèse d'Etat de l'Université Paris 6, 1984.

[27]   MURAKAMI S., "Notion of continuum damage mechanics and its application to anisotrope creep damage theory", J. Engng Mechanics ASCE, in press, 1986.

[28]   RABOTNOV Y.N., "Creep rupture", Proc. XII, Int. Cong. Appl., Mech., Stanford-Springer, 1968.

[29]   REIFFNIDER K., "Stiffness reduction mechanism in composite materials",ASTM-STP 775, novembre 1980.

[30]   REMOND Y., "Sur la reconstitution du comportement mécanique d'un matériau composite 3D à partir de ses constituants", Thèse de 3e cycle de l'Université Paris 6 / ENS Cachan, 1984.

[31]   POSS M., "Endommagement et rupture des matériaux composites Carbone-Carbone", Thèse de 3e cycle de l'Université Paris 6 / ENS Cachan, 1982.
[32]   TALREJA R., "Transverse cracking and stiffness reduction in composite laminates", Journal of composite materials, Vol. 19, July 1985.

IDENTIFICATION OF TEMPERATURE DEPENDENCE
FOR ORTHOTROPIC MATERIAL MODULI

PER S. FREDERIKSEN
Department of Solid Mechanics
The Technical University of Denmark, Lyngby, Denmark

ABSTRACT

The present paper deals with the identification of orthotropic material moduli by a combined numerical and experimental approach. Though static material properties are our goal, the identification is based on eigenfrequencies for a free rectangular plate, because excellent agreement between measured and calculated eigenfrequencies can be obtained.

The numerical identification problem can be formulated as an optimization problem where an error functional is minimized. This error functional expresses the difference between model analysis response and experimental response.

The method is fast, non-destructive and requires only one experiment to obtain all four moduli. In addition, the method is suitable for the study of material moduli in different environments since the test plate is removed from the test equipment.

Examples for the temperature dependence of the material moduli for plates made from aluminium and glass/epoxy are presented.

INTRODUCTION

Traditional laboratory tests are designed to give the desired quantities in the most direct way. As an example, determination of material parameters of constitutive laws leads to special design of test samples and choice of applied load in an attempt to obtain homogeneous stress-strain fields. These idealized tests are not easy to perform however, and especially for composite systems, problems often arise. Furthermore, these tests have a local nature at the point of the strain gauge, for example, and thus more general material information requires series of tests.

The identification approach is an experimental strategy from an opposite point of view. The experiment is chosen to be as simple as possible to give reliable results. On the other hand this often leads to complex non-homogeneous stress-strain fields where the desired quantities are not among the directly measured quantities. This causes a complicated interpretation, but nowadays this is comfortably taken care of by computer calculations.

Though our aim is to find static material data, we decide to measure eigenfrequencies of a free rectangular plate, because excellent agreement between measured and calculated eigenfrequencies can be obtained. A structural eigenfrequency is an integrated quantity and we thus obtain material quantities that are valid in the mean for the entire structure.

The interpretation of the measurement (i.e. the numerical identification problem) can be formulated as an optimization problem where we minimize an error functional that expresses the difference between model analysis response and experimental response. The optimization is an iterative procedure which is based on analytical sensitivity analysis.

Eigenfrequency measurements are fast and simple to perform using modern equipment and there is nothing to prevent these measurements from being performed under different conditions. This means that the present identification method is suitable for the study of the moduli dependence due to, say, moisture or temperature.

The intention of the paper is to focus on the general technique as well as the practical application of the method illustrated by examples of an aluminium plate and a glass/epoxy plate.

The present work is based on the pilot project by Markworth & Petersen [1] and related work can be found in the doctoral thesis by Sol [2].

ORTHOTROPIC MATERIAL PARAMETERS

The quantities of interest are the engineering parameters E_L, E_T, G_{LT} and V_{LT}, i.e. Young's modulus in the fiber direction, in the transverse direction, the shear modulus and the Poisson's ratio.
With reference to Pedersen [3] we define the following practical parameters based on these four moduli :

$$\alpha_2 = 4 - 4(E_T/E_L)$$
$$\alpha_3 = 1 + (E_T/E_L)(1 - 2V_{LT}) - 4(G_{LT}/E_L)\alpha_0$$
$$\alpha_4 = 1 + (E_T/E_L)(1 + 6V_{LT}) - 4(G_{LT}/E_L)\alpha_0 \tag{1}$$
$$\alpha_0 = 1 - V_{LT}^2\, E_T/E_L$$

Corresponding parameters in dimensional form can be found in Vinson & Sierakowski [4]. The non–dimensional form (1) is chosen because our main interest is to identify the relative quantities E_T/E_L, G_{LT}/E_L and V_{LT}.

The identification is performed on $(\alpha_2, \alpha_3, \alpha_4)$ and the inverse relations of (1) then gives

$$E_T/E_L = (4 - \alpha_2)/4$$
$$V_{LT} = (\alpha_4 - \alpha_3)/(8 - 2\alpha_2)$$
$$\alpha_0 = 1 - V_{LT}^2 \, E_T/E_L \tag{2}$$
$$G_{LT}/E_L = (8 - \alpha_2 - 3\alpha_3 - \alpha_4)/(16\alpha_0)$$

The constitutive equation for a single ply of the laminated plate relative to a fixed coordinate system may be written as

$$\begin{Bmatrix} \sigma_{11} \\ \sigma_{22} \\ \sigma_{12} \end{Bmatrix} = \frac{E_L}{8\alpha_0} [Q] \begin{Bmatrix} \varepsilon_{11} \\ \varepsilon_{22} \\ \varepsilon_{12} \end{Bmatrix} \tag{3}$$

The matrix $[Q]$ only depends on the non–dimensional parameter α_m and the ply angle θ.

The laminate analysis is not covered here but can be found in textbooks eg. Vinson & Sierakowski [4].

IDENTIFICATION PROBLEM

The identification technique is based on the assumption that for a sufficient number of eigenfrequencies only one set of realistic material parameters α_m for the plate in question will provide agreement between the eigenfrequencies obtained through the numerical model and by the experimental measurements.

For the free plate considered numerical discretization is necessary for the model analysis. The eigenvalues are then obtained by solving a matrix eigenvalue problem

$$[S]\{\Delta_i\} = \lambda_i[M]\{\Delta_i\} \tag{4}$$

where the eigenvalue λ_i is related to eigenfrequency ω_i by $\lambda_i = \omega_i^2$.

In the following, I is the number of eigenvalues involved in the identification. Typical values for I are 7 - 10.

Now let the experimentally obtained eigenvalues be designated

$$\bar{\lambda}_1, \ \bar{\lambda}_2, \ \bar{\lambda}_3, \ \ldots, \ \bar{\lambda}_I \tag{5}$$

The numerical calculated eigenvalues for a given plate model with assumed material data α_m are designated

$$C\lambda_1 \, , \, C\lambda_2 \, , \, C\lambda_3 \, , \, \ldots \, , \, C\lambda_I \tag{6}$$

where the scaling factor C is defined by

$$C = \frac{1}{I} \sum_{i=1}^{I} \bar{\lambda}_i/\lambda_i \tag{7}$$

whereby the influence of quantities which just scale the eigenvalue spectrum is eliminated.

According to (5)–(7) we can define the error functional Φ. We choose the often–used square form

$$\Phi = \frac{1}{I} \sum_{i=1}^{I} (\bar{\lambda}_i - C\lambda_i)^2 / \bar{\lambda}_i^2 \tag{8}$$

The identification problem can now be formulated as an optimization problem, i.e. the identification of the set of material parameters α_m that minimizes the error functional Φ :

$$\text{Minimize } \Phi(\alpha_2, \alpha_3, \alpha_4) \geq 0 \tag{9}$$

From the optimal solution α_m the relative material moduli are evaluated through the relation (2). E_L and α_0 contribute to the scaling factor of (3); thus E_L can be found since C and α_0 are known.

Sensitivity Analysis

The optimization problem is solved by a first order gradient method, where we need the error functional sensitivities

$$\frac{\partial \Phi}{\partial \alpha_m} = -2 \sum_{i=1}^{I} \frac{1}{\bar{\lambda}_i^2} (\bar{\lambda}_i - C\lambda_i) \frac{\partial (C\lambda_i)}{\partial \alpha_m} \qquad \text{for } m=2,3,4 \tag{10}$$

This requires the knowledge of the sensitivities for the scaled eigenvalues:

$$\frac{\partial(C\lambda_i)}{\partial\alpha_m} = C\frac{\partial\lambda_i}{\partial\alpha_m} + \lambda_i\frac{\partial C}{\partial\alpha_m} \qquad \text{for } i=1,2,...,I \text{ and } m=2,3,4 \tag{11}$$

From the definition (7) we obtain the scaling factor sensitivity

$$\frac{\partial C}{\partial\alpha_m} = -\frac{1}{I}\sum_{i=1}^{I}\frac{\bar{\lambda}_i}{\lambda_i^2}\frac{\partial\lambda_i}{\partial\alpha_m} \tag{12}$$

The eigenvalue sensitivities $\frac{\partial\lambda_i}{\partial\alpha_m}$ are calculated using the numerical model. Differentiating (4), taking into account that the mass matrix is independent of the material parameters, gives the known result

$$\frac{\partial\lambda_i}{\partial\alpha_m} = \frac{\{\Delta_i\}^T [S]_{,\alpha_m} \{\Delta_i\}}{\{\Delta_i\}^T [M] \{\Delta_i\}} \tag{13}$$

Gradients of the stiffness matrix are determined by

$$[S]_{,\alpha_m} = \sum_{ij} [S]_{,d_{ij}} d_{ij,\alpha_m} \qquad \text{for } ij=11,22,12,33 \text{ and } m=2,3,4 \tag{14}$$

where the sensitivities of the bending stiffnesses d_{ij} are obtained from laminate analysis.

EXPERIMENTAL SETUP

Figure 1 shows the integrated setup. The plate is hung on two rubber bands to simulate the free boundaries condition. The rubber bands ensure simple rigid body motions with frequencies far below the first plate eigenfrequency.

The signal is amplified and subjected to Fast Fourier Transformation (FFT), using a high performance Brüel & Kjær frequency analyzer. The result is the discrete frequency response function. This data set is transferred to the personal computer and curvefitted to give precise values for the plate eigenfrequencies. These frequencies, along with the plate geometry, mass and stacking sequence for laminated plates, are the basis for the identification program running on the PC.

The possibility of isolating the test plate in different environments is obvious since the only connection between the test plate and the test equipment are the two small electrical wires. As indicated by the dashed box in figure 1 the test plate was isolated in a heater and identification

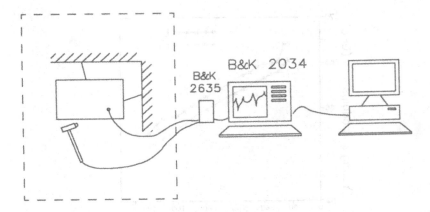

Figure 1. Scheme of the experimental setup

was performed at different temperatures. The plate excitation was remotely controlled simply by dragging and releasing a nylon wire connected to the hammer through a small hole in the heater wall. The hammer itself was fixed to a blade spring and a rubber damper was mounted to prevent double hints.
Valuable information regarding practical modal testing can be found in Ewins [5].

RESULTS

Results are shown for a rolled aluminium plate and a glass/epoxy plate. The temperature was increased from room temperature (approximately 20°C) in step of 5°C. In each step the temperature was raised over a period of 5 minutes and held in 25 minutes to stabilize the conditions. At the end of each 30 minutes period plate excitation and identification was performed.
On the curves shown in figure 2 and figure 3 identification points are connected with strait lines.

Example with Rolled Aluminium Plate

Figure 2 shows the temperature dependence for the four moduli. Ending temperature was 120°C. At all temperatures the relative differences between measured and calculated frequencies were within ±0.4% after the identification.

Figure 2 upper shows the weak anisotropic behavior with a ratio E_L/E_T between 1.01 and 1.02 over the hole temperature range. The two Young's moduli and the shear modulus have a fairly linear decrease with temperature. The relative decrease over the range of 100°C is 5.6% for E_L and E_T 5.8% for G_{LT}. The Poisson's ratio is rather insensitive to temperature.

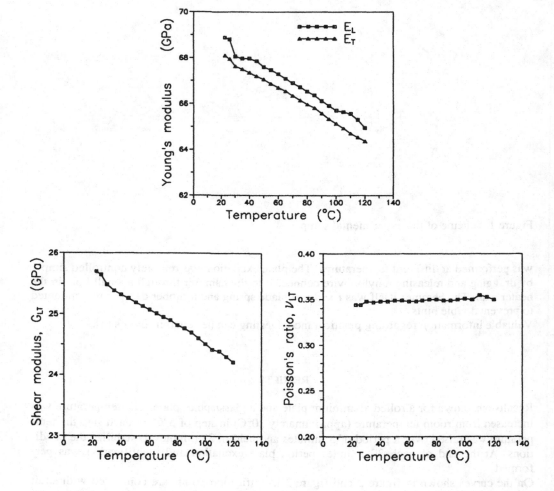

Figure 2. Temperature dependence for Young's modulus in main direction E_L, in transverse direction E_T, in-plane shear modulus G_{LT} and Poisson's ratio ν_{LT}.

Rolled aluminium plate.

Example with glass/epoxy plate

The second material was glass/epoxy with moderate anisotropy. The ratio $E_L/E_T = 2.8$ at room temperature. The ending temperature was 100°C. For this plate agreements within ±0.5% between measured and calculated frequencies were obtained at all temperatures.

Young's modulus E_L is dominated by the glass fiber and has low sensitivity to temperature.

The relative decrease over the range of 80°C is only 1.6% compared to 22% for E_T and 27% for G_{LT}. For E_T and G_{LT} the decrease with temperature is faster as the temperature approaches the transient temperature which for the epoxy used is 120°C.
The Poisson's ratio is the most difficult parameter to identify but nevertheless we get a clear indication of a strong increase with temperature.

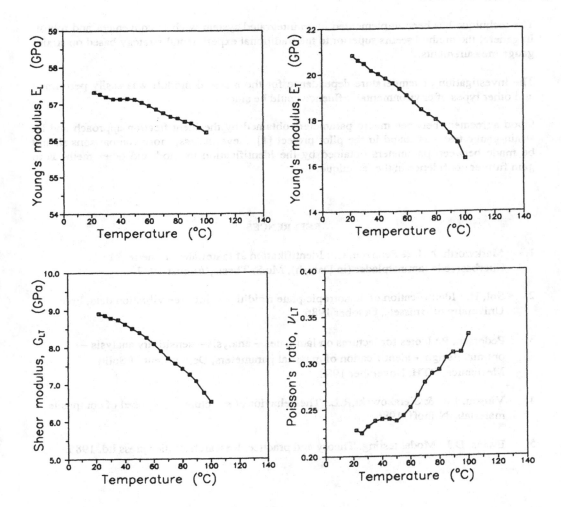

Figure 3. Temperature dependence for Young's modulus in main direction E_L, in transverse direction E_T, in-plane shear modulus G_{LT} and Poisson's ratio ν_{LT}.
Glass/epoxy plate.

CONCLUSION

The identification technique is a fascinating approach involving very close cooperation between experimental work and analytical/numerical work.

For the present method, analytical sensitivity analysis is found to be very important in the numerical procedure and extensive results can be obtained even though we are dealing with composite structures.

The technique has been implemented in an integrated system wich is convenient and robust. In general the method seems superior to the traditional experimental strategy based on strain-gauge measurements.

The investigation of temperature dependence for the material moduli was easily performed and other types of environmental influence could be studied.

Good agreement between macro parameters obtained by the identification approach and by strain-gauge test was found in the pilot project [1]. Nevertheless, more comparisons should be made between parameters obtained by the identification method and other methods to gain further confidence in the technique.

REFERENCES

1. Markworth, N.J. & Petersen, C. : Identifikation af materialeparametre for fiberbaserede laminatplader (in Danish), Mc.S. Thesis, August 1987.

2. Sol, H. : Identification of anisotropic plate rigidities using free vibration data, Free University of Brussels, October 1986.

3. Pedersen, P : Notes for lectures on laminates – analysis – sensitivity analysis – optimal design – identification of material parameters, Department of Solid Mechanics, DTH, November 1988.

4. Vinson, J.R. & Sierakowski,R.L.: The behavior of structures composed of composite materials, Nijhoff, 1986.

5. Ewins, D.J : Modal testing: Theory and practice, Research studies press ltd. 1985.

EXPERIMENTAL IDENTIFICATION OF COMPLEX STIFFNESSES OF COMPOSITE MATERIALS BY ULTRASONIC WAVE PROPAGATION

Yvon CHEVALIER
Rheology and structures laboratory
Institut Supérieur des Matériaux et de la Construction Mécanique (ISMCM)
3, Rue Fernand Hainaut
93407 Saint Ouen - Cedex (FRANCE)

ABSTRACT

Ultrasonic testing has become the method of choice for the determination of viscoelastic proper-ties of anisotropic media. This procedure works better than the usual way of testing materials because of classical difficulties (the need to have a lot of samples, off axis test,...) which are weeded out. The method is based on an analysis of the perturbation of a plane harmonic wave through a composite slab. The transit time (different wave velocities between the composite and a coupling medium) gives the storage moduli of the composite material, wave attenuation gives the loss moduli. The ultrasonic tests, required to completely characterize all nine complex moduli of orthotropic materials, are performed. The tests are conducted using contact and immersion trans-ducers at normal incidence to identify the dilatational and shear complex moduli (C_{ii} , i=1,...,6). To determine coupling complex moduli (C_{ij} , i≠j) we use immersion transducers with mode conversion to generate the required waves (oblique incidence and quasi-transverse waves). First and second invariant components of the complex stiffness matrix are helpful and necessary in determining in practice the coupling complex moduli of viscoelastic composites.

The method is rapid, accurate and sufficiently compact to be implemented on a laboratory so that it would be useful for viscoelastic testing purpose.

INTRODUCTION

Wave propagation in anisotropic media has been extensively studied and used in non destructive testing and has become an outstanding method for the quality assurance of advanced materials. However one rarely uses wave propagation to determine elastic or viscoelastic behavior of ani-sotropic material and to determine storage and loss moduli of materials. The fundamental pro-perties of wave propagation in composite materials have been analysed by Dieulesaint et Royer [1], Vinh [2], Chevalier [3], Auld [4], Kline [5], whereas Krautkramer et al. [6], Schreiber et al. [7], Hosten et al. [8], Bucur [9], Fideler [10], Veidt et al. [11] describe experimental procedures.

The common acceptance of an orthotropic model of composite viscoelastic behavior means that nine independant complex components of the stiffness matrix are required for the complete determination of the material : six diagonal terms which can be easily obtained from on-axis wave velocity measurements and three off-diagonal terms which involve off-axis bulk wave velocity. For many obliquely directions of wave propagation in a coordinate plane, the first and second invariant components of the complex stiffness tensor are computed and the best fit of the complex stiffness tensor is found by choosing the minimum of variation between the invar-iants and their mean values. A new stiffness tensor is calculated and the one starts again. The process stops when the complex stiffness tensor does not change.

THEORY

In material characterization ultrasonic techniques are introduced to obtain the complex stiffness terms of orthotropic media from velocity and attenuation measurements. Before discussing the optimization process, it is first useful to review quickly the concept of velocity and attenuation. A phase harmonic displacement wave $U(x, t)$, propagating in the direction ν ($\|\nu\| = 1$), is defined by

$$U(x, t) = AP \exp [i(\omega t - \tilde{k} .\nu x)] \quad , \quad (i^2 = - 1) \tag{1}$$

where $x = (x_i)$ are the spatial coordinates , A the magnitude of the wave , P the polarization vector ($\|P\| = 1$) and ω the circular frequency. The complex Wave number $\tilde{k}$ may be represented as

$$\tilde{k} = k - i\alpha \tag{2}$$

α is the attenuation of the wave and $v = \omega/k$ the phase velocity. Relationships between stiffnesses and ultrasonic velocity are obtained by replacing the displacements given by Eq.(1) in the equations of motion (see ref. [2] , [3]). This operation leads to equations

$$\{ \tilde{k}^2 \tilde{c}_{mnpq} (\omega) \nu_n \nu_q - \rho\omega^2\delta_{mp} \} P_p = 0 \quad , \quad m = 1, 2, 3 \tag{3}$$

where $\tilde{c} (\omega) = (\tilde{c}_{mnpq} (\omega))$ is the complex stiffness tensor of the composite material

$$\tilde{c}_{mnpq} (\omega) = c'_{mnpq} (\omega) + c''_{mnpq} (\omega)$$

$$\uparrow \qquad\qquad \uparrow$$

storage modulus   loss modulus

ρ the mass density and $\delta = (\delta_{pq})$ the kronecker tensor. Equation (3) may be rewritten as

$$[\tilde{\Gamma}_{pr} - \tilde{\lambda}_{pr}] P_r = 0 \quad , \quad p = 1, 2, 3 \tag{4}$$

which is known as **Christoffel equations** .

$$\tilde{\Gamma}_{pr} (\omega) = \tilde{c}_{pnrq} (\omega) \nu_n \nu_q \tag{5}$$

is the acoustical (or Christoffel) tensor whose eigenvalues $\tilde{\lambda}$ depend on phase velocity and on attenuation α :

$$\tilde{\lambda} = \frac{\rho v^2}{\left(1 - \dfrac{i\alpha v}{\omega}\right)^2} \quad , \quad i^2 = -1 \tag{6}$$

The Christoffel tensor is developed in various studies, (see [2], [3]). The general test principle is summerized on table 1.

The experimental procedure constits :

- First to note the output signal corresponding to an impluse excitation of the emission tranducer (a coupling medium whose characteristics v_0 and α_0 are known remains between the transducers).

- Secondly to note the output signal again when the composite material is introduced between the tranducers.

The transit time Δt give the velocity and the ratio between the amplitudes of the input and output signals leads the attenuation α in the composite material, see [3].

TABLE 1
Scheme for testing composite materials by ultrasonic method

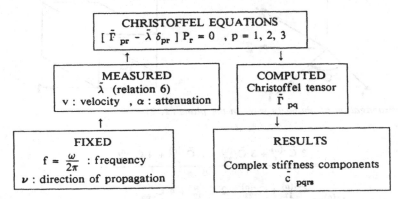

Diagonal terms : The diagonal complex moduli are obtained by measuring the bulk velocities of propagation waves along the symetry axes.

$$\tilde{C}_{qq}(f) = \tilde{\lambda}_{qq}(f) \quad , \quad q = 1, 2 ..., 6 \tag{7}$$

f is the frequency and the, 6x6 stiffness matrix $(\tilde{C}) = (\tilde{C}_{pq})$ is linked to the Stiffness tensor by the classical relations (see [1], [3], [9]). This procedure is not dicussed here as it is very simple. **Off – diagonal terms :** The numerical values of the components $\tilde{C}_{pq}$ of the complex stiffness matrix $(p \neq q)$ are obtained via the propagation of quasi-longitudinal and quasi-transverse waves (cf. [2], [3], [10], [11] by the relationship :

$$\tilde{C}_{pq} = \frac{1}{\nu_p \nu_q} \sqrt{\tilde{C}_{pp}\,\nu_p^2 - \tilde{G}_{pq}\,\nu_q^2 - \tilde{\lambda}_{pq}} \quad \sqrt{\tilde{C}_{qq}\,\nu_q^2 - \tilde{G}_{pq}\,\nu_p^2 - \tilde{\lambda}_{pq}} \quad - \tilde{G}_{pq}$$

$$p = 1, 2, 3 \quad , q = 1, 2, 3 \quad , p \neq q \tag{8}$$

when the propagation takes place at an angle from the elastic symetry axes.

$\tilde{G}_{pq}$ is the complex shear modulus in the (p, q) plane. The directional dependence of the off-diagonal complex stiffnesses $\tilde{C}_{pq}$ (Eq. 8) renders conventional averaging techniques inappropriate.

The approach consist of selecting the maximum $\tilde{\lambda}_{pq}$ values $(p = 1, 2, 3, \quad q = 1, 2, 3, \, p \neq q)$. For p and q fixed, $p = 1, \, q = 2$ for example cf. Fig 1) the off-diagonal components $\tilde{C}_{pq}^{(n)}$ are obtained from relation (8), $n = 1, 2,...,N_3$ denotes the number of measurements in obliquely direction α $(\alpha = \alpha_1, \alpha_2,..., \alpha_{N_3})$. Then for each value of α_n $(n = 1, 2,..., N_3)$ first and second scalar invariants $\tilde{I}_1^{(n)}, \tilde{I}_2^{(n)}, \tilde{J}_1^{(n)}, \tilde{J}_2^{(n)}$ are determined from the complex components $\tilde{C}_{pq}'^{(n)}$ expressd in the $(1', 2', 3')$ coordinate system (cf. Fig 1 & Eq. 9).

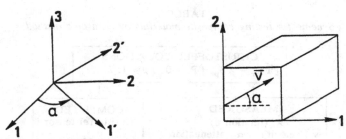

Figure 1 . *Propagation of obliquely wave in the plane (1, 2).*

$$\tilde{I}_1^{(n)} = \frac{3\,\tilde{C}\,'^{(n)}_{11} + 3\,\tilde{C}\,'^{(n)}_{22} + 2\,\tilde{C}\,'^{(n)}_{12} + 4\,\tilde{C}\,'^{(n)}_{66}}{8}$$

$$\tilde{I}_2^{(n)} = \frac{\tilde{C}\,'^{(n)}_{11} + \tilde{C}\,'^{(n)}_{22} - 2\,\tilde{C}\,'^{(n)}_{12} + 4\,\tilde{C}\,'^{(n)}_{66}}{8}$$

$$\tilde{J}_1^{(n)} = \frac{\sqrt{\left[\tilde{C}\,'^{(n)}_{11} - \tilde{C}\,'^{(n)}_{22}\right]^2 + 4\left[\tilde{C}\,'^{(n)}_{16} + \tilde{C}\,'^{(n)}_{26}\right]^2}}{2} \qquad (9)$$

$$\tilde{J}_2^{(n)} = \frac{\sqrt{\left[\tilde{C}\,'^{(n)}_{11} + \tilde{C}\,'^{(n)}_{22} - 2\,\tilde{C}\,'^{(n)}_{12} - 4\,\tilde{C}\,'^{(n)}_{66}\right]^2 + 16\left[\tilde{C}\,'^{(n)}_{26} - \tilde{C}\,'^{(n)}_{16}\right]^2}}{8}$$

The complex stiffness components can be calculated (Eq. 9) from the **mean values** $\bar{\tilde{I}}_1, \bar{\tilde{I}}_2,$ $\bar{\tilde{J}}_1, \bar{\tilde{J}}_2$ of the four invariants

$$\begin{aligned}
\tilde{C}_{11} &= \bar{\tilde{I}}_1 + \bar{\tilde{J}}_1 + \bar{\tilde{J}}_2 \\
\tilde{C}_{22} &= \bar{\tilde{I}}_1 - \bar{\tilde{J}}_1 + \bar{\tilde{J}}_2 \\
\tilde{C}_{12} &= \bar{\tilde{I}}_1 - 2\,\bar{\tilde{I}}_2 - \bar{\tilde{J}}_2 \\
\tilde{C}_{66} &= \bar{\tilde{I}}_2 - \bar{\tilde{J}}_2
\end{aligned} \qquad (10)$$

with these new values of stiffnesses one starts again to compute the invariants (Eq. 9) and the process stops when the R.M.S (Root mean square) between the invariants and their mean value is minimum. From this values of the complex stiffness components one begins a new optimization process with another values of p and q (p = 1, q = 3 then p = 2, q = 3, then p = 1, q = 2 etc). The process stops when the complex stiffness matrix does not change , ie. the variation must not exceed 2% (cf. Fig 2).

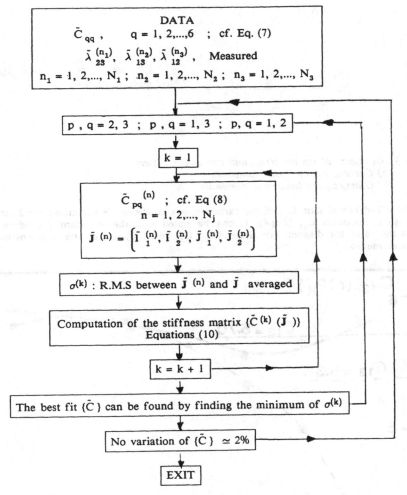

Figure 2 . *Optimization of the off-diagonal components of the complex stiffness matrix.*

RESULTS AND DISCUSSION

The set of experimental results is concerned with ensuring the validity of the method of optimization. Diagonal and off diagonal terms are found from a series of measurements on 2 composite materials : a carbon/Epoxy plain weave material and a glass/Epoxy laminated composite ($\pm$ 35°). Each constituant of these materials are supposed to be elastic (loss moduli of stiffness vanish) and the geometry is illustrated in figure 3.

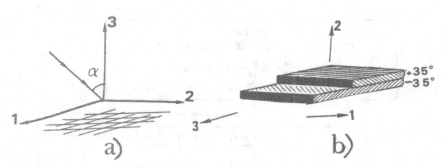

Figure 3. *Geometry of the materials and coordinate systems*
a) Carbon/Epoxy woven composite
b) Glass/Epoxy laminated composite

The off-diagonal term C_{13} of the carbon/epoxy composite is optimized for 2 starting values of the shear modulus C_{55} (Fig.4). It can be noted that the invariant procedure does note change the value the diagonal sriffnesses (C_{11}, C_{33} and C_{55}) so that the invariant method is not a sensitive method.

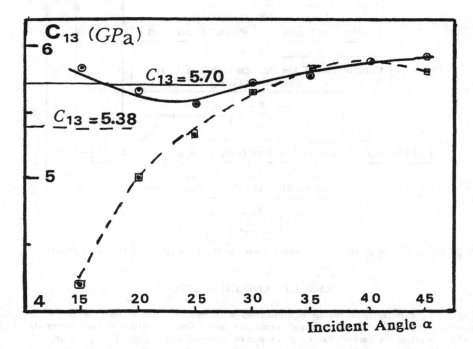

Figure 4. *Optimization of C_{13} for a carbon/Epoxy woven composite (C_{11}=18.55 GPa C_{33}=14.0 GPa).*

———— C_{55} = 4.80 GPa (1 iteration, variation of C_{13} : 0.8%)
— — — C_{55} = 4.50 GPa (10 iterations, variation of C_{13} : 4%)

The glass/Epoxy composite is used to build cylindrical containers for deep water and the specimen of material is not plane : it is a very difficult case. A rather good agreement can be observed between this method and other approach (Table 2) for the Young moduli E_1, E_2 and for the 3 shear moduli.

TABLE 2
Elastic moduli of an elastic glass/epoxy laminated composite ($\pm$ 35 ^{0})

GPa	Measurements	Optimization [14]	tensile test [14]	Predictions Finite Elem. [14]	Halpin.Tsai
E_1	31.73	29.56	27.00	26.59	
E_2	20.97	19.81	18.96	16.78	
E_3	21.53	22.48	17.41	15.21	
ν_{21}	0.121	0.267	0.139	0.086	0.35
ν_{32}	0.306	0.249	0.275	0.254	-
ν_{31}	0.297	0.275	0.240	0.344	-
G_{23}	7.32	7.32	-	5.88	-
G_{13}	11.60	11.63	-	13.4	12.90
G_{12}	8.14	7.49	-	5.83	-

CONCLUSION

The invariant technique is an effective method for optimizing the stiffness matrix of an orthotropic material, especially for dispersive data. This method allows to optimize, **Simultaneously** the storage and loss moduli of viscoelastic composite materials (viscoelastic data must be available for the composite material and also for the coupling medium).

The optimization process will be improved by testing many values of the shear moduli for one material (10% for the variation of G_{ij}). This process is rapid and sufficiently compact to be implemented in a laboratory.

REFERENCES

1. Dieulesaint, E. & Royer, D., **Ondes élastiques dans les solides**, Masson & Cie, Paris, 1973. (Elastic waves in solids).

2. Vinh, T., Mesures ultrasonores de constantes élastiques des matériaux composites, **Sciences et Techniques de l'Armement**, 54, 2ème fascicule, 1981, pp. 89-139 . (Ultrasonic measurement of elastic moduli of composite materials).

3. Chevalier, Y., Caracterisation dynamique des materiaux composites par propagation d'ondes ultrasonores. In **Caractérisation mécanique des composites - Extensiométrie appliquée aux composites**, Ed. A. Vautrin, Editions pluralis, Paris, 1990, pp.177-194 .(Testing composites materials by propagation of ultrasonic waves).

4. Auld, B.A., **Acoustic fields and waves in solids**, Vol I & II, John Wiley & sons, New York, 1973.

5. Kline, R.A., Wave propagation in fiber reinforced composites for oblique incidence, **J. Composite Materials**, 1988, 22, pp. 287-305.

6. Krautkramer, J. & Krautkramer, H., **Ultrasonic testing of materials**, Springer verlag, Hei-

delberg, 1977.

7. Schreiber, E. & Anderson, O., **Elastic constants and their measurement**, Mc Graw Hill Book Company, New York , 1973.

8. Hosten, B. & Deschamps, M., Mesure des coefficients élastiques complexes de matériaux anisotrope à l'aide d'un interféromètre ultrasonore assisté par ordinateur, **Cahiers du groupe français de rhéologie**, 1983, tome VI, **4**, pp. 145-154. (Determination of complex moduli of anisotropic materials by ultrasonic interferometer helped by a computer).

9. Bucur, V., **Ondes ultrasonores dans le bois - Caracterisation mécanique et qualité de certaines essences de bois**, thèse de docteur ingénieur, ISMCM, Saint Ouen, 1984. (Ultrasonic waves in wood - Mechanical Characterization and quality of various species)

10. Fideler, B.; **Caractérisation mécanique du bois, méthodes dynamiques de mesure des constantes élastiques**, mémoire de diplôme Ingénieur CNAM, Paris, 1983. (Mechanical characterization of wood, dynamical measurement to determine elastic constants).

11. Veidt M., Sayir M., Experimental evaluation of global composite laminate stiffnesses by structural wave propagation, **J.Composite Materials**, 1990, 24, pp.688-706

12. Bucur, V. & Rocaboy, F., Surface wave propagation in wood : prospective method for determination of wood off-diagonal terms of stiffness matrix, **Ultrasonic**, 1988,**28**, pp. 344-347

13. Yinh, T., Vibration des corps viscoélastiques. In **La rhéologie** ,Ed. B. Persoz , Masson, Paris 1969, pp. 89-139. (Vibrations of viscoelastic bodies).

14. Chauchot, P., Guillemin, O., Hassim, A., Léné, F., Caractérisation des propriétés mécaniques des matériaux composites pour enceintes sous marines, **Mécanique Matériaux Electricité - J. du GAMI**, 1990, 433, pp. 10-13.(Characterization of mechanical properties of composite materials for deep water containers).

15. Chauchot, P., Guillemin, O., Hassim, A., Léné, F., Caractérisation mécanique des structures bobinées par homogénéisation et techniques expérimentales. In **Caractérisation mécanique des composites - Extensiomètrie appliquée aux composites**, Ed. A. Vautrin, Editions Pluralis, Paris, 1990, pp. 129-141. (Mechanical characterization of Wound structures by homogenization and experimental methods).

A HYBRID METHOD TO DETERMINE
MATERIAL PARAMETERS OF COMPOSITES

M. HENDRIKS, C. OOMENS, J. JANSSEN
Department of Engineering Fundamentals
Eindhoven University of Technology
P/O Box 513, 5600 MB Eindhoven, The Netherlands

ABSTRACT

This paper presents a method to determine parameters in constitutive equations, used for the description of the mechanical behaviour of composites. The method is based on: numerical analysis, strain distribution measurements and systems identification techniques. The method is especially suitable to study the behaviour of inhomogeneous materials. By means of a simulation it will be shown that for a solid with a varying fibre direction it is possible to estimate the stiffness parameters as well as the local fibre directions from one single test.

INTRODUCTION

With traditional experimental methods to study the mechanical behaviour of composites it is common practice to manufacture samples which have to be representative for the mechanical properties and can be loaded in a way that a homogenous stress and strain distribution occurs in part of the sample. Examples are bars in tension and circular rods in torsion. The homogeneous stress and strain distribution is essential in these tests for analytical as well as technical reasons. The procedure has a number of disadvantages: homogeneous strains in an experimental set—up cannot be obtained if properties are inhomogeneous, the internal structure is disrupted when test samples are manufactured and many experiments are necessary to measure all parameters when material models are complex.

In the present paper an alternative approach is proposed, which is based on the idea, that an inhomogeneous strain distribution, obtained by loading objects of arbitrary shape, may contain enough information to uniquely determine all material parameters from one experiment, provided that the strain distribution can be measured and the numerical analysis as well as the identification can be performed in a reasonable amount of time.

At some point in the process estimations for the parameters $\underline{x}_{k-1}$ are available based on the first $k-1$ observations. A finite element calculation leads to model observations $h(\underline{x}_{k-1})$, which can be compared to the newly added observation y_k. The difference is used to improve the estimation and find $\underline{x}_k$. In Hendriks [1] the procedure to derive a recursive scheme is given. The scheme can be obtained by minimizing the following quadratic expression:

$$S_k = [y_k - h(x)]^T R^{-1} [y_k - h(x)] + [\underline{x}_{k-1} - x]^T \underline{P}_{k-1} [\underline{x}_{k-1} - x]$$

$$(2)$$

R represents the covariance matrix of the observation error in $\underline{x}_k$. $\underline{P}_{k-1}$ represents the covariance matrix of the estimation error in $\underline{x}_{k-1}$. (For more details, see [2]).

Hendriks et al. [1] have applied the above method to determine the mechanical properties of an orthotropic, linear-elastic, homogeneous textile. From one single experiment two Youngs moduli, one Poisson ratio, one shear modulus and the direction of material symmetry were estimated. In the next section the procedure will be applied to an inhomogeneous material. In this case the experiment is replaced by a computer simulation with known parameters. It was possible to simulate measurement errors by disturbing the data from the simulation. Three different simulatinos were used. A simulation were no noise was generated and simulations with normal divided noise with standard deviations σ of 0.001 and 0.01 respectively, while the average displacements were 0.1.

IDENTIFICATION OF INHOMOGENEOUS MATERIALS

This section will focus on an orthotropic sheet of which the material symmetry varies with position. The "experiment" will be replaced by a numerical simulation with known parameters. Two approaches will be distinguished. In the first approach the inhomogeneous properties of the sample are modeled with help of a continuous function over the sample domain. The parameters in this function will be identified together with the stiffness parameters. We will call this the *global* approach. In the second approach, which we call *local*, in each analysis a small part of the sample is modeled and confronted with "experimental" data. The properties of each part are assumed to be homogeneous. The approach has advantages with regard to boundary conditions and sample geometry.

Figure 2 shows the loaded finite element model of the sample used for the artificial generation of measured data. The membrane is symmetrically loaded with two equal forces working in the plane of the sample. It will be clear that the deformation is not symmetrical, which is caused by the variing fibre direction. The material is considered to be orthotropic elastic (fig.3). The main fibre directions are along concentric circles with its centroid in the point with coordinates $\xi = 3$, $\eta = 3$. The nondimensional material parameters are: $E_1 = 1$, $E_2 = 0.2$, $\nu_{12} = 0.3$, $G_{12} = 0.2$.

In the next section the method is presented. After that a simulation on a orthotropic, elastic sheet will be discussed. Because of a fibre direction that changes with position, the properties are inhomogeneous. The stiffness parameters as well as the local fibre direction will be estimated. Two approaches will be used. The first approach will be called *global* and uses a functional relationship between the fibre directions and the material coordinates and estimates the fibre directions in the whole sheet at once. The second *local* approach uses no a priori assumptions concerning the fibre directions but estimates the fibre directions in only a small part of the sheet.

METHOD

The method is based on imposing inhomogeneous stress and strain distributions in the specimens that are studied. The strain distribution on the surface can be measured. This leads to columns y of observational data; strains at a large number of places and at different times. Assuming that a constitutive law for the material is available a column of material parameters x can be defined (elastic moduli, poisson ratio's, time constants, but also fibre directions etc.). It is also assumed that some algoritm is available to calculate y when x is known. This algoritm, usually based on the Finite Element Method is symbolized by a functional $h(x)$. The observation model then becomes:

$$y = h(x) + v \tag{1}$$

, with v an observation error.

The problem is, to find x when y is known. Because of the usually very large number of observations (strain distributions at different times) a sequential identification method is used. (fig. 1). This means that not all observational data are used at once. The set of data is divided in parts. In each iteration step new observational data are used to improve earlier estimations of the parameters x.

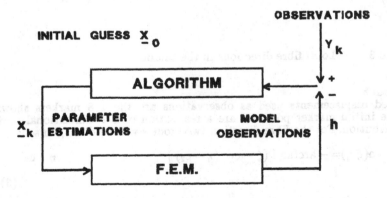

Figure 1: Schematic drawing of the sequential identification method

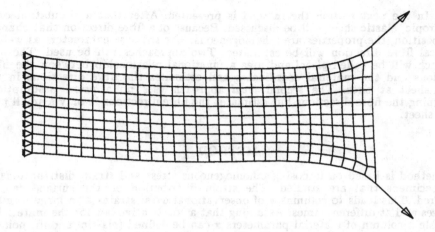

Figure 2   Finite Element Model for the generation of
"experimental" data.

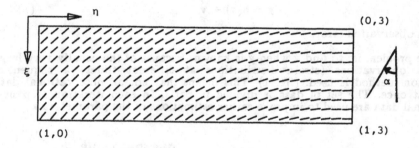

Figure 3   Local fibre directions in the sample

Global approach

The measured displacements used as observations are the 128 markers shown in figure 4. The initial marker positions are a realization of a 2–dimensional uniform random distribution. For the fibre direction two models will be distinghuised:

model 1: $\alpha(\xi,\eta) = -\arctan\left[\,(\xi - c_1)/(\eta - c_2)\,\right]$ $\eta \neq c_2$

                                                                         (3)

model 2: $\alpha(\xi,\eta) = b_0 + b_1\,\xi + b_2\,\eta$ (4)

In these equations α denotes the positive rotation of the material 1–direction from the model ξ–axis. This rotation is a function of the position coordinates ξ and η. The parameters c_1 and c_2 in model 1 can be interpreted as the coordinates of the centroid of the concentric circles.

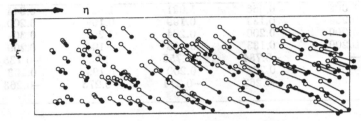

o position before deformation

● position after deformation

Figure 4   Measured displacements that have been used as observational data.

The third column of table 1 shows the estimation of the parameters with the use of model 1 after 10 iterations and without observation errors. The fourth and fifth column show the estimation results when observation data are disturbed with errors with a zero mean normal distribution. It can be observed that the identification approach works well, even with a noise to signal ratio of 10% ($\sigma = 0.01$).

TABLE 1
Results of estimations of properties with model 1

Param. x_i	Exact values	Initial guess	Estimations No noise	$\sigma=0.001$	$\sigma=0.01$
E_1	1.000	0.666	1.000	0.993	0.931
E_2	0.200	0.133	0.200	0.200	0.198
ν_{12}	0.300	0.200	0.300	0.301	0.305
G_{12}	0.200	0.133	0.200	0.201	0.211
c_1	3.000	2.000	2.998	3.004	3.055
c_2	3.000	2.000	2.999	3.010	3.106

Table 2 shows the estimation results of the estimations with model 2. Here similar estimation results are presented as in table 1, but now 7 parameters are estimated. The table shows that in this case the obvious model errors hardly effect the estimations for E_1, E_2, ν_{12} and G_{12}.

A comparison of the fibre directions calculated with model 1 and model 2 show hardly any difference. Evidently, in this case, the estimation results are neither very sensitive to this type of model errors, nor are they very sensitive for the combination of model errors and random observation errors.

TABLE 2
Results of estimations of properties with model 2

Param. x_i	Exact values	Initial guess	Estimations No noise	$\sigma=0.001$	$\sigma=0.01$
E_1	1.000	0.666	1.031	1.022	0.944
E_2	0.200	0.133	0.199	0.199	0.200
ν_{12}	0.300	0.200	0.294	0.296	0.304
G_{12}	0.200	0.133	0.193	0.194	0.209
b_0	—	−0.784	−0.755	−0.753	−0.735
b_1	—	0.262	0.200	0.199	0.193
b_2		−0.262	−0.274	−0.273	−0.263

Local approach

In this approach only a part of te sample is modeled (fig. 5). Within this part the material is considered to be homogeneous and only the kinematic boundary conditions are used. This has experimental advantages. An additional advantage is that the finite element model can be small with few elements. Also it is not necessary in this approach to make a priori model assumptions for the fibre direction. A disadvantage is that no complete characterization is possible. In the present example the following parameters will be estimated:

$$x^T = \{ E^*, \nu_{12}, G^*, \cotan(\alpha) \} \tag{5}$$

with, $E^* = E_2/E_1$; $G^* = G_{12}/E_1$.

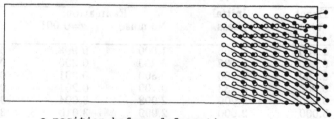

o position before deformation

• position after deformation

Figure 5 Area definition and displacements that have been used for the local approach.

Table 3 shows the estimation results. Some discussion on the exact values is worthwhile in this case. In the model it is assumed that the considered sample part has homogeneous properties. This model error makes the use of the term exact misleading. The slightly biased parameters of column 3 may give better results in the homogeneous model than the exact parameters. It is also clear from the last column that a large error in the displacements results in bad estimates for the material properties.

TABLE 3
Results of estimations of properties in local approach

Param. x_i	exact values	Initial guess	Estimations no noise	$\sigma=0.001$	$\sigma=0.01$
E^*	0.200	0.133	0.216	0.221	0.497
ν_{12}	0.300	0.200	0.322	0.330	0.000
G^*	0.200	0.133	0.202	0.220	0.848
$COTAN(\alpha)$	–	−0.100	−0.093	−0.090	−0.056

DISCUSSION

For the characterization of inhomogeneous materials a hybrid numerical–experimental approach is favorable, compared to traditional methods. Via two approaches it is shown that it is possible to find the properties even when the material is inhomogeneous.

The advantages of the global approach are: the procedure leads to a complete quantification of the entire sample, the identification is not very much affected by the observation noise and the models errors used in the example.

The advantages of the local approach are: that a priori model assumptions for the fibre directions are not necessary, in general the models contain less parameters, the geometry and boundary conditions of the considered part are accurately known and the finite element models are small and simple. However it seems that the local approach is more sensitive to observation noise than the global approach. This needs to be investigated more extensively

REFERENCES

1. Hendriks, M.A.N., Oomens, C.W.J., Jans, H.W.J., Janssen, J.D., Kok, J.J., A numerical experimental approach for the mechanical characterization of composites. Proceedings of the 9th Int. Confer. on Experimental Mechanics. Part 2. Aaby Tryk, Copenhagen, 1990, pp. 552 – 561.
2. Norton, J.P., An introduction to identification, Academic Press, New York and London, 1986.

NON DETERMINED TESTS
AS A WAY TO IDENTIFY WOOD ELASTIC PARAMETERS
THE FINITE ELEMENT APPROACH

F. Rouger , M. Khebibeche, C. Le Govic

Centre Technique du Bois et de l'Ameublement
10 av de St Mandé, 75012 PARIS, FRANCE

ABSTRACT

Wood is an orthotropic material whose elastic constants need a large number of tests to be identified. This paper proposes an alternative method of testing combined with a finite element optimization method. In the case of non determined tests, the identification procedure is an inverse problem. The solution is based on a numerical minimization technique. An example is given on a modified bending test. The influence of experimental noise is analysed through a Monte-Carlo simulation. An experimental validation is reported.

INTRODUCTION

The elastic behavior of wood [3] involves nine constants which might be identified by different techniques. The simplest one is the use of statically or cinematically determined tests. But a minimum of six tests is required. Other non destructive techniques have been used. The use of non determined tests reduces the experiments because much more mechanical information from the specimen is taken into account. This method has been already applied for composite plates [4] and adapted for wood material [1]. This paper attempts to extend the approach to a general two dimensionnal case.

WOOD ELASTIC BEHAVIOR

Due to its structure, wood material has a strong orthotropic behavior, which is observed in the elastic domain. This orthotropy is cylindrical :
- the longitudinal direction (L) is the axis of the tree. This is also the main direction of the constitutive cells.
- the radial (r) and tangential (t) directions are defined locally according to the growth rings.

To simplify the equilibrium equations, this orthotropy is considered as cartesian. This assumption remains valid if the size of the specimen is small compared with the size of the tree. Therefore, the equations are formulated in global axis (L,T,R).

The elastic behavior is described by the following equation :

$$\sigma_{ij} = C_{ijkl} \cdot \varepsilon_{kl} \quad \Leftrightarrow \quad \varepsilon_{kl} = S_{klij} \cdot \sigma_{ij}$$
$$i,j,k,l \in \{1,2,3\} \qquad \text{(Einstein notation)}$$

The stiffness [C] and the compliance [S] are fourth order tensors. A simplified notation (VOIGT) is usually preferred in which stresses [σ] and strains [ε] are expressed as vectors. In the case of an orthotropic material, 9 terms are independant and non zero :

$$S_{11} = \frac{1}{E_x} \qquad S_{22} = \frac{1}{E_y} \qquad S_{33} = \frac{1}{E_z} \qquad S_{44} = \frac{1}{G_{xy}} \qquad S_{55} = \frac{1}{G_{yz}} \qquad S_{66} = \frac{1}{G_{xz}}$$

$$S_{12} = -\frac{\nu_{xy}}{E_x} \qquad S_{13} = -\frac{\nu_{xz}}{E_x} \qquad S_{23} = -\frac{\nu_{yz}}{E_y}$$

The elastic parameters to be identified are :

3 Young's moduli	3 Poisson's ratios	3 Shear moduli
E_L, E_R, E_T	$\nu_{LR}, \nu_{RT}, \nu_{LT}$	G_{LR}, G_{RT}, G_{LT}

As an example, average values are reported below for two species :

Species	Spruce	Beech
E_L (MPa)	17000	14450
E_R (MPa)	830	1650
E_T (MPa)	650	750
ν_{LR}	0.37	0.45
ν_{RT}	0.56	0.47
ν_{LT}	0.43	0.83
G_{LR} (MPa)	640	1400
G_{LT} (MPa)	870	1100
G_{RT} (MPa)	40	290

In the case of a specimen whose main directions are not the orthotropic directions, the tensor notation is preferred :

$$S'_{i'j'k'l'} = S_{ijkl} \; \alpha_{i'i} \; \alpha_{j'j} \; \alpha_{k'k} \; \alpha_{l'l}$$

with   S   orthotropic tensor   (L,R,T)
           S'   specimen tensor (X,Y,Z)
           $\alpha_{t't}$ rotation matrix (from LRT to XYZ).

NUMERICAL METHODS FOR MATERIAL IDENTIFICATION

Basic equations.

A constitutive equation is a particular case of a physical relation between stresses and strains for a given volume considered as continuous :

$$\sigma(M, t) = \Re\left[M ; t ; F(M, \tau) ; V_i(M, \tau) \right]_{\tau=-\infty}^{t}$$

F is the gradient of the geometrical transformation T which associates to a reference configuration S_a an actual configuration at time t noted S_t :

$$M\ (a_1, a_2, a_3) \qquad \xrightarrow{\quad * \quad} \qquad M\ (x_1, x_2, x_3) \qquad F_{ij} = \partial x_i / \partial a_j$$
$$S_a \qquad\qquad\qquad\qquad S_t$$

F is the product of a rotation tensor (R) and a pure strain tensor (S). Vi are physical quantities as temperature, moisture content,...

Under certain assumptions, this relationship is simplified :

- objectivity principle : $\Re$ does not depend on R
- small deformations : $R_{ij} = \delta_{ij}$
- small deformations : $S_{ij} = \varepsilon_{ij} = 1/2\ (u_{i,j} + u_{j,i})$
- homogeneous material : $\Re$ does not depend on M.

$$\sigma(M, t) = \Re\left[t ; \varepsilon(M, \tau) ; V_i(M, \tau) \right]_{\tau=-\infty}^{t}$$

This relation is valid when the unknown is the stress tensor, given from the strain history. If the strain is the unknown, the previous equation is inverted.

To identify a constitutive equation, we have to do qualitative experiments and write the analytical expression of the functionnal. Furthermore, we must give numerical values to the constitutive parameters. To fit these values, we use the following procedure :

- In a test, assign values to the excitation and to the physical variables.
- measure the response.
- evaluate parameters such as the calculated response is "as close as possible" to the measured value. This might be expressed by the minimum of a least squares norm between responses.

This simple scheme is only partially valid because we never know the excitation and the response everywhere in the specimen.

(1) If the excitation is easily known where the response is measured, the test is statically or cinematically determined. The equation to be solved is :

Find c_j such as :

$$\sum_{\substack{pts \\ experimentaux}} \left|\left| R_i - R_i^* \right|\right| \to 0$$

with R_i^* : measured response
$R_i = F(S_i; V_i; c_j)$: calculated response

(2) The excitation is unknown where the response is measured, but these variables are related through the equilibrium equations. This problem is an inverse problem, which might be solved numerically. If we use a kinematic formulation, stresses and strains are replaced with displacements :

If $\{u\}^*$ are the measured displacements, we have to fit c_j such as the calculated displacement field $\{u\}$ gets close to $\{u\}^*$. Once $\{u\}$ calculated, we are facing to a type (1) problem.

Numerical methods.

The displacements u_i are calculated from the parameters c_j using th finite element method :

$$\left\{ \begin{array}{l} \text{Equilibrium Equations} \\ \text{Boundary Conditions} \\ \text{Constitutive Equations} \end{array} \right\} \quad \rightarrow \quad [K]\{U\} = \{F\} \text{ to solve}$$

The cost function is given on the nodes of the mesh :

$$J = \frac{1}{2} \sum_k (u_k - u_k^*)^2 \, \delta_{ik}$$

k : degrees of freedom $\delta_{ik} = 1$ if DOF corresponds to a measurement point
                                $= 0$ if not.

In a matrix notation, this equation is equivalent to :

$$J = 1/2 \ <U-U^*> . [\delta] . \{U-U^*\}$$

The cost function reaches a minimum if the gradients are zero. These gradients have the following expression :

$$\frac{\partial J}{\partial c_j} = <U-U^*> . [\delta] . \left\{ \frac{\partial U}{\partial c_j} \right\} = <U-U^*> . [\delta] . \left\{ \frac{\partial \, [K]^{-1} . \{F\}}{\partial c_j} \right\}$$

Because this equation requires the inversion of $[K]$, we use a adjoint state formulation. The generalized functionnal is given by :

$$\pi^*(U,c,P) = 1/2 \ <U-U^*>.[\delta].\{U-U^*\} + <P>.(\ [K].\{U\}-\{F\} \ )$$

We have to find the stationnarity of π^*, given a constraint on $\{U\}$, which satisfies the finite element equations :

$$[K].\{U\}-\{F\} = 0$$

The stationnarity principle applied to π^* gives :

$$\delta \, \pi^* = \partial \, \pi^*/\partial P . \, \delta P + \partial \, \pi^*/\partial U . \, \delta U + \partial \, \pi^*/\partial c . \, \delta c \quad \rightarrow \quad 0$$

$$\partial \pi^*/\partial P = [K].\{U\} - \{F\} = 0 \quad \Rightarrow \quad \{U\}$$

$$\partial \pi^*/\partial U = <U-U^*>.[\delta] + <P>.[K] = [K]\{P\} + [\delta].\{U-U^*\} \quad \Rightarrow \quad \{P\}$$

$$\partial \pi^*/\partial c = <P>.[\partial K/\partial c].\{U\} \quad \Rightarrow \quad \{c\}$$

The solution of the problem gives a zero value for each of these gradients. Therefore, the algorythm is :

Initial estimation of cj

Calculation of {U}

Calculation of {P}

Calculation of J

Calculation of J gradients.

New parameters evaluation
using a gradient method (2).

Convergence test.

End

Implementation of the method in a code.

The previous algorythm has been implemented in a object oriented finite element program, named SIC3 and developed by the University of Technology of Compiègne (FRANCE). This program is an interactive code, made with commands which use objects. The minimization of a cost function belongs to a wide range of optimization problems. Therefore, we used two general objects to describe different variables.

* the OPTIM object.

It contains the cost function, its gradients with respect to the optimization variables, the limitations (linear or not) and their gradients, and some input parameters for the line search routine. Instead of direct values, some zones can contain an identifier to a macro-commad which performs the calculation.
* the VARCO object.

It contains the optimization variables : initial values, current values, lower and upper bounds.

A set of basic commands has been written to solve the problem :

RECUPERER EXPERIENCE
INITIALISER OPTIMISATION /VARIABLES = n
DEFINIR VARIABLE-CONCEPTION
CALCULER FONCTION-COUT (activates a macro)
CALCULER LIMITATIONS
CALCULER GRADIENTS-COUT VARIABLES

CALCULER GRADIENTS-COUT DDL
EXECUTER VF02AD
NORMALISER
DENORMALISER
MULTIPLIER

An additionnal set of macro-commands has been written :

LIN solves the finite element system. GRAD to compare different methods for $\partial J/\partial c$
COUT calculates cost function after LIN ITER performs an iteration.
ADJOINT calculates <P> (eq. to LIN with a different 2nd member)

VALIDATION OF THE THEORY

An example has been derived using a modified bending test [6]. In order to get more information about the elastic parameters, the specimen has a span to depth ratio of 5-8. Several measurement points have been located according to the following procedure:

- a finite element calculation is performed with a given set of parameters. Displacements are recorded.
- One of the parameters is modified. A new finite element calculation is done. Modifications on the displacements are recorded. The most sensible points are noted.
- This procedure is done for each of the parameters.
- The final choice of the locations for measurements is done considering practical considerations (possibility to use LVDT's on the edge of the specimen, calculate displacements from strain gages,...).

The designed specimen is illustrated in figure below :

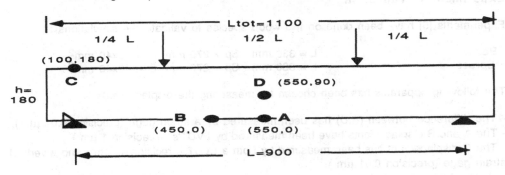

Monte-Carlo Simulations.

A first example has been treated using " simulated experimental displacements" , i.e. the experimental values come from a direct finite element simulation whose parameters are given. The results are given below :

Parameter	Initial Value	Lower Bound	Upper Bound	Final Value
E_x (MPa)	15000	14000	20000	17000
E_y (MPa)	700	650	900	830
G_{xy} (MPa)	800	500	1000	640
ν_{xy}	0.25	0.2	0.45	0.37

This result is obtained after 45 iterations. Because this example does not take into account any experimental noise, a Monte-Carlo simulation has been carried out, introducing a random noise (uniform distribution between -10% and +10% on each of the displacements. The number of replications is 20. The results are reported below :

	E_x	E_y	ν_{xy}	G_{xy}
Expected mean	17000	830	0.37	640
Calculated mean	17136	836	0.369	654
Error on mean	0.8%	0.67%	0.2%	2.2%
95% , lower	16662	776	0.357	605
95% , upper	17610	895	0.381	703
Minimum	15727	640	0.33	491
Maximum	19138	1087	0.435	850
Standard deviation	1014	127	0.026	105
Coefficient of var.	5.9%	15.2%	7.04%	16.1%
Amplification	1.025	2.63	1.22	2.8

This simulation leads to the following conclusions :
- the calculated mean is close to the expected one.
- the 95% confidence interval contains the expected mean.
- the amplification factors are low for E_x and ν_{xy} but high for E_y and G_{xy} .

Experimental Validation

Experiments [5] have been done on two wood species to validate the simulations :

- Beech   L = 330 mm   Sp = 270 mm   S = 40x40 mm^2
- Spruce   L = 330 mm   Sp = 270 mm   S = 40x40 mm^2

The following apparatus has been chosen for measuring the displacements :

- The difference between (A,D) has been measured by a strain gage (precision 0.01 μm).
- The A and B displacements have been measured by LVDT's (precision 1 μm).
- The C displacement has been measured by both a LVDT (precision 1 μm) and a vertical strain gage (precision 0.01 μm).
- A supplementary strain gage has been placed horizontally on the lower edge on the specimen between A and B (precision 0.01 μm).

The following results have been recorded :

Specimen	P daN	V_A mm	V_B mm	V_D mm	V_C mm	U_B-U_A mm
Beech	250	0.412	0.397	0.415	0.045	0.0202
Spruce	240	0.409	0.394	0.412	0.047	0.0202

An approximation is used to get an estimation on E_x , G_{xy} :

- E_x is calculated using the circular bending assumption in the central zone :

$$E_x = \frac{3\,P\,\frac{L}{3}\,(x_A - x_B)^2}{b\,h^3\,(v_A - v_B)}$$

- G_{xy} is calculated using theory on bending of beams with shear effect :

$$G_{xy} = \frac{\alpha\,P\,a}{2\,b\,h\left(v_A - \dfrac{P\,a\,(3\,L^2 - 4\,a^2)}{4\,b\,h^3\,E_x}\right)} \qquad \text{with } a = \frac{L}{3}$$

These formulae applied to the three specimens give the following results :

Specimen	E_x GPa	G_{xy} GPa
Beech	16.7	0.7
Spruce	16.0	0.69

The optimization has been done on four parameters simultaneously. It gave the following results :

		Initial Value	Lower Bound	Upper Bound	Final Value
Beech	E_x GPa	10	5	25	20.4
	E_y GPa	1.5	0.1	3	1.01
	v_{xy}	0.45	0.05	0.5	0.262
	G_{xy} GPa	1	0.1	3	0.52
Spruce	E_x GPa	10	10	30	19.6
	E_y GPa	0.8	0.1	3	0.89
	v_{xy}	0.5	0.1	0.7	0.196
	G_{xy} GPa	1	0.1	3	0.52

These values are close to the expected ones. The corresponding calculated displacements are given below :

Specimen	P daN	V_A mm	V_B mm	V_D mm	V_C mm	U_B-U_A mm
Beech	250	0.416	0.398	0.417	0.047	0.0194
Spruce	240	0.413	0.395	0.414	0.049	0.0193

CONCLUSION

The numerical method which has been developed might find a wide range of applications. The example derived in this paper leads to the following remarks :
- the Monte-Carlo simulations give stable results.
- the experiment has been done with success but the metrology is difficult to achieve because of small displacements.
- the calculated elastic parameters are slightly different from the approximations, even if the calculated displacements are close.

This method could be improved in further work using for example optical devices on the whole specimen, which would provide more measurement points to be used in the analysis.

Anyway, the method is available for any kind of two dimensionnal specimens and can be easily extended to three dimensions or to plates and shells. Another direction of investigation would be to take into account other kinds of behaviors.

AKNOWLEDGEMENTS

The authors would like to congratulate Paul Nyault, technician in the Mechanics laboratory, who helped to carry out the experiments.

REFERENCES

[1] FOUDJET A. & al. , 1982 - Indirect identification methods for the elastic constants of orthotropic materials and their application to wood. Wood Sc. & Technol. , n.16 , p:215-222

[2] GILL P.E., MURRAY W. , 1974 - Numerical methods for constrained optimization, Academic Press , N.Y.

[3] GUITARD D. , 1987 - Mécanique du matériau bois et composites, Cepadues Ed., collec. Nabla

[4] KERNEVEZ J.P. & al. , 1978 - An identification method applied to an orthotropic plate bending experiment. Int. Journal for numerical methods in engineering, n.12 , p:129-139

[5] NYAULT P., 1990 - Mise au point d'une méthodologie d'essai pour la détermination de cinq complaisances de la matrice de souplesse du bois, IUT Clermont-Ferrand.

[6] ROUGER F., 1988 - Application des méthodes numériques aux problèmes d'identification des lois de comportement du matériau bois. PhD dissertation , University of Technology of Compiègne.

MEASUREMENT OF LAMINATE BENDING ELASTIC PARAMETERS FROM NON-UNIFORM STRAIN FIELDS.

M. GREDIAC* & A. VAUTRIN**

* Ecole Nationale d'Ingénieurs
de Saint-Etienne
56 rue Jean Parot
42023 Saint-Etienne Cedex 2
FRANCE

** Ecole Nationale Supérieure des
Mines de Saint-Etienne
Département Mécanique et Matériaux
158 cours Fauriel
42023 Saint-Etienne Cedex 2
FRANCE

ABSTRACT

The assumption of uniform states of strain in most conventional tests carried out on anisotropic specimens is questionable. The present paper deals with a method allowing the determination of elastic bending compliances of laminates from tests giving rise to non-uniform states of strain. Data are presented from experiments performed on two laminates of two different stacking sequences.

INTRODUCTION

The experimental characterization of the mechanical properties of composite materials reveals specific difficulties due to the anisotropy of these materials. Conventional mechanical tests are performed and interpreted under the assumption of uniform states of strain. However, strictly speaking, such states of strain cannot be obtained on anisotropic materials. For instance, Pagano and Halpin [1] pointed out the influence of end constraint on the strain field in off-axis tensile tests, which is non-uniform because of the shear coupling. They proposed a corrective formula that provides the actual Young's modulus as a function of the measured one. Pindera and Herakovich [2] used the same test in order to determine the intralaminar shear modulus with a rotating-grip fixture described at first by Wu and Thomas [3]. Such a grip arrangement can noticeably reduce perturbations in the strain field, but setting up the sample seems to be a difficult operation.

On the other hand just a few papers deal with the proper performing of bending tests on anisotropic specimens. Nevertheless, the laminated plate theory provides in-plane and bending stiffness matrices which are independent in the general case. Therefore, bending stiffnesses cannot be deduced from any tensile test, because the order of the plies in a stacking sequence is critical for the bending properties, unlike for the in-plane ones. Any checking of the predicted elastic parameters demands both tensile and bending tests.

Some well-known bending tests on beams are available in order to measure the elastic parameters. Three- and four-point bending tests allow the measurement of such parameters but it is worth noting that those tests can only be carried out on samples cut in orthotropy directions of an orthotropic in bending plate; otherwise, the shear coupling gives rises to severe discrepancies between actual and apparent moduli [4]. Note that in-plane and bending stiffnesses being different, many industrial stacking sequences are isotropic or orthotropic in stretching while they are completely anisotropic in bending. Any bending test can therefore be considered as an off-axes one and the bending elastic parameters cannot be properly measured. For instance, the well-known quasi-isotropic $\frac{\pi}{4}$ laminates are isotropic in stretching and completely anisotropic in bending. The polar representations of both in-plane and bending rigidity moduli of a graphite/epoxy [0,45,90,135]s in Figure 1 clearly point out how significative the difference may be.

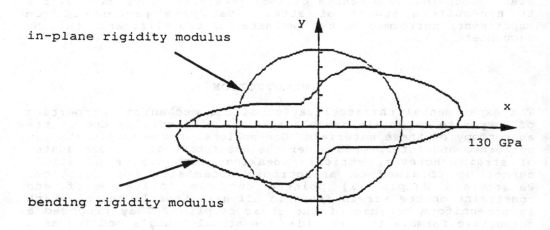

Figure 1. Polar representation of in-plane and bending rigidity moduli of a quasi-isotropic graphite/epoxy laminate.

ANALYSIS

The experimental procedure presented herein clearly departs from the usual ones. Instead of considering local measurements on strain fields whose uniformity is questionable, we suggest to get the bending compliances from the **whole strain field** even if it is **non-uniform**. Moreover, in order to avoid any sampling variability, we use only **one sample** of plate instead of many samples considered as beams cut in different directions of the plate. The shape of the sample is **circular**, in such a way that each rotation of this sample in the mechanical device provides a new mechanical configuration. The six unknown independent compliances are derived from two different static tests using relationships involving the applied loading, some geometrical parameters, the mean value of the strain on the top surface of the sample and the compliances to be determined. Those two tests are presented below.

Three-point bending test
This test is an extension of the three-point bending test carried out on beams. In the latter case, the state of strain is linear. On the other hand, the same test carried out on a circular plate (Figure 2) gives rise to a non-uniform strain field, but three out of the six unknown compliances can be determined.

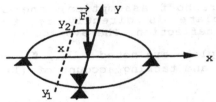

Figure 2 . Three-point bending test.

Consider a section of the plate located at x (-R<x<0). F is the algebraic value of the loading (F<0 in the present case). The equilibrium of the plate leads to :

$$\int_{y_1}^{y_2} M_{xx}dy = \frac{F}{2}(R+x)$$

Integrating on the left half part S' of the plate, we get :

$$[\frac{F}{2}(Rx+\frac{x^2}{2})]_{-R}^0 = \frac{FR^2}{4} = \iint_{S'} M_{xx}dxdy$$

If we consider now the whole surface S of the specimen, we obtain :

$$\frac{FR^2}{2} = \iint_{S} M_{xx}dxdy$$

(1)

The same argument leads to :

$$\iint_S M_{yy} \, dxdy = 0 \qquad \iint_S M_{xy} \, dxdy = 0$$

(2)

Consider now the constitutive law provided by the classical lamination theory [5] :

$$\begin{pmatrix} w_{,xx} \\ w_{,yy} \\ 2w_{,xy} \end{pmatrix} = - \begin{pmatrix} d_{11} & d_{12} & d_{16} \\ d_{12} & d_{22} & d_{26} \\ d_{16} & d_{26} & d_{66} \end{pmatrix} \begin{pmatrix} M_{xx} \\ M_{yy} \\ M_{xy} \end{pmatrix}$$

(3)

where w, deflection of the plate,
         M_{ij}, $(i = x, y)$ bending or twisting moment per unit
            length, and
         d_{ij}, $(i = 1,2,6)$ flexural compliance.

Under the Love-Kirchhoff assumption, the strain on the top surface of the plate is directly related to the second derivatives of the deflection. The average of the strains on the top surface of the plate are noted : $\overline{\varepsilon_{xx}}$, $\overline{\varepsilon_{yy}}$ and $\overline{\varepsilon_{xy}}$. Integrating equation (3) over S and tacking account of (1) and (2) leads to:

$$\begin{pmatrix} \overline{\varepsilon_{xx}} \\ \overline{\varepsilon_{yy}} \\ 2\overline{\varepsilon_{xy}} \end{pmatrix} = \frac{h}{4S} \begin{pmatrix} d_{11} & d_{12} & d_{16} \\ d_{12} & d_{22} & d_{26} \\ d_{16} & d_{26} & d_{66} \end{pmatrix} \begin{pmatrix} FR^2 \\ 0 \\ 0 \end{pmatrix}$$

Finally, introducing the value of the surface $S = \pi R^2$,

$$\overline{\varepsilon_{xx}} = \frac{hF}{4\pi} d_{11} \quad \overline{\varepsilon_{yy}} = \frac{hF}{4\pi} d_{12} \quad 2\overline{\varepsilon_{xy}} = \frac{hF}{4\pi} d_{16}$$

(4)

Those relationships allow the determination of three out of the six unknowns. The plate being rotated through 90 deg., we have the same relationships involving d_{22}, d_{12} and d_{26}; therefore, an other test is required to determine the last compliance : d_{66}.

Twisting test
This test is presented in Figure 3. It is an extension of the well-known twisting or anticlastic test carried out on square plates. In the latter case, the state of strain is uniform and three compliances can be determined : d_{16}, d_{26} and d_{66}. With our approach, the plate is circular and the state of strain is non-uniform.

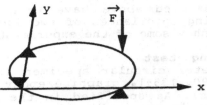

Figure 3. Twisting test.

The same reasoning as above provides the three following relationships :

$$\overline{\varepsilon_{xx}} = \frac{-hF}{2\pi} d_{16} \qquad \overline{\varepsilon_{yy}} = \frac{-hF}{2\pi} d_{26} \qquad 2\overline{\varepsilon_{xy}} = \frac{-hF}{2\pi} d_{66} \qquad (5)$$

Using the two present tests, the six flexural compliances can be determined.

EXPERIMENTS AND RESULTS

Experimental concept and set-up
Figure 4 shows the experimental set-up with its main components. Plate **1** is attached in the mechanical device **2**. The load is produced by small hydraulic jacks. A moiré optical device **3** provides the slope contour $\frac{\partial w}{\partial x}$ on the whole top surface of the plate. Note that the plate is coated with a thin reflective resin layer (thickness = 20μm). The moiré pattern **4** is taken by a video camera **5** and digitized by the digitizing board of the personal computer **6**. The mean value of the strain is then computed using a numerical processing of the pattern based on the finite elements method [6]. Note that the mechanical device **2** can be rotated through 90 deg in order to obtain both $\frac{\partial w}{\partial x}$ and $\frac{\partial w}{\partial y}$ moiré patterns.

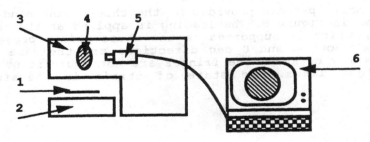

Figure 4. Experimental set-up.

The two tests presented above have been used in order to determine the bending compliances of two laminated plates. The following examples show some of the experimental results.

Three-point bending test

The 150mm in diameter circular specimen is cut in a quasi-isotropic $[0_2, 45_2, 90_2, 135_2]_s$ graphite/epoxy stacking sequence (thickness = 2.08mm). As specified in the first part of the paper, in-plane elastic properties of such a stacking sequence are isotropic, while the bending ones are completely anisotropic. Hence the elastic compliances cannot be measured using conventional three- or four-point bending tests because of the shear coupling. On the other hand, the three-point bending test on a circular plate allows the measurement of the longitudinal compliance in any base from a non-uniform strain field.

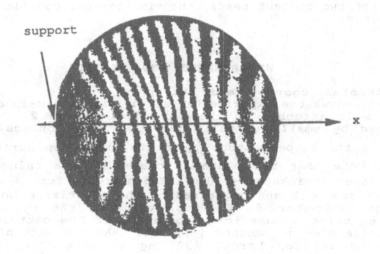

Figure 5. Typical moiré pattern.

A typical moiré pattern provided by the three-point bending test is depicted in Figure 5. The loading is applied at the centre of the plate, which is supported on the three points presented in the figure. Both x and 0 deg directions are the same in this figure. As may be seen, the fringes are not equidistant and are not straight lines. The state of strain is therefore not uniform.

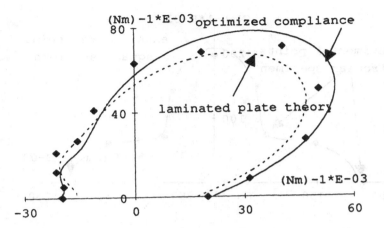

Figure 6. Three-point bending test on a "quasi-isotropic"
$[0_2, 45_2, 90_2, 135_2]_s$ graphite/epoxy plate. Experimental results.

The results for the longitudinal bending compliance d_{11} are presented in Figure 6. It is important to mention that **all the experimental points** are obtained from **only one sample**. Also presented are both optimized and expected evolutions of d_{11}. The optimized one is obtained applying the least square method on the whole set of the six compliances measured in 13 different bases. The second one is obtained using predicted values provided by the classical lamination theory, the four independent parameters of a ply being determined by in- and off-axis tensile tests carried out on samples cut in a unidirectional plate. It is worth noting that the shapes of both curves are quite similar, but the laminated plate theory provides rather lower values. That point shows that the present method allows the checking of laminate elastic parameters.

Twisting test
The 150mm in diameter circular specimen is cut in a 16-ply unidirectional glass/epoxy laminated plate (thickness = 2.60 mm). The twisting test has been used in order to measure the shear compliance in 9 different bases by rotation of the plate in the mechanical device. Experimental points and optimized curve are shown in Figure 7. It can be noted that the scatter of the experimental result is rather low. It is certainly due to the fact that only one specimen of plate provides the whole set of the 9 experimental points. The same twisting test carried out on a square sample cut in the 0 deg - 90 deg base enables us to measure the shear compliance in only one base. The experimental point is reported in Figure 7. The agreement between experimental results obtained with both square and circular samples can be observed, but a square shape provides only one experimental point while the circular one provides 9 points in this case.

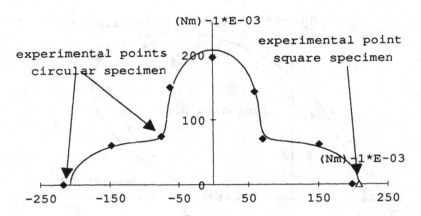

Figure 7. Twisting test on a unidirectional plate.
Experimental results.

CONCLUSION

The present approach of the mechanical tests on composites takes
directly into account the non-uniformity of the strain field
provided by suitable tests performed on plates. The whole field
is considered by averaging the strain on the whole surface of
the specimen. This contributes to avoid local variabilities of
the material and to give a global mechanical response of the
plate to be characterized.

ACKNOWLEDGEMENT

The *Ministère de la Défense* is gratefully acknowledged for the
financial support provided for this work (DRET grant 88/1208).

REFERENCES

1. Pagano, N.J. and Halpin, J.C., <u>Journal of Composite
   Materials</u>, 1968, **2**, 1, pp. 18-31.

2. Pindera, M.-J. and Herakovitch, C.T., <u>Experimental Mechanics</u>,
   1986, **26**, 1, pp. 103-112.

3. Wu, E.and Thomas, R., <u>Journal of Composite Materials</u>, 1968,
   2, pp. 523.

4. Whitney, J.M. and Dauksys, R.J., <u>Journal of Composite
   Materials</u>, 1970, **4**, pp. 135-137.

5. Jones, R.M., <u>Mechanics of Composite Materials</u>, McGraw-Hill,
   New-York, 1975.

6. Grédiac, M. and Vautrin, A., <u>Journal of Applied Mechanics</u>,
   paper 8516, to be published.

IDENTIFICATION OF THE SHEAR PROPERTIES OF THE LAMINA IN THE FRAMEWORK OF PLASTICITY.

Y. SURREL, A. VAUTRIN
Dpt. Mécanique et Matériaux
École Nationale Supérieure des Mines de Saint-Étienne
158, cours Fauriel
42023 ST ÉTIENNE CEDEX 2, F

ABSTRACT

The topic of the present work is twofold. We present first a modelling of the shear response of the unidirectional lamina in the framework of plasticity, and also we present a novel process of identifying the model parameters, based on the minimizing of an appropriate "distance" between the experimental and the modelized responses. By response, we mean *the whole plane strain response, i.e. we take into account the three components of the plane strain tensor.*

INTRODUCTION

Shear properties are prominent mechanical features of fibre-reinforced composite materials. They are important both in rigidity-oriented design and in strength-oriented design, due to the poor shear properties of the matrix and of the interfaces. The extensive nonlinearity observed in the shear stress/strain response is also another aspect of primary importance for the complete characterization of the mechanical behaviour.

Some models have already been proposed to describe the material nonlinearity. SURREL & VAUTRIN [1] have shown that viscoelasticity could not explain the shape of the shear stress / shear strain response observed in monotonic tensile tests performed on off-axis coupons. SUN & CHEN [2] develop a one parameter plastic model and introduce an equivalent stress and an equivalent strain linked by a power-type flow-rule. GILLETTA [3] and LADEVEZE [4] consider the coupling between damage and plasticity.

We will present here a model derived from the one proposed in

[1]. In this paper, we have chosen to focus on the *identification procedure* rather than on the model itself. A discussion of the relevancy of that model as well as an extension to the case of non monotonic loadings are presented in [5].

MATERIAL AND TESTS

The present work concerns tensile tests performed on graphite-epoxy T300/914 laminates in on-axis ($[0]_{4S}$, $[90]_{4S}$), off-axis ($[10]_{4S}$, $[10]_{4S}$) and cross-ply ($[\pm45]_{2S}$) configurations. The coupons were 160/200 x 13 x 1.1 mm^3 and the fiber volume fraction was found $V_f = 62 \pm 3$ %. Special emphasis was given to conditioning: all specimens were carefully conditioned during at least two months at 60 °C in an environmental chamber at various humidities: dry, 60 % r.h., 80 % r.h. and 96 % r.h.. Some specimens were also immersed in distilled water. Complete results concerning the influence of moisture content are given in [5].

The test was a monotonic tensile test with a load speed adjusted for each configuration to have a test duration around 1000 s. The specimens were equipped with a 2D rosette and a strain gauge 45° off the tensile axis, in order to record the whole plane strain response. Load and strains were logged on a PC-type computer in order to make the unavoidable corrections on rough data due to the nonlinearity of the quarter bridge setup, the transverse effect of the strain gauges and the reduction of the cross section of the specimens.

MODEL

The classical lamination theory is used, and we propose the following model ([1],[5]) for the individual ply seen as a homogeneous anisotropic material (1 and 2 are the material axes, with 1 along the fibers, and E_1 and v_{12} are assumed constant):

$$\begin{cases} G = (G_0 - G_1)\exp\left(-\dfrac{R}{R_6}\right) + G_1 \\ E_2 = E_{20}\exp\left(-\dfrac{R}{R_2}\right) \end{cases} \tag{1}$$

where R is the maximum in-plane engineering shear strain, i.e. twice the radius of Mohr's circle:

$$R = \sqrt{(\varepsilon_1 - \varepsilon_2)^2 + \varepsilon_6^2}$$

which plays the role of a hardening parameter.

The total number of parameters to identify is 7 (4 moduli, 2 decrements and 1 asymptotical parameter). The problem is now to get the set of parameters allowing the best fit between the three experimental responses obtained during a single test (three strains are recorded) and the three corresponding "modelized" responses computed incrementally using (1) and the classical lamination theory.

IDENTIFICATION

Scalar product and distance
In order to carry on the identification procedure on sound mathematical bases, we define first an appropriate distance between the modelized and experimental responses. For a given test, we will call "response" the in-plane tensorial function $\varepsilon(\sigma_x)$ where x is along the axis of the sample and σ_x varies between 0 and σ_{xMax}. Between two real functions f and g defined on an interval $[a,b]$, the following scalar product is well known:

$$<f \mid g> = \int_a^b f(x) \cdot g(x)\,dx \qquad (2)$$

Between two mechanical responses $\varepsilon(\sigma_x)$ and $\varepsilon'(\sigma_x)$, the following expression defines also a scalar product:

$$<\varepsilon \mid \varepsilon'> = \int_0^{\sigma_{xMax}} \varepsilon_{ij}(\sigma_x) \cdot \varepsilon_{ij}'(\sigma_x)\,d\sigma_x \qquad (3)$$

In the space of the mechanical responses to a given test, a distance may be defined by:

$$D(\varepsilon,\varepsilon') = \sqrt{<(\varepsilon - \varepsilon')\mid(\varepsilon - \varepsilon')>} \qquad (4)$$

which is the "length" of the vector joining ε and ε'.

Let us mention here that we will use not only the distance (4) as an error function, but also the scalar product defined in (3) in an operative manner during the optimization procedure, as will be seen hereafter.

Optimizing one parameter
Let M_0 be the representative point of the experimental response in the space where the scalar product (3) and the distance (4) have been defined, giving it an Euclidean structure. Suppose a model has been proposed, depending on only one parameter λ. Let C_λ be the curve which is the locus of the different responses M_λ for varying parameter λ. We will define as the best value λ_∞ of the parameter λ the value for which the distance $M_\lambda M_0$ is minimum (Figure 1).

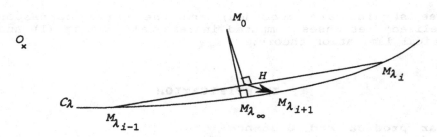

Figure 1. Optimization scheme with one parameter.

It is easy to show that this condition implies that $M_0M_{\lambda_\infty}$ be orthogonal to the tangent to C_λ at M_{λ_∞}. The corresponding equation has to be solved numerically through an iterative process. We have used the following procedure to get an estimate of λ_∞:

–a first gross value λ_0 of the parameter is needed, as well as a first value for an increment $\Delta_1\lambda$;

–being known the two last estimates λ_{i-1} and λ_i, the next increment $\Delta_{i+1}\lambda$ is so that $\lambda_{i+1} = \lambda_i + \Delta_{i+1}\lambda$ divides the numerical interval $[\lambda_{i-1}, \lambda_i]$ with the same ratio as H divides the geometrical segment $M_{\lambda_{i-1}}M_{\lambda_i}$ (Figure 1). With the tool provided by the scalar product (3), it is easy to find the following expression:

$$\Delta_{i+1}\lambda = \Delta_i\lambda \cdot \frac{\overrightarrow{M_{\lambda_i}M_0} \cdot \overrightarrow{M_{\lambda_{i-1}}M_{\lambda_i}}}{\overrightarrow{M_{\lambda_{i-1}}M_{\lambda_i}}^2} \tag{5}$$

So, in that expression, the scalar product defined has been used for its geometrical significance when orthogonal projections are involved and not only as a basis to define a distance.

The point $M_{\lambda_{i+1}}$ does not coincide with M_{λ_∞} because of the curvature of C_λ and because of the nonlinear scaling of C_λ with respect to λ. Obviously, these two effects become less and less important as the estimates λ_i get closer to λ_∞. So, the speed of convergence increases very rapidly, as will be seen on the example given in a further section.

It is easy to quantify the accuracy of the fit by the ratio of the norm of the difference between the experimental and modelized responses over the norm of the experimental response. We will call that ratio the *precision* of the fit and denote it by p:

$$p = \frac{|\overrightarrow{M_0M_{\lambda_\infty}}|}{|\overrightarrow{OM_0}|} \tag{6}$$

Optimizing many parameters

For many parameters, one has to proceed through the optimization process for each parameter, and then loop back to the first parameter and go on optimizing and looping until all the parameters reach a stable value.

The modulus E_1 and the Poisson's ratio ν_{12} are identified from the results on $[0]_{4S}$ on-axis tests. Of course, on those tests, the parameters linked to the shear properties cannot be determined. We will focus here on the identification carried on off-axis and cross-ply configurations. Fortunately, the various parameters involved in the model have not the same action on the response curves. The moduli E_{20} and G_0 govern the initial slopes of the curves. It is wise for those parameters to work only on a small part of those responses, say the first quarter, in order to discard the effect of the curvature. On the other hand, the two decrements R_6 and R_2 act upon the curvature only, and the whole length of the curves has to be taken into account. Finally, the refinement parameter G_1 is identified independently after the others. The flow chart is represented on Figure 2. Let us mention that there is a graphical monitoring possible, as the experimental and the modelized curves can be plotted simultaneously at any step in the procedure.

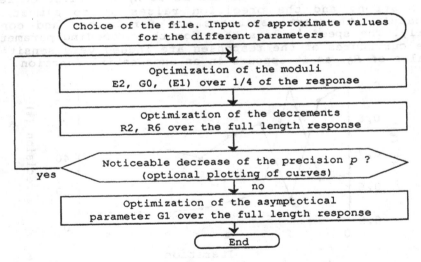

Figure 2. Flow chart of the optimization procedure.

Example

On Figure 3 are represented the evolutions of the initial shear modulus G_0 and of the related precision for a $[10]_{4S}$ sample. The first values input by the user are $G_0 = 5$ GPa and $\Delta_1 G = .1$ GPa. It is apparent on Figure 3 that this increment had the wrong sign. Nevertheless, the second computed value around 4.6 GPa is already very near the asymptotic value (iterations are stopped when the relative evolution of the parameter is below 10^{-5}).

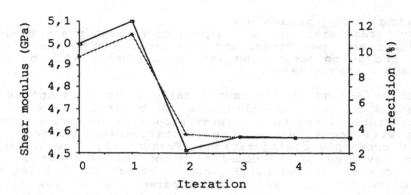

Figure 3. First identification of the initial shear modulus.
(First quarter of the responses, [10]$_{4s}$.)
Black: parameter ; grey: precision.

On Figure 4 are represented the principal decrement, R_6 and the related precision. It is interesting since it shows the stability of the procedure. The values input by the user are R_6 = .9 % and ΔR_6 = -.05 %. As can be seen, the first correction is too strong and the precision raises up to almost 80 %. Fortunately, the next estimates are far better and converge steadily. The speed is less than for the preceding parameter G_0 as the curvatures of the responses are indeed very sensitive to the value of R_6, as it appears in an exponential function.

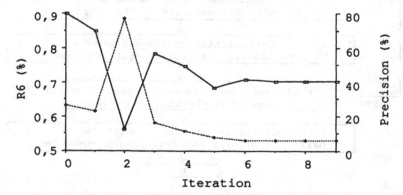

Figure 4. First identification of the shear decrement R_6.
(Whole of the responses, [10]$_{4s}$.)
Black: parameter ; grey: precision.

Experience have shown that in general only two passes through the loop represented on Figure 2 were necessary, the second pass introducing refinements on the parameters below 5 % in relative value and a third pass, variations below 1 %. For instance, concerning the sample corresponding to Figures 3 and 4, a third pass to optimize G_0 made the best value vary from 4.699 GPa to 4.716 GPa (.35 % relative variation).

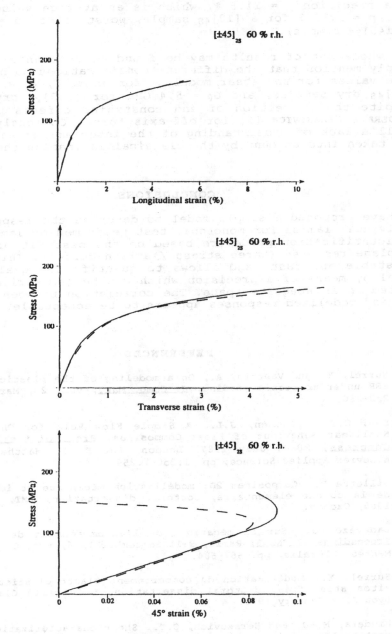

Figure 5. Comparison between experimental and modelized
responses ([±45]2s, precision = 11.5 %)
Solid line: experimental; dashed line: model

On Figure 5 are presented the comparisons between the experimental and modelized responses for a $[\pm45]_{2S}$ sample, corresponding to a precision $p = 11.5$ %, which is an average value (best value: $p = 1.32$ % for a $[10]_{4S}$ sample, worst value: $p = 22.7$ % for a $[\pm45]_{2S}$ sample).

The whole set of results may be found in reference [5]. Let us simply mention that the different configurations do not give the same values for the shear modulus (for example, $G_0 = 6.5$ GPa for $[10]_{4S}$ dry samples, and $G_0 = 5.4$ GPa for $[\pm45]_{2S}$ dry samples), despite the correction of end constraint effects proposed by PINDERA & HERAKOVICH [6] for off-axis tests. Obviously, there is still a lack of understanding of the interlaminar shear effects not taken into account by the classical lamination theory.

CONCLUSIONS

We have proposed a simple model to describe the response of the individual lamina for monotonic test performed on laminates, and an identification procedure based on the best fit of the whole in-plane response (three stress / strain curves). This procedure is stable and fast, and allows to quantify the quality of the model by means of a precision which can be geometrically interpreted in an adequate space. The correlation between experimental and modelized responses appears to be acceptable.

REFERENCES

1. Surrel, Y. and Vautrin, A., On a modeling of the plastic response of FRP under monotonic loading, J. Comp. Mat., vol. 23, March 1989, pp. 232-250.

2. Sun, C.T. and Chen, J.L., A Simple Flow Rule for Characterizing Nonlinear Behaviour of Fiber Composites, Proc. of the ICCMVI/ECCM2 Congress, 20-24 July 1987, London, Ed. F. L. Matthews and al., Elsevier Applied Science, pp. 1.250-1.259.

3. Gilletta D., Composites 2D: modélisation mécanique et identification de la couche élémentaire, doctoral dissertation, LMT, 25 October 1985, Cachan.

4. Ladevèze, J., Sur la mécanique de l'endommagement des composites, Proceedings of JNC 5, Paris, 9-11 September 1986, Eds. C. Bathias & D. Menkès, Pluralis, pp. 667-684.

5. Surrel, Y., Modélisation du comportement élasto-plastique de composites stratifiés, doctoral dissertation, Université Claude Bernard Lyon I, 9 January 1990.

6. Pindera, M.-J. and Herakovich, C.T., Shear characterization of unidirectional composites with the off-axis tension test, Exp. Mechs., vol. 26, n°1, March 1986, pp. 103-112.

PARAMETRIC IDENTIFICATION OF MECHANICAL STRUCTURES : GENERAL ASPECTS OF THE METHODS USED AT THE LMA

G. LALLEMENT, S. COGAN
Laboratoire de Mécanique Appliquée, Associé au CNRS,
UFRST, Université de Franche-Comté, BESANCON

ABSTRACT

The accent is placed on the localization of the dominant modelling errors by a sub-space method. Other points also discussed concern the preparation of tests for a parametric identification : optimal selection of the pick-up positions with respect to : the reconstitution of the unobserved degrees of freedom ; the orthogonality and the numerical conditioning of the Jacobian matrix.

OBJECTIVES

This paper investigates the problem of the improvement in the quality of predictive models used in calculating the dynamic behavior of mechanical structures. The following assumptions are made : the structures are linear elastodynamic ; they lead to mathematical models which are "simple" in the sense that they are governed by state matrices which are strictly diagonalizable (no Jordan blocks ; no principal vectors). The case of non self-adjoint and "non simple" structures can be treated with analogous although more complicated methods. The proposed methods exploit data which are stationary with respect to time : eigensolutions and frequency responses.

It is sought to correct the state matrices, M^a ; B^a ; $K^a \in R^{C,C}$ symmetric positive definite, representing the initial estimation (Finite Elements) using the measured dynamic behavior of the real physical structure. This parametric identification problem is reduced to a linear or non-linear parametric optimization problem with constraints.

The technical applications envisaged are :
* The improvement of the quality of predictive models for the global (deplacement field) or local (stress field) behavior in the presence of exterior excitations or new structural configurations (Ex : dynamic sub-structuring) ;
* Improvement of the data bases in the case of sub-structures or of boundary conditions which at present are difficult to modelize (Ex : ground-structure interaction) ;
* Detection of the modelization errors and aid in a partial analytical remodelization ;
* Surveillance and localization of structural failures (Civil engineering examples : Dams, Cooling towers).

The aspect studied here is : The parametric correction of the initial estimation. In the case where this leads to incoherent results, a complete reformulation of the initial estimation must be performed. This aspect is not studied here.

The "localization of dominant errors" is to be understood in the following sense. The reduction of the "distances" between the predicted behavior of the model and the partially observed behavior of the structure is often reduced to a problem of minimizing a cost function constituted of quadratic forms. The necessary condition for the minimization of this function with or whithout local linearization is thus reduced to the resolution of a local problem which is linear and non homogeneous, either under or over determined and having real matrices of the form :   Ax = b   (1)
where : b is the representation of the distances,
         x the local parametric correction vector (in the simplest case).
     The resolution of (1) by pseudo-inverse generally does not lead to a physically satisfying solution. Schematically :
- if (1) is under-determined, a solution $\hat{x}$ of minimal norm privileges the columns of A having dominant norms. This solution has no physical significance ;
- if (1) is over-determined, a solution $\hat{x}$ of minimal error norm leads to dominant components $\hat{x}_i$ of $\hat{x}$ which are associated with columns of A having small norms.
- a solution $\hat{x}$ obtained form a regularization (for example of the type Tikhonov [1]) leads to a intermediate solution between the 2 preceeding cases.
     In all these cases, the solution $\hat{x}$ does not allow the poorly modeled columns of A (and ultimately the sub-domains of the structure) to be detected (except if A is an orthogonal matrix). In terms of the concept of attempting to localize the physical modelization errors it is proposed to solve (1) in a qualitative manner with the objective of detecting the smallest number of parameters contributiong in a dominant way to the distance "b".

LOCALIZATION BY THE SUB-SPACE METHOD [2;3]
     The localization of the dominant modelling errors in the estimation M^a ; K^a (and eventually B^a) is reduced to a selecting among the columns of A those which are the closest to the direction defined by b. The global quality of this localization depends on the projection of the vector b in the sub-space whose basis vectors are the columns of A.
     The first useful operation consists thus in evaluating the rank of the matrix A and that of the augmented matrix $\tilde{A}$ = [A;b]. If the ranks of these 2 matrices are the same, then the necessary condition for the localization is satisfed. If $R(\tilde{A}) > R(A)$, a perfect localization will be impossible.
     From a numerical point of view, the following prodedure is followed. Let σ_i be the ith singular value of A and : $r_i \overset{\Delta}{=} \sigma_i/\sigma_1$. R(A) is defined to be equal to the number of quantities r_i superior to a given tolerance T. After determination of the singular values of $\tilde{A}$ and A, the evolution of r_i as a function of i can be plotted for $\tilde{A}$ and A (Fig. 1).

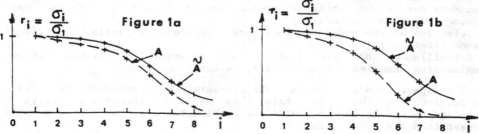

If the configuration is of type (1a), then it is assumed that $R(\tilde{A}) = R(A)$: the localization is potentially possible. If the configuration is of type (1b), $R(\tilde{A}) > R(A)$: the localization is probably imperfect.

The second operation consists in the localization properly speaking. The best sub-space of dimension d is sought (d = 1,2...) in the space of the parameters x which best reduces, in the least squares sense, the error vector ε_d of the over-determined problem :

$$\varepsilon_d = \frac{1}{||b||} [b - A_d x_d],$$

hence : $x_b = A_d^+ b$, where A_d is a sub-matrix of A formed by d columns ; A_d^+ is the Moore-Penrose pseudo-inverse of A_d ; $x_p \in R^{p,1}$ is the sub-vector of x associated with the p columns of A_d. The norm $||\varepsilon_d||$ is a non-increasing function of d. Its evolution as a function of d is represented in figure 2.

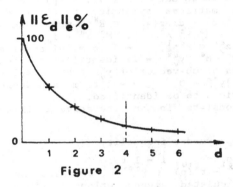

Figure 2

In this example, $||\varepsilon_d||$ is practically stationary for d > 4. The dominant localized parameters which must be taken into account in priority are those corresponding to the first 4 selected columns.

In practice :
a) Before localization, the following operations are performed : α) a regrouping (or elimination) of the columns of A having small norms ; β) a regrouping of columns which are quasi-linearly dependent if and only if they correspond to adjacent sub-domains.
b) The localization starts with a search for the best combination of d = 1,2 to 3 columns among the possible combinations of the columns of A. Retaining these 3 columns, the best fourth column among those remaining is selected, and so on, with the objective of limiting the number of combinations.
c) The evolution of $||\varepsilon_d||$ depends on the definition of the sub-domain partitioning and the parameterization chosen in these domains. Several localizations are performed associated with different partitionings, then comparisons and intersections between the localized parameters are made.
d) The quality of the localization in the sense of selecting the smallest number of poorly modelized parameter in the finite element model conditions the quality of the ulterior parametric adjustment. The selection of the equations forming the lines of A is thus considered to be very important. An optimal selection is indicated : In what follows, several formulations allowing the construction of the matrix A are examined.

MINIMIZATION OF THE DISTANCES BETWEEN THE EIGENSOLUTIONS
INTRODUCES AS OUTPUT QUANTITIES

Hypothesis : The considered eigensolutions are those of the conservative structure associated (Abr. ACS) to the physical dissipative structure. The partially observed complex eigensolutions are thus initially transformed into those of the ACS by a numerical method [4;5].

This restriction is made for the following reasons :
a) Decoupling between the identification of the conservative parameters of $M^a(p)$; $K^a(p)$ and the dissipative parameters of $B^a(p')$
b) Poor initial estimation of $B^a(p')$ and poor quality and limited quantity of measured data with respect to the effect of $B^a(p')$.

Formulation

The localization and parametric correction are performed using the following data : calculated data : $K^a(p)$; $M^a(p) \in R^{C,C}$, fi-nite element mass and stiffness matrices symetric, positive definite ; $Y^a(p) = [\ldots y^a_v \ldots] \in R^{C,N}$, $\Lambda^a(p) = \text{diag}\{\lambda^a_v\} \in R^{N,N}$ modal and spectral matrices such that : $K^a Y^a = M^a Y^a \Lambda^a$; $^T Y^a M^a Y^a = I_N$.

Measured data : λ^{ex}_v ; $z^{ex}_v \in R^{c,1}$, $v = 1$ to m, eigenvalues and normalized sub-eigenvectors ($^T y^{ex}_v M^{ex} y^{ex}_v = 1$) identified on the c pickup dof of the structure. z^{ex}_v is a sub-vector of $y^{ex}_v \in R^{c,1}$, solution of :

$$[K^{ex} - \lambda^{ex}_v M^{ex}] y^{ex}_v = 0, \qquad v = 1 \text{ to } m \text{ and where } K^{ex} ; M^{ex} \in R^{C,C}$$

are the unknown matrices to be identified.

The procedure consists in an iterative minimization of the functionnal $f(p)$:

$$f(p) = \frac{1}{2} {}^T r(p) \; P_r \; r(p), \text{ under inequality}$$

constraints : $p_{iL} \leq p_i \leq p_{iU}$, $i = 1$ to q (4)
where : * $r(p) = r^{ex} - r^a(p) \in R^{cm+m,1}$ contains the distances between the measured and calculated eigensolutions ; $r^{ex} = {}^T({}^T z^{ex} ; {}^T \lambda^{ex})$; $r^a(p) = {}^T({}^T z^a(p) ; {}^T \lambda^a(p))$ where z^{ex} ; $z^a(p) \in R^{cm,1}$; $\lambda^{ex}, \lambda^a(p) \in R^{m,1}$
 * p ; $p^a \in R^{q,1}$ vectors of the design variables, p unknown vector to be sought, p^a its estimation
 * $P_r \in R^{m(c+1), m(c+1)}$ diagonal, positive definite, covariance matrix of the uncertainties in the measured quantities.

According to Newton's method, the minimization of (4) without constraints is written : $[{}^T S(p^a) P_r S(p^a) + V(p^a)] x = {}^T S(p^a) P_r r(p^a)$ (5)

with : $S(p^a) \overset{\Delta}{=} \left[\dfrac{\partial r}{\partial p} \right]_{p^a} \in R^{m(c+1),q}$;

$\Delta p = p - p^a \overset{\Delta}{=} x$;

$V(p^a) \in R^{q,q}$, having the general element :

$$V_{ij} = \left[\frac{\partial^2 r_1}{\partial p_i \partial p_j} ; \ldots ; \frac{\partial^2 r_{cm+m}}{\partial p_i \partial p_j} \right]_{p^a} P_r \; r(p^a).$$

In practice :
a) The approximation $f(p)$ by a quadratic model does not allow convergence in a single step. An iterative procedure is thus adopted. In addition, to reduce the cost of the calculation, the influence of the second derivatives is neglected (matrix V) and a Gauss-Newton type resolution is frequently employed.
b) For the localization, a Gauss-Newton type approach has been used until now. Equation (5) is thus equivalent to the minimization of the residual error vector ε :

$$\varepsilon = P_r^{1/2} [r(p^a) - S(p^a)] x \overset{\Delta}{=} b - Ax \text{ (cf 1) (6)}$$

c) The proposed procedure has a fundamental fault : it is not based on a minimum variance estimator for the unknown parameters x (H.G Natke [6;7] ; M.I Friswell [8]). This type of approach must thus be modified at the level fo the cost functions (4). The accent will be placed here on other aspects in a deterministic, thus imperfect, approach.
d) The details concerning the numerical eva-luation of the Jacobian matrix S and its updating during the iterations are given in [9].

Special aspects
★ Definition of the vector x of design variable. The unknown matrices K^{ex} ; M^{ex} are written :

$$K^{ex} = K^a + \sum_{r}^{} k_i K_i^a \; ; \; M^{ex} = M^a + \sum_{s}^{} m_i M_i^a \quad (7)$$

where K_i^a ; M_i^a are the matrices associated with the ith finite macro-element ; k_i, m_i optimal values of the correction parameters, hence $x = {}^T(...k_i...m_i...) \in R^{q,1}$, $q = r+s$.

The partitioning into macro-elements must : α) allow the independent representation of the different types of kinetic and potential energies which are dominant in the observed frequency band ; β) allow the use of evolving macro-elements sizes and definitions : localizations with several macro-element de-finitions ; following the localization of the sub-domains presenting the dominant errors rise of smaller macro-elements definitions.
★ Normalization errors in the measured eigenvectors. The uncertainties in the identified generalized masses are frequently very important [10]. In r(p), the sub-vector $(z^a(p) - z^{ex})$ is directly affected by these uncertainties. They are taken into account by the intermediary of m additional unknown, each one associated with an identified sub-eigenvector. Let : $z_v^{ex} = (1 + \alpha_v)z_v$, where : $z_v^{ex} \in R^{c,1}$, exact sub-eigenvector, z_v observed sub-eigenvector, α_v real scalar. In order to simplify the formulation, it is supposed that $P_v = I$ in equation (6). For $\alpha_v = 0$, the c equations relatives to the vth identified sub-eigenvector intervenes in (6) in the form : $z_v^{ex} - z_v^a = S_z \Delta p$.

For $\alpha_v \neq 0$: $z_v - z_v^a = [S_{zv} \mid -z_v] \begin{bmatrix} \Delta p \\ -- \\ \alpha_v \end{bmatrix}$,

$v = 1$ to m.
(8)

The m correctors α_v are introduced in this way.

Selection of the pickup dof (Test preparation in view of the parametric correction [11]) :
The Jacobian matrix $S \in R^{(c+1)m,q}$ has the general form : ${}^TS = [{}^TS_z \mid {}^TS_\lambda]$. In a deterministic approach, the 2 sub-matrices of S are considered to have the same relative importances. A weighting coefficient is thus introduced for the equations relative to S_λ in such a way as to obtain matrices S_z and S_λ with the same norms. In a stochastic approach, the weightings defined by P_r will then be introduced. The optima pickup mesh is defined by one of the following two procedures.
★ Best orthogonality of S. S_z is rewritten in the form :
${}^TS_z = [s_1 \mid ... \mid {}^Ts_i \mid ... \mid {}^Ts_n]$ where $s_i \in R^{m,q}$ is the Jacobian matrix of the ith dof. The number n corresponds to the number of pickup dof possible after having eliminated form among the C model dof : the inaccessible dof ; the rotational dof ; the dof having small displacements for the m identified modes... The procedure consists in selecting one by one the dof in such a way as to ob-tain a matrix ${}^TSS$ which tends toward a diagonal matrix.
a) Choice of the first dof. Starting with the n potential dof, n

matrices $S_1^i = W_i \begin{bmatrix} s_i \\ -- \\ S_\lambda \end{bmatrix} \in R^{2m,q}$, $i = 1$ to n, are constructed, where W_i

allows an identical norm to be obtained in the sensitivity blocks for the eigenvalues and eigenvectors. The dof i which best diagonalizes $^T S_1^i$ S_1^i is retained.
b) Choice of the second dof. The n-1

$$\text{matrices } S_2^j = W_j \begin{bmatrix} s_i \\ s_j \\ S_\lambda \end{bmatrix} \in R^{3m,q}, \ j = 1 \text{ to } n,$$

j ≠ i, are constructed where W_j allows as before, to obtain the same norms in the sensitivity blocks for the eigenvalues and eigenvectors. The dof j is selected which best diagonalizes $^T S_2^j$ S_2^j.
c) By proceeding in this way, step by step, a local minimum of the diagonality is obtained for a given number c of pickups. The global minimum is obtained by evaluating the diagonalities of the C_n^c combinations. This number is, in general, too large to envisage such a procedure. Thus a sub-optimum of the optimum pickup mesh is obtained.
d) Choice of the number c. Figure 3 represents the typical evolution of the diagonality δ(S) as a function of the number of selected pickups. The number c which is selected corresponds to one of the first local minima which are obtained.

★ Best conditioning of S. This alternative method consists in selecting one by one the dof yielding the best possible conditioning of S. The rank of S is defined as before . The condition number is defined as equal to the inverse of the smallest value r_i superior to a given threshold. Having chosen the n possible dof, a matrix S is constructed by a technique analogous to the preceeding one : selection one by one of the dof allowing to simultaneously obtain the maximum rang and the minimum condition number for the normalized sensitivity matrix. This normalization is performed by normalizing each column of S to 1. Figure 4 represents the typical evolution of the rang and the condition number as a function of the number of selected dof. The optimal number c is that which yields a maximum rank with a minimum of dof and a minimum condition number.

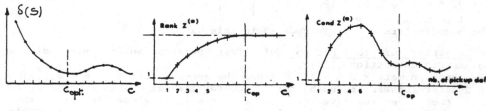

Figure 3   **Figure 4**

<u>Critical analysis of this type of parametric identification</u>
Inconvenients :
a) In the context of an industrial modal identification, the uncertainties in the generalized masses (errors from 20 to 100 %) impose the introduction of the additional unknowns α_ν, $\nu = 1$ to m.
b) The matching between the calculated and measured eigensolutions is necessary. The matching can lose all meaning in the case of multiple and quasi-multiple eigensolutions. Some solutions have been proposed [(12);(13)]. In addition, in the case of dense spectra, an automatic control of the matching during the updating iterations is indispensable ("crossing" problems of eigen-vectors).
c) The localization depends on the linearization of the residual vector r(p) with respect to the parameter vector. Even though in the case of a

fairly sparse spectrum, the convergence radius of the Taylor series expansion of r(p) is relatively large, several localizations during the parametric updating are recommended in the case of significant structure-model distances.

Advantages :
a) No reconstitution of the unobserved sub-eigenvectors.
b) Good robustness with respect to the uncertainties in the measured data.
c) Domains of application :
   *) case where only a small number c of pickups are available ;
   *) case where the model-structure distances are small or moderate.
The experience gained form numerical and experimental test cases has confirmed the ef-ficacity of the optimal selection of pickup dof ;
   *) Improvement of the quality of the dominant error localization;
   *) Improved robustness with respect to the uncertainties in the measured data.

EXPLOITATION OF THE DYNAMIC STIFFNESS
AND FLEXIBILITY RELATIONS [11]

Objective
These methods are complementary to the preceeding ones. They do not exploit the direct minimization of a residual constructed from the outputs. They eliminate the matching constraint, the generalized masses and the linearization of the distances, but require the reconstitution of the unobserved sub-eigenvectors. The formulation leads to an equation analogous in form to (1) in which the localization is performed by the sub-space method.

Formulation
The localization and identification exploit three blocks of equations which are regrouped in a single system after having been weighted with respect to the others.
★ Equation relative to the eigenvalues :
The model sought must satisfy, for each one of the m measured eigenvectors, the dynamic stiffness relation :
$$Z_v \, y_v^{ex} = f_v(p)$$
where : $Z_v = (K^a - \lambda_v^{ex} M^a) y_v^{ex}$ (9)
$$f_v(p) = (\lambda_v^{ex} \Delta M - \Delta K) y_v^{ex}, \quad v = 1 \text{ to } m$$

Prémultiplying by $^T y_\sigma^a$, $\sigma = 1,2\ldots$ leads to :

$$d_{\sigma v} \, \Delta \lambda_{\sigma v} = {}^T y_\sigma^a [\Delta K - \lambda_v^{ex} \Delta M] y_v^{ex}, \quad v = 1 \text{ to } m ;$$
$$\sigma \cong 1 \text{ to } m + 2 \qquad (10)$$

with : $\Delta \lambda_{\sigma v} \overset{\Delta}{=} \lambda_v^{ex} - \lambda_\sigma^a$; $d_{\sigma v} \overset{\Delta}{=} {}^T e_\sigma d_v$, is
the σth component of the linear combination vector defined by : $y_v^{ex} = Y^a d_v$ where $Y^a \in R^{c,p}$ is the modal sub-base formed by the p first eigenvectors of $(M^a ; K^a)$.

Following the parameterization of ΔM ; ΔK analogous to (7), the equations (10) are regrouped in the form :
$A_1 x = b_1$, $A_1 \in R^{a_1, q}$. The number a_1 is defined by the condition : $d_v \geq 0,5$ and in practice : $m \leq a_1 \leq 2m$ (11)

★Orthogonality relations :
They are defined by :
$$^{T}y_{\sigma}^{ex}[M^{a} + \Delta M]y_{v}^{ex} = 0,$$
$^{T}y_{\sigma}^{ex}[K^{a} + \Delta K]y_{v}^{ex} = 0,$ where $v \neq \sigma$; $v,\sigma = 1$ to m and regrouped in the
form : $A_{2} x = b_{2}$ where : $A_{2} \in R^{a2,q}$, $a_{2} = m(m-1)$ (12)

★Dynamic flexibility relations :
They have the form : $y_{v}^{e} = [K^{a} - \lambda_{v}^{ex} M^{a}]^{-1} [\lambda_{v}^{ex} \Delta M - \Delta K]y_{v}^{ex}$, $v = 1$ to
m, and are regrouped in the form :
$$A_{3}x = b_{3}, \text{ where } A_{3} \in R^{a3,q} \text{ with : } cm \leq a_{3} \leq Cm \quad (13)$$

The difficulties related to the numerical conditioning of the ma-
trix : $[K^{a} - \lambda_{v}^{ex} M^{a}]$ are avoided by the introduction of shifts in the
eigenvalues of the model M^{a} ; K^{a}. The totality of equations (11 to 13)
form the linear non homogeneous system : $Ax = b$ which is successively
exploited for the localization of the dominant errors followed by the
parametric correction.

Reconstitution of the unobserved sub-eigenvectors

★Direct reconstitution :
Among all the reconstitution procedure envisaged at the AML, the most
acceptable solutions in terms of the precision of the vector $y_{v}^{ex} \in R^{c,1}$
reconstituted form $z_{v}^{ex} \in R^{c,1}$ are obtained by projection on the sub-
basis $Y^{a} \in R^{c,p}$ of the finite element estimation : $y_{v}^{ex} = Y^{a} d_{v}$, where
$d_{v} = Z^{a+} z_{v}^{ex}$
$Z^{a} \in R^{c,p}$ (p < c, $R(Z^{a}) = p$), sub-matrix of Y^{a} corresponding to the c
observed dof ;
$Z^{a+} \in R^{p,c}$ Moore-Penrose pseudo-inverse of Z^{a}.

★Optimal repartition of the pickups in terms of the reconstitution :
The optimal pickup mesh is defined here as that one which leads to
the best numerical conditioning of the modal sub-basis Z^{a}. It is
constructed line by line

a) Choice of the first dof. Given the n observable dof, the n norms
$||z_{i}^{a}||_{n} = ||^{T}(z_{i1}^{a} ; \ldots, z_{ip}^{a})||$ are formed, where $z_{ij}^{a} =$ displacement of
the ith dof for the jth mode. The first dof retained is the one which
maximizes this norm. Let i be this dof.
b) Choice of the second dof. All the
matrices $Z_{2j}^{a} = \begin{bmatrix} ^{T}z_{i}^{a} \\ \overline{^{T}z_{z}^{a}} \end{bmatrix}$, $j \neq i$ are formed. The
2nd dof retained is the one which maximizes the rank while minimizing
the condition number of Z_{2j}^{a}.
c) Choice of the optimum number c of dof. By exploiting the graphs of
the variations of the rank and condition number of the retained matrices
Z^{a} as a function of the number of dof, the number c corresponds to the
minimum of the conditioning of Z^{a} with a maximum rank of Z^{a}.

Note : * If certain dof are imposed for technical reasons, they are
introduced from the start in this sub-optimal construction process.
     * The sub-optimal matrix Z^{a} thus constructed, leads to a
significant improvement of the quality of the reconstitution and an
improved robustness with respect to the uncertainties in z_{v}^{ex}.

★Iterative reconstruction of the unobserved sub-eigenvector :
The quality of the reconstruction conditions the quality of the
localization and parametric correction. This quality improves as the
projection basis approachs the eigenbasis of the structure, hence the
following interative procedure.

1st iteration :   . A fairly
crude localization is performed and the localized parameters are
corrected. Let ΔK^1 ; ΔM^1 the obtained correction. The eigensolutions of
the problem are then evaluated :
$[(K^a + \Delta K^1) - \lambda_v^{(1)}(M^a + \Delta M^1)]y_v^{(1)} = 0$, giving the construction of $Y^{a(1)}$.
At the $k^{i\grave{e}me}$ iteration : the reconstitution is performed from :
$y_v^{ex(k)} = Y^{a(k-1)} d_v^{(k)}$, where : $d_v^{(k)} = [Z^{a(k-1)}]^+ z_v^{ex}$.

These iterations are performed in view of improving the quality of
the Ritz basis. The localization and adjustment are performed at each
iteration in terms of the M^a ; K^a initial estimation.

Critical analysis
Inconvenients :
* require s the reconstitution of the unobserved dof, requirement which
implies that significant number c of dof have been observed (c ≃ 3 to
5 m, c beeing a function of the types of observed modes).
* less robust   with respect to the uncertainties in the measured
data. The developments concerning the relative robustness of : the
stiffness and dynamic flexibility relations ; the orthogonality
relations ; the differences between the eigenvalues, with respect to the
uncertainties and modal troncation errors, have been established.

Advantages :
* no mode matching but "recognition" of the observed modes is
  recommended.
* generalized mass errors have no influence.
* no linearization, except the parametrization (7).

CONCLUSION
Two complementary parametric correction strategies have been proposed. This complementary is a function : of the significance of the
distance between the model and the structure ; of the number of observed
degrees of freedom ; of a minimization of residuals constructed form the
outputs and inputs. The key point controlling the convergence of the
finite element model toward the discrete model of the same order which
is closest to the physical structure lies in the localization of the
poorly modelized regions. The optimal selection of the observed degrees
of freedom contribute considerably toward this goal. The present
developments are made for the ESA-ESTEC, Noordwijk, in the framework of
a research contract and we would like to thank Mrs. E. FISSETTE and S.
STRAVRINIDIS for their support.

REFERENCES
1. Tihkonov, A., Arsenine, V., Méthodes de résolution de problèmes mal
   posés. Editions Mir, Moscou, 1974.
2. Fillod, R., Lallement, G., Piranda, J., Zhang, Q., Parametric
   correction of regular non-dissipative finite element models, 11th
   Int. Modal Analysis Seminar, Leuven, 22-26th Sept., 1986.
3. Lallement, G., Localization techniques, Proc. Worshop on Structural
   Safety Evaluation based on System Identification Approches, June 29-
   July 1, 1988, Lambrecht, FRG, In Vieweg Int. Sci. Book Series,
   Braunschweig, 213-233.
4. Zhang, Q., Lallement, G., Estimation of the generalized damping
   matrix from identified complex eigensolutions, Mech. Struct. and
   Mach. 197, 15(4), 543-557.
5. Zhang, Q., Lallement, G., Simultaneous determination of normal
   eigenmodes and generalized damping matrix from complex eigenmodes,
   Proc. 2nd Int. Conf. Aeroelasticity, 1985, Aachen, FRG, 529-535.

6. Natke, H.G., Einführung in Theorie und Praxis der Zeitreihen-und Modal-analyse- Identifikation schwingungs-fähiger elastomechanischer Systeme. Friedr. Vieweg & Sohn, 1983, Braunschweig, Wiesbaden.
7. Natke, H.G., Survey on the identification of mechanical systems, Proc. 2nd Workshop on Road-Vehicle systems and related mathematics, June 20-25, 1987, Torino.
8. Friswell, M.I., The adjustment of structural parameters using a minimum variance estimator, Mech. Syst. and Signal Processing, 1989, 3(2), 143-155.
9. Lallement, G., Piranda, J., Recalage paramétrique par une méthode de sensibilité, Actes de Strucome, Paris, 2-4 nov., 1988.
10. Fillod, R., Piranda, J., Bonnecase, D., Taking non linearities into account in modal analysis by curve fitting of transfer functions, Proc. 3rd IMAC, Orlando, USA, 28-31 Jan., 88-95, 1985.
11. Andriambololona, R., Identifications modale et paramétrique, Thèse Doctorat, 1990, Université de Franche-Comté, Besançon, 25030 Cedex, France.
12. Zhang, Q., Identifications modale et paramétrique de structures mécaniques auto-adjointes et non auto-adjointes, Thèse de Doctorat es Sciences Physiques, Mention Mécanique, 1987, Université de Franche-Comté, Besançon, 25030 Cedex, France.
13. Lallement, G., Zhang, Q., Inverse sensitivity based on the eigen-solutions : Analysis fo some difficulties encountered in the problem of parametric correction of finite element models, Proc. 13th Int. Sem. Modal Analysis, 1988, Leuven, Belgium, Part 1, 20 p.

MATERIAL PARAMETERS IN ANISOTROPIC PLATES

KARL-EVERT FÄLLSTRÖM
Division of Experimental Mechanics
Luleå University of Technology
S-95187 Luleå, Sweden

ABSTRACT

Material parameters, that is, the two Youngs moduli, the in-plane shear modulus and the Poisson's ratio, in anisotropic rectangular plates are determined in three different ways.

1. Using modes of vibration

Real-time, TV-holography is used, to determine frequencies and shapes of the first five modes of vibration of plates with free-free boundary conditions. By comparing experimental results with FE-calculated ones the material parameters are determined in a non-destructive way.

2. Using Rayleigh's Method

The plates are tuned, by changing the quotient between the length of the sides, so that the second and the third mode of vibration, for free-free boundary conditions, degenerates into the well-known cross- respectively ring-mode. The first three modes of vibrations for these plates are determined by optical methods.
It is found that bending waves generated in the middle of the tuned plates will reach the boundaries of the plates simultaneously. This gives a relationship between main material parameters. Using this relation and Rayleigh's method for the first three modes of vibrations, the main material parameters for the plates are determined.

3. Using transient bending waves

Bending waves are generated by the impact of a ballistic pendulum. Hologram interferometry, with a double pulsed ruby laser as light source, is used to record the out of plane motion of the waves.
Studying the propagating bending waves in the plates makes it possible to determine the material parameters in a non-destructive way.

INTRODUCTION

 Methods for the testing of materials are an area of central interest in material evaluation. Testing methods are needed both during the material development process, production and the use of the final product. The problem of finding adequate testing methods has become a field of current interest in recent years. One reason for this is that a lot of new materials have been introduced. Many of these "modern materials" are collectively termed "composite materials". Composite materials represent a wide range of materials, and a problem with many of these materials is the lack of methods for non-destructive testing (NDT).

 NDT is the detection of material or manufacturing imperfections by procedures which do not require destruction or significant alteration of the component being tested. Non-destructive testing employs a variety of techniques such as ultrasonic methods, thermography, acoustic emission, visual inspection, vibration measurements and holographic interferometry.

 In our testing of materials we use holographic interferometry. TV-holography or speckle interferometry (ESPI) is used when we study the modes of vibrations and holographic interferometry with a double pulsed laser as light source when transient bending waves are studying.

 Useful characteristics of holographic non-destructive testing techniques include simple, whole-field visual display, applicability to inspection of components having fairly complicated shapes, and lack of special requirements for surface preparation of the test object. Limitations of the technique are stringent requirements for mechanical stability and a restricted range of sensitivity.

TEST SAMPLES

Six different glass-fibre reinforced composite plates with orthotropic and anisotropic properties were produced for the investigation. The first five plates (Plate 1 - 5) were wound on a flat tool in a filament winding machine and then

TABLE 1
Properties of test plates used in the calculations

Plate	Laminate sequence	Thickness (mm)	Density (kg/m³)	Properties	Size (mm*mm)
1	$(2.5/-2.5)_3$	3.34	1940	aniso, antisym	216*164
2	$(4.5/-3.5/4.5)_S$	3.10	1993	aniso, sym	229*171
3	$(2.0/-2.0/2.0)_S$	3.15	2027	aniso, sym	214*166
4	$(30.5/-29.5/30.5)_S$	4.10	1957	aniso, sym	208*172
4a	$(25/-35/25)_S$	4.10	1957	aniso, sym	172*141
5	$(29/-29)_3$	4.03	1938	aniso, antisym	204*175
6	4 plies, weave	2.87	1736	orthotropic	182*140
7	6 plies, weave	4.70	1660	orthotropic	225*181

pressed between two steel-plates. This gave an even thickness over the surface. The plates were cured for 4 hours in 140°C in a convection oven. Plate number 6 and 7 were produced with hand layup and cured at room temperature.

EXPERIMENTAL SETUPS

In our experiments, two different kinds of experimental set ups are used. In the experiments in which the material parameters are estimated by studying modes of vibrations, these modes are recorded by means of TV-holography. When the material parameters are estimated studying propagating transient bending waves, the interferograms of the waves are recorded with holographic interferometry with a double pulsed laser as light source.

Set-ups for recording modes of vibrations in plates
If a plate is excited into harmonic vibrations it comes in resonance for certain frequencies. At each such frequency the plate vibrates in a certain manner, called a mode. With TV-holography it is possible to record the shape of such a mode, see figure 1a. The broad white lines in the figure indicate the parts of the plate which do not move. These lines are called nodal-lines. Figure 1c illustrate the vibration of the plate. The dashed lines in this figure represent the nodal-lines. Figure 1 shows the second mode of vibration, called the (2,0)-mode. (2,0)-mode means that there is, two nodal-lines parallel with the longer and none parallel with the shorter side of the plate.

  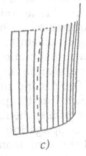

a)   b)   c)

Figure 1. a) shows the (2,0)- mode of vibration recorded with TV-holography, b) the same mode recorded with double pulsed interferogram technique and c) the vibrating plate. The nodal-lines are indicated in a) with the broad white lines and in b) and c) with dashed lines.

TV-holography is based on image-plane, time average holographic recordings on a TV-vidicon. Since the holograms are recorded on a TV-vidicon, they are updated with 25 Hz. This means that the instrument works as in real time. TV-holography is a non-contact and non-destructive measuring technique. The TV-holography system we used is limited to a lower frequency of about 20 Hz and in object size up to 1 m^2.

As seen in figure 1a the picture is very blurry. The reason is that the TV-holographic measurement gives a speckled image on the monitor. This is a disadvantage for documentation purposes. In our experiments it is necessary to use an instrument with real time presentation to get the correct shape of the modes, as the process is iterative. To get better interferograms for documentation we use holographic interferometry with a doubled pulsed ruby laser as light source to record the interferograms, when a mode had been identified by TV-holography. Figures 1a and 1b show a comparison between pictures obtained with TV-holography and the double pulsed interferogram technique. It is seen that the picture obtained with the double pulsed laser has much better resolution. The black curves in figure 2b can be considered as iso-amplitude curves, that is, the curves connecting points vibrating with equal amplitude. With less coarse speckles in the TV-image it is not necessary to use the double pulsed interferogram technique. Today such instruments exist, which reduce the speckle noise in the pictures from TV-holography.

Set-up for recording transient bending waves in plates

If transient phenomena are to be studied it is not possible to use TV-holography, as this technique is based on time average holographic recordings. Therefore, we employ holographic interferometry with a double pulsed ruby laser as light source to study such phenomena. With the doubled pulsed laser it is possible to get two pulses with a duration between them from 1 to 800 µs. Since each laser-pulse duration is as short as 25 ns, almost all mechanical movements are recorded "frozen" in the double exposed holograms.

We use this technique to study transient bending waves in plates [1,2]. These are impacted with a small pendulum. The interferograms of the created propagating transient bending waves are recorded with a double pulsed interferogram technique. The first pulse from the laser records the object just before the pendulum impacts on the plate and gives the first exposure of the hologram. The second pulse is launched at a preset time after the first one. As the pendulum hits the plate in the time interval between the first and the second laser pulse, the latter gives a second exposure of the hologram, which records the propagating wave. In the experiments we also measure the time of impact and the time from the initiation of the impact to the second laser pulse. Figure 2 shows the propagating bending wave in two plates. The left one is isotropic and the right anisotropic. The left figure shows that the wave travels at

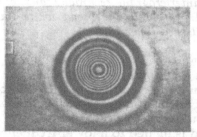 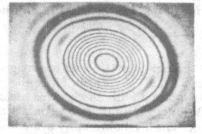

Figure 2. The propagating bending waves in two plates. The left plate is isotropic and the right one is anisotropic.

the same speed in all directions, which is naturally because the plate has the same properties everywhere. The pattern for the anisotropic plate is, however, more complicated. As such a plate does not have the same properties in different directions, the speed of the wave varies in different directions.

One of the advantages of the double pulsed interferogram technique is that it is fast. An experiment sequence for detecting defects in a plate takes about one hour. A sequence normally contains about 30 interferograms, showing the propagating waves at different times after the start of the impact. The impact time is normally about 30-50 μs.

DETERMINING MATERIAL PARAMETERS IN ANISOTROPIC PLATES

1. Comparing experimental results with FE-calculated ones
A more detailed description of this method is found in ref [3].

Modes of vibration of the test plates, both the shape and frequency, are determined using TV-holography. Rectangular plates with free-free boundary conditions are used. For orthotropic plates, it is shown by the finite element method (FEM), that the first mode of vibration depends strongly upon the in-plane shear modulus, the second mode upon the Young's modulus E_1 and the third upon the Young's modulus E_2. This is a consequence of the shapes of the modes of vibration, mainly twisting for mode no. 1, mainly bending across the main fibre direction for mode no. 2 and mainly bending at a right angle to this direction for mode no. 3, see figure 3. It is also shown that a change in Poisson's ratio has very little influence on the frequency of a mode but quite large influence on the shape of the mode. The first three modes of vibration for an anisotropic plate do not show as strong a dependence upon the main material parameters, one at a time, as the orthotropic ones. For anisotropic plates it is possible, for example, that the first mode depends upon both the shear modulus and the two Young's moduli. To determine the material parameters in orthotropic plates we studied the first three modes of vibration, but for the anisotropic plates we studied the first five modes. The procedure to determine the four elastics constants is to vary the material parameters in the FE-

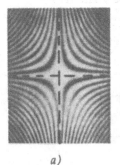  

a)   b)   c)

Figure 3. The first three modes of vibration for an orthotropic plate. The first mode is a twisting mode and the other bending modes. The second mode bends across the fibre direction and the third bends along the fibre direction.

calculations, so that the frequencies and the shapes of the modes of vibration coincide with the experimentally obtained ones.

The advantages with this method is that the material parameters, that is, the two effective Young's moduli, the in-plane shear modulus and the Poisson's ratio, are determined with the use of only one plate. Furthermore, it is fast (the experiments for one plate take about thee hours), and the instrument we use is moveable (the experiments can be performed not only in a laboratory).

2.Rayleigh's method

A more detailed description of this method is found in ref [4].

With this method we also study the frequency and shape of the first three modes of vibration of rectangular plates with TV-holography. The plates are tuned, by changing the quotient between the length of the sides, so that the second and the third modes of vibration degenerate into the well-known cross and ring mode, respectively. A cross-mode is characterized by the nodal-lines crossing each other. For the ring-mode we have only one nodal-line which describes a closed curve, see figure 4. The first three modes of vibrations can be approximately describe with three simple functions.

A good approximation for the (1,1)-mode, that is the first mode, is

$$w = \sin \frac{\pi}{2}\left(\frac{y}{b} + m\frac{x}{a}\right) \cdot \sin \frac{\pi}{2}\left(p\frac{y}{b} + \frac{x}{a}\right) \qquad (1)$$

where m is the slope of one of the nodal lines relative the x-axis and p is the slope of the other nodal line relative the y-axis. The x- and y-axis are parallel to the longer respectively the shorter side of the plate.

The cross mode, the (0,2)-mode, is described by the function

$$w = \sin \frac{\pi}{2}\left(\frac{y}{b} + k\frac{x}{a}\right) \cdot \sin \frac{\pi}{2}\left(\frac{y}{b} + q\frac{x}{a}\right) \qquad (2)$$

where k and q are the slopes of the nodal lines relative the x-axis.

The ring mode, (2,0) mode, can with good accuracy be written

Figure 4. The first three modes of vibration for an orthotropic plate which has tuned so that the second and third modes degenerate to a cross- respective a ring-mode. The dashed lines represent the nodal-lines.

$$w = \cos\left(\frac{\pi}{2}\left(\frac{y}{b} + k\,\frac{x}{a}\right)\right) \cdot \cos\left(\frac{\pi}{2}\left(\frac{y}{b} + q\,\frac{x}{a}\right)\right) - \left(\frac{b}{a}\right)^2 \tag{3}$$

The same values of k and q as in equation (2) are used in this equation.

We have also shown that transient bending waves in plates which are cut in this way, reach the boundaries at the same time if the plates are impacted in the centre. Using this fact and Rayleigh's method for the first three modes of vibration, it is possible to determine the two Young's moduli, the in-plane shear modulus and the Poisson's ratio for anisotropic plates.

The advantages of this method are the same as for the previous method; all parameters can be estimated with the use of only one plate, it is fast and the instrument is moveable. Another advantage is that we need very little computer capacity. The disadvantage is that the sides of the plate have to be cut so that the second mode of vibration degenerates to a cross-mode.

3. Using bending waves
This method is described in more detail in ref [2,5].

We have studied the propagation of bending waves in isotropic and anisotropic plates. The plate equation has been solved for the isotropic case. The plate is assumed to be of uniform thickness and composed of a linear elastic material. If the starting conditions are modelled as a Dirac pulse in space and time, a similarity variable is found. This variable depends upon the plate thickness, the density and the material parameters. The existence of this variable brings new understanding to the importance of specific parameters for wave propagation in plates.

In isotropic plates the transient bending waves travel at the same speed in all directions. For anisotropic plates the travelling wave pattern is more complicated, see figure 2. In this case the plate equation is more complicated than for the isotropic case.

We have not yet found the equivalent solution to the plate equation for the orthotropic or the anisotropic case. However, we have shown that it is possible to expand the isotropic theory to be also valid for anisotropic plates if we introduce effective material parameters, which describe the material properties in two mutually orthogonal directions [2]. Thereby we substitute the isotropic material parameters with the effective material parameters valid for the anisotropic plates. That is, for example the isotropic Young's modulus is substituted with the effective Young's moduli in the anisotropic case. This expansion makes it possible to interpret the behaviour of the propagating bending waves in anisotropic plates in a better way. It also makes it possible to determine the effective Young's moduli in the main directions for anisotropic plates, based on the results from the experiments.

RESULTS

In table 2 the results from these three ways to determine material parameters in anisotropic plates are compared between themselves and with static and dynamic tests. The static test is a tensile test of bars cut from the plates and the dynamic test is a vibration test of the same bars. Static or dynamic measurements have not been done for the in-plane shear modulus. The agreement is quite good except for E_2 for plate 2. One reason, why especially the E_2-values for the static and dynamic tests are uncertain, is; when the bars are cut from the plates cracks often appear, especially since the main fibre direction is across the bar. In the tensile test we used a strain gauge glued to the bar, which is sensitive to cracks and gives locally determined Young's moduli. In the dynamic test of bars initial cracks also influence on the result. Other reasons for deviating results are that in the tensile test we get very "short signals" for the bars with fibre directions across the bar, giving large errors and that in the FE-calculus no considerations have been taken for damping or non-linear effects.

TABLE 2

A comparison between Young's modulii, shear modulus and Poisson's ratio obtained with different methods. Ra = Rayleigh's method, FE = FE-calculation, Bw = Bending waves, St = Static test and Dy = Dynamic test.

Plate	E_x (GPa)					E_y (GPa)					E_s (GPa)			v_{xy}		
	Ra	FE	Bw	St	Dy	Ra	FE	Bw	St	Dy	Ra	FE	Bw	Ra	FE	St
1	39.5	42.1	-	47	42	13.1	13.3	-	12	12	6.1	6.4	-.	0.28	0.27	0.30
2	44.3	43.8	45.5	48	45	13.9	14.5	13,5	18	12	6.0	6.5	7,6	0.2	0.26	0.30
3	44.7	46.2		49	41	16.2	16.6	-	16	14	6.8	6.9	-	0.23	0.22	0.30
4	24.9	25.2	24.4	22	21	12.1	12.8	11.7	14	11	9.7	9.6	8,8	0.45	0.43	0.45
4a	26.9	26.9	-	-		12.6	12.7	-	-	-	9.3	9.3	-	0.38	0.44	-
5	23.0	24.6	24.6	23	24	12.6	13.1	7,6	12	11	9.0	9.6	9.3	0.43	0.42	0.50
6	30.0	31.8	-	32	-	10.6	10.7	-	10	-	4.6	4.8	-	0.21	0.19	0.13
7	19.7	20.5	-	20	23	8.3	8.4	-	8	7	2.8	2.8		0.22	0.27	0.26

REFERENCES

1. K-E Fällström, H Gustavsson, N-E Molin and A Wåhlin. Transient Bending Waves in Plates Studied by Hologram Interferometry. Accepted for publication in Experimental Mechanics. (1989)
2. K-E Fällström, L-E Lindgren, N-E Molin and A. Wåhlin. Transient Bending waves in anisotropic Plates Studied by Hologram Interferometry. Accepted for publication in Experimental Mechanics. (1989)
3. K-E Fällström and M Jonsson. A Nondestructive Method to Determine Material Properties in Anisotropic Material. Accepted for publication in Polymer Composites. (April 1990).
4. K-E Fällström. Determining Material Properties in Anisotropic Plates using Rayleigh's Method. Accepted for publication in Polymer Composites. (April 1990).
5. K-E Fällström. A Nondestructive Method to Determining Material Parameters through Studying Bending Waves. To be submitted to Polymer Composites.

EVALUATION OF GLOBAL COMPOSITE LAMINATE STIFFNESSES BY STRUCTURAL WAVE PROPAGATION EXPERIMENTS

MARTIN VEIDT and MAHIR SAYIR
Institute of Mechanics
Swiss Federal Institute of Technology, ETH-Zurich
CH-8092 Zurich

ABSTRACT

An experimental method based on the propagation of structural waves is presented which allows to determine the five most important elastic stiffnesses of an orthotropic plate. In the first part flexural waves are studied. A consistent second order theory which includes shear deformation perpendicular to the middle-plane of the plate is derived by means of the asymptotic expansion of the three dimensional, linear-elastic equations. Phase velocities of flexural waves which are induced centrally with a piezoceramic transducer are determined from the phase spectra of the time signals which were measured with a heterodyne interferometer along straight lines through the loading point. The flexural wave experiments allow to determine four elastic constants with very satisfactory accuracy by only one single measurement. In the second part in-plane shear waves are studied. The measurement of the phase velocity in the principal direction of anisotropy allows to determine the in-plane shear modulus. Thus the five most important moduli of the orthotropic plate are determined.

INTRODUCTION

Several kinds of static and dynamic test methods which can be used to determine the global elastic moduli of laminates have been reported in the literature, eg. [1]-[6]. As in [3] and [4] the propagation of structural waves is investigated here. The use of piezoceramic transducers allows to generate reproducible, narrow-band input signals and the highly accurate and sensitive heterodyne interferometer can measure displacements in the range of some nm. In contrast to other methods the anaysis of the experimental data is based on a second order approximation. Thus some of the elastic properties perpendicular to the middle plane of the laminate can be determined.

In the first part the propagation of flexural waves is studied. The results of an asymptotic expansion of the full three dimensional dynamic equations as described in detail in [7] are summarized. The analytical calculations lead to dispersion relations of the second order for plane flexural waves propagating in various directions.

These theoretical results are used to determine experimentally the tensile moduli along the principal axes of anisotropy and the shear moduli perpendicular to the middle plane of a cross-ply carbon fiber-reinforced plate. The plate response due to a reproducible input signal generated by a piezoceramic transducer is measured at distinct points of the plate by a heterodyne interferometer. With the aid of a fast Fourier transform the phase velocity as a function of the wave number for the direction of propagation is obtained. Using a two parameter nonlinear regression procedure, the two tensile moduli along the principal axes of anisotropy and the two shear moduli across the plate thickness can be calculated with sufficient accuracy from the experimental data.

In the second part the same experimental concept and measurement technique is used for in-plane shear waves. The most simple model of plane waves which travel through the material allows to determine the in-plane shear modulus using the phase velocity in the principal directions. The presented experimental data shows that the plane wave model can not be used to explain all physical effects which occur if in-plane waves are centrally induced in the plate.

FLEXURAL WAVES IN AN ORTHOTROPIC PLATE

Results of a consistent second order theory

The physical system under consideration consists of a thin fiber reinforced laminate. According to the Cartesian coordinates defined in figure 1 the axes x_2 and x_3 coincide with the principal axes of anisotropy. The fibres are thin compared to the plate thickness and their volume is about equal to that of the matrix. The structure is regarded as an orthotropic continuum with 9 independent material parameters.

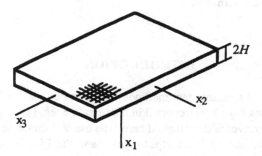

Figure 1. Cross-ply, fiber-reinforced plate, reinforcement in the principal directions of orthotropy, x_2 and x_3.

The calculation scheme for flexural waves in a thin orthotropic plate is described in full detail in [7]. The asymptotic approach which is used here had formerly been applied in [8] to a transversely isotropic plate and in [9] to a transversely isotropic beam. With the application of the two variable perturbation technique, it is possible to derive a uniformly valid solution for the second approximation.

The phase behaviour of centrally induced flexural waves can be approximated quite accurately by the corresponding behaviour of plane waves at distances of as little as one or two wavelengths from the source. This was confirmed for example in [8].

The phase velocity c for a plane bending wave

$$U = U_0 \, exp \, [i \, k \, (x_\alpha \, n_\alpha - c \, t)] \qquad\qquad (\alpha = 2, 3) \qquad\qquad (1)$$

is calculated up to the second order of approximation. The vector $n = n_2 \, e_2 + n_3 \, e_3$ is the unit vector of the propagation direction and k is the wave number. The dispersion relation becomes

$$c^2 = \frac{E_{22} \, (H \, k)^2}{3 \, \rho \, (1 - v_{23}^2 \frac{E_{33}}{E_{22}})} \, [n_2^4 + 2 \, v_{23} \frac{E_{33}}{E_{22}} \, n_2^2 n_3^2 + \frac{E_{33}}{E_{22}} \, n_3^4] \, [1 + \frac{\beta_1}{2 \, \pi} \, (H \, k)] \qquad (2)$$

This relation looks very similar to the one which can be calculated from the classical plate theory. The only difference occurs in the expression with the mixed derivatives. Usual laminates are weak for in-plane shear deformation. In contrast to the the classical plate theory, the perturbation scheme takes into account that in a usual cross-ply laminate the shear stresses σ_{23} (connected with torsional moments) are one order of magnitude smaller than the normal stresses σ_{22} and σ_{33} (connected with flexural moments). Therefore no shear term which is connected with the shear moduli G_{23} appears in the expression with the mixed derivatives.

In the first approximation with $\beta_1 = 0$ the phase velocity is a linear function of the wave number. The slope of the straight line is a function of the propagation direction n. Measuring the phase velocity for very long wavelengths the three material parameters E_{22}, E_{33} and v_{23} can be determined. The tensile moduli follow from considering the propagation along the principal axes $n_2 = 1$, $n_3 = 0$ or $n_2 = 0$, $n_3 = 1$ and the Poisson's number v_{23} from directions in-between, provided a suitable iteration procedure is used.
For smaller wavelengths, which still remain large compared with the plate thickness second order effects become important. They are taken into account through the time scaling factor β_1 which is a function of k, n and the various material parameters. It is determined from the second approximation and includes shear effects perpendicular to the plate.

For propagation in the principal directions x_2 and x_3 the dispersion relation (2) for plane flexural waves become

$$c = (H \, k) \, (\frac{E_{22}}{3 \, \rho})^{\frac{1}{2}} \, [1 + (H \, k)^2 \frac{2 \, E_{22}}{5 \, G_{12}}]^{-\frac{1}{2}} \qquad \text{and} \qquad (3)$$

$$c = (H \, k) \, (\frac{E_{33}}{3 \, \rho})^{\frac{1}{2}} \, [1 + (H \, k)^2 \frac{2 \, E_{33}}{5 \, G_{13}}]^{-\frac{1}{2}}$$

These two equations show that the second order dispersion curves in the principal directions can be used to determine the shear moduli perpendicular to the plate. For in-between directions the expression for β_1 is much more involved and no further information is available.

Experimental concept

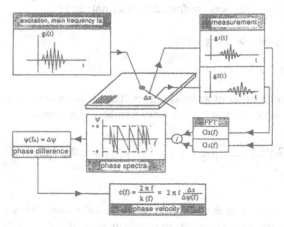

Figure 2. Experimental concept and arrangement

The phase velocity as a function of frequency or wave number is experimentally determined from the phase spectra of time-dependent signals, measured at points along radial straight lines that are starting from the point of excitation. The flexural excitation of the orthotropic plate is produced by a piezoceramic transducer (diameter 8mm). The reproducible narrow-band input signal is digitally generated by a personal computer (Olivetti XP5). The response to the loading is measured by a heterodyne interferometer which is described in detail in [10]. With the aid of a digital storage oscillograph (LeCroy 9400) the responses to 500 input signals are averaged. The resulting time signal at one point of the plate is transferred to the PC where the data is processed by the use of fast Fourier transform. If Δx is the distance between two measuring points on the radial line and $\Delta \psi (f)$ is the measured phase difference for a frequency f as determined from the Fourier transformed time-dependent signals, then the $c(f)$ is calculated as mentioned in figure 2.

Representative dispersion curve

Figure 3 shows a representative dispersion curve which was measured for waves propagating in the principal direction x_2.
The experimental points lie almost perfectly on a slightly concave smooth curve in the whole input frequency range from about 10kHz to 100kHz.
With a piezoceramic transducer as exciter exhibiting a lower cutoff frequency of about 5kHz, the linear range of the dispersion curve is not accessible. Second order effects become too important even for wavelengths of more then thirty times the plate thickness. Thus the first order expression with $\beta_1 = 0$ is not adapted to the range of measured frequencies. The data must

therefore be analyzed using the second order relation with $\beta_1 \neq 0$. Only the two dispersion curves along the principal axes of anisotropy are needed. In the frequency range corresponding to the second step of approximation, the information obtained from in-between propagation is not significant enough to obtain v_{23}.

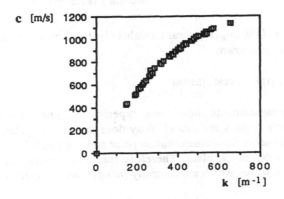

Figure 3. Dispersion curve for the propagation in the principal direction x_2.

From the experimental data together with a nonlinear regression procedure for the theoretical dispersion curves (3) the in-plane tensile stiffnesses E_{22} and E_{33} and the shear moduli G_{12} and G_{13} can be determined. An estimation of the accuracy of the measurement technique shows that for the tensile stiffnesses it is better than 2.5% and for the shear moduli better than 5%.

IN-PLANE SHEAR WAVES

The same experimental concept with little changes in some of the components of the measuring system is used to examine centrally induced in-plane shear waves.
In figure 4 the time signals of the displacement u_3 at 4 distinct points along the principal axis x_2 are shown. The excitation signal consists of 3 cycles of a sinusoidal which are multiplied with a Hanning window.

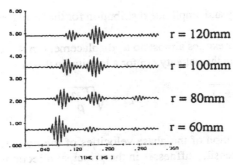

Figure 4. Displacement u_3 along the principal direction x_2, distances from the excitation point 60, 80, 100, 120mm.

A plane wave model allows to determine the in-plane shear modulus G_{23} from the measured phase velocity c_2 of the non dispersive shear waves in the principal directions x_2 and x_3 respectively.

$$G_{23} = (c_2)^2 \rho \qquad \text{accuracy better than 2.5\%} \qquad (4)$$

With the determination of the in-plane shear modulus G_{23} the 5 most important elastic constants of the orthotropic plate are known.

Extended experimental investigation

The in-plane wave measurements show some experimental phenomena which can not be explained with a simple plane wave model. Why does the second pulse in figure 4 become larger for increasing distances from the excitation point and run much faster than the first pulse? Where is its origin located in the plate? Therefore further experimental investigations were carried out for in-plane waves which are centrally induced with a special torsional transducer [10].

In figure 5 the group velocity and the amplitude distribution of pulses in which only the displacement u_3 are considered is shown.

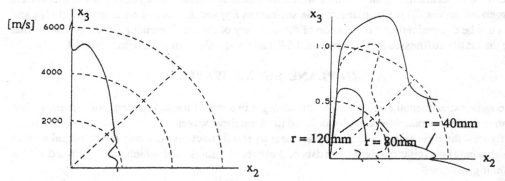

Figure 5. Group velocity and amplitude distribution for the displacement u_3.

The torsional transducer excites almost no u_3 displacement in the x_3 direction. In that direction the pulse propagates with the velocity c_{L3} for a longitudinal wave.

$$c_{L3} = \sqrt{\frac{E_{33}}{\rho}} \qquad\qquad c_{L2} = \sqrt{\frac{E_{22}}{\rho}}$$

$$(4)$$

Therefore the determination of the phase velocities of the non dispersive longitudinal waves allows to calculate the tensile stiffnesses in the principal directions. The results of the flexural wave experiments can be controlled.

Figure 6 gives an experimental explanation for the second pulse phenomenon. The time signals at different points along a ray which includes an angle of 10.8° to the axis x_3 and along the reflected ray are plotted. These time signals can be explained if the experimental results of figure 5 are taken into account.

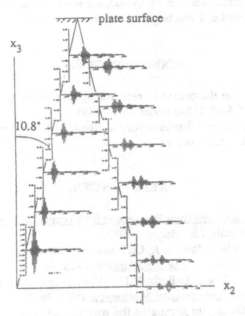

Figure 6. Time signals at different points on a ray under 10.8° to the x_3 axis and on the reflected ray.

The amplitude attenuation and the group velocity of the pulse varies for different propagation directions. The pulse which propagates along the ray has a high velocity and undergoes small attenuation. Therefore it may happen that the reflected pulse arrives at a comparable time and with a comparable amplitude as the pulse which comes directly from the excitation point because this pulse travels much slower and its attenuation is much higher.

CONCLUSIONS

Structural wave experiments can be used to determine in-situ the five most important elastic constants of an orthotropic plate. One single experiment of flexural wave propagation together with a corresponding data analysis based on the results of a consistent second order perturbation approximation of the three dimensional dynamic equations leads to four material parameters. The accuracy of the tensile moduli in the principal directions is better than 2.5% and of the shear moduli perpendicular to the mid-plane of the plate better than 5%.

The experiment is relatively easy to perform and can be devised to be extensively computer controlled. The heterodyne interferometer is positioned with the aid of two step-motors. The personal computer can be used in addition to change the input signal frequency and to analyze the data automatically.

The experimental data of centrally induced in-plane shear waves shows that the model of planar waves is insufficient to describe all observed experimental phenomena. These are explicable with the aid of a perturbation approach for cylindrical waves in an orthotropic plate. The theory has recently been developed and will be published soon.

ACKNOWLEDGEMENT

This paper was prepared in the course of research sponsored by the Swiss National Science Foundation (NF 2.394-0.84/2.275-0.86/20-255.56.88).

The author wish to express their thanks to Mr. A. Eisenhut for performing all experiments for the propagation of in-plane shear waves.

REFERENCES

[1]   ASTM, Standards and Literature References for Composite Materials, American Society of Testing and Materials, Philadelphia, 1987.

[2]   McIntyre, M.E. and Woodhouse, J., On measuring the elastic and damping constants of orthotropic sheet materials, Acta Metallurgica, 1988, **36**, 1397.

[3]   Daniel, I.M., Liber, T. and LaBedz, R.H., Wave propagation in transversely impacted composite laminates, Experimental Mechanics, 1979, **19**, 9.

[4]   Doyle, J.F., Toward in situ testing of the mechanical properties of composite panels, Journal of Composite Materials, 1988, **22**, 416.

[5]   Castagnede, B. and Sachse, W., Optimized determination of elastic constants of anisotropic solids from wavespeed measuremendts, in Review of Progress in Quantitative Nondestructive Evaluation, Vol. 8B, ed. D.O. Thompson and D.E. Chimenti, Plenum Press, New York, 1988, 1855.

[6]   Kline, R.A. and Chen, Z.T., Ultrasonic technique for global anisotropic property measurement in composite materials, Materials Evaluation, 1988, **46**, 986.

[7]   Veidt, M., and Sayir, M., Experimental evaluation of global composite laminate stiffnesses by structural wave propagation, J. Comp. Mat., 1990, **24**, 688.

[8]   Kreis, A. and Sayir, M., Propagation of flexural waves in a thin transversely isotropic plate, Z. angew. Math. Phys., 1983, **34**, 816.

[9]   Sayir, M., Flexural vibrations of strongly anisotropic beams, Ing. Archiv, 1980, **49**, 309.

[10] Dual, J., Experimental methods in wave propagation in solids and dynamic viscometry, Diss. ETH 8659, Zurich, 1988.

DYNAMIC MEASUREMENTS OF ELASTIC PROPERTIES OF FILAMENT - WOUND CYLINDRICAL SHELLS

JURG DUAL and MAHIR SAYIR
Institute of Mechanics
Swiss Federal Institute of Technology
ETH Zentrum
8092 Zürich
Switzerland

ABSTRACT

The present paper deals with a theoretical and experimental investigation of the first two axisymmetric modes of waves in filament - wound circular cylindrical shells. The specimens were manufactured from carbon fibres embedded in an epoxy matrix keeping the winding angle constant during the winding process. Five different winding angles were considered, covering a range from 0° to 82°. The dispersion relation was measured over a large frequency range using a combination of resonance and transient pulse experiments. It was found, that the behavior of filament wound shells can be modelled by homogeneous orthotropic constitutive equations. Four of the nine independent constants were determined from the measurements.

INTRODUCTION

There is no need to point out the importance of filament - wound circular cylindrical shells as structural elements in various applications. However, *very little experimental work* has been reported on the subject of dynamic analysis of such shells, although for quantitative nondestructive evaluation, a thorough knowledge of the propagation characteristics of waves *both from a theoretical and an experimental viewpoint* is essential.

It will be shown, that filament - wound cylindrical shells can be modelled by *homogeneous orthotropic constitutive equations*. Various approximations have been used to derive the dispersion relation for waves in orthotropic cylindrical shells. Shul`ga [ 1 ] and Ramskaya and Shul`ga [2] have solved the problem of axisymmetric waves in an orthotropic hollow cylinder. The analysis is limited to the case of specially orthotropic materials, where the axes of orthotropy coincide with the coordinate lines of a cylindrical coordinate system with its axis of symmetry along the geometrical axis of the tube. The solution of Shul´ga was used for comparison of the experimental data with theoretical results in the present investigation.

Another approach to derive an approximate theory uses *asymptotic analysis*, where quantities are considered according to their orders of magnitude and the physical behavior that dominates various frequency ranges is immediately evident . Such a theory is available only for the case of transversely isotropic cylindrical shells and weak to moderate anisotropy and was given by Gasser. [3]

Very little is reported on *experimental* work regarding dynamics of anisotropic cylindrical shells. Egle and Bray [4] have studied free vibrations of cylindrical shells with discrete longitudinal stiffening.

Only the *first two axisymmetric modes of waves* will be considered here. It is well known, that for wavelengths, which are small with respect to the radius of curvature, shells behave very much like plates. [5] However, for larger wavelengths the geometry of the cylindrical shell produces a dispersion behavior, which is characteristic for the cylindrical shell and qualitatively independent of its material properties. A thorough discussion of this aspect is given in [6].

The *first mode* (mode 1) starts at the bar wave speed c_0 with a uniaxial state of stress and primarily longitudinal motion. For higher frequencies, the waves are slowed down by lateral inertia effects due to nonvanishing lateral contraction and reach a minimum wavespeed, where the shell behaves like a plate in bending on an elastic foundation. The strength of the elastic foundation is influenced by the curvature of the shell and by the plate modulus in the circumferential direction. The motion in this range is primarily radial. If the frequency is further increased, the effect of the elastic foundation diminishes, while shear effects in plate bending become increasingly important.

The *second mode* (mode 2) has a cut-off frequency f_c, whose inverse is given by the time it takes a longitudinal plate wave to travel around the shell in the circumferential direction. It then approaches the wavespeed for longitudinal plate waves in the axial direction, with primarily axial motion.

The various physical models that govern the modes in various frequency ranges make dispersion measurements on the shell suitable for the determination of material constants.

MATERIALS AND METHODS

Specimens

Carbon-fibre reinforced shells with an Epoxy matrix produced by filament - winding were used for the experiments. The tubes had a nominal inner and outer diameter of 2.8 10^{-2} m and 3.1 10^{-2} m, respectively, and a length of 2.0 m. Variations in the radial dimensions amount to about $\pm$ 0.3$\cdot10^{-3}$ m. The density was calculated from the measured weight and the dimensions of the tube. The resulting average density is equal to 1.549$\cdot10^3$ kg/m^3.

The carbon fibres were of the HS type (Young´s modulus in the fiber direction: 2.37$\cdot10^{11}$ N/m^2, density 1.81$\cdot$ 10^3 kg/m^3, volume fraction V_f = 50 % , number of filaments per roving: 12000), while the matrix consisted of R31 Epoxy.

Five different winding angles were considered (α = 0°, 22.5°,45.0°,67.5°,82.0°) for each of which two tubes were examined (called A and B, respectively). The angle of winding α was kept constant for a single tube, such that the principal material direction was alternately oriented at an angle of $\pm$ α with respect to the axis of the tube. The composite configuration is not constant along the length and the circumference. Owing to the winding process, where the rovings follow a helix path along the tube, the top layer is alternately oriented at an angle of + α or - α . Viewed from the outside, the tubes look like a distorted chessboard.

Resonance Experiments

Dispersion data was obtained by measuring resonance frequencies for the low frequency regime. For these frequencies, only one propagating modes exists and waves propagate in tubes as they do in bars. Dispersion data can therefore be obtained simply by observing longitudinal resonance frequencies. The corresponding wavenumbers have to be obtained from a transcendental characteristic equation. [6]

$$\varepsilon k L + tg(k L) = 0 \tag{1}$$

L is the length of the tube, k the wavenumber and ε the ratio of the mass of the transducer to the mass of the tube. The wavespeed c_{0i} at the frequency f_i is then given by

$$c_{0i} = \frac{2 \pi f_i}{k_i} \tag{2}$$

where k_i is the ith solution of Eq. 1 and f_i the ith resonance frequency.

The vibrations were excited using a large piezoelectric transducer and measured with a heterodyne laser interferometer. Dependent on the damping and the specific

dispersion behavior, between 12 and 67 resonance frequencies could be measured, all
belonging to the first mode. Due to high damping of the tubes, a phase criterion was
used to ascertain the correct resonance frequencies. The resulting dispersion curves
are shown in Fig. 1

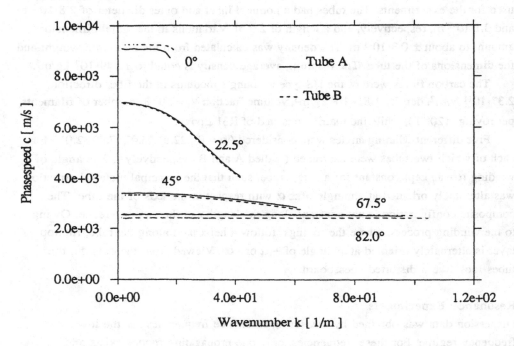

Figure 1. Dispersion Curves for Mode 1 in the Low Frequency Regime, Measured
Using Resonance Experiments.

The largest deviations (about 2 %) between supposedly identical tubes occur for
the ones with a winding angle of 0° and 45°. For the other tubes, the agreement is
better than 1%.

Transient Pulse Fourier Analysis

It is well known, that transient pulse Fourier analysis can be used to determine
dispersion relations. However, to the author's knowledge, it has not been applied to
dispersion in cylindrical shells. This is probably due to the complex nature of pulse
propagation in tubes above the cut-off frequency f_c of the second mode: Two strongly
dispersive modes propagate, which both involve coupled radial and axial motion.
Precisely in this range, however, data is sought, for the determination of material
constants.

The dispersion relation for a wave propagating in the +z direction is obtained from the Fourier transform of two displacement signals measured at say two locations z_1 and z_2.

$$k(f) = \frac{1}{z_2 - z_1} \text{Arg} \left\{ \frac{u^*(z_1,f)}{u^*(z_2,f)} \right\} \tag{3}$$

where u is any non - zero displacement component and * denotes its Fourier transform. For the measurement of mode 1, u was the radial displacement in accordance with the predominant motion for the frequencies considered. On the other hand, for mode 2, which is primarily longitudinal, the displacement at an angle of 45° with respect to the axis of the tube was measured.

Eq. 3 applies only, if a single mode contributes to the measured displacement and if no reflections are superimposed on the outgoing wave. Both conditions are difficult to satisfy for waves in carbon-fibre reinforced shells. Above the cut-off frequency f_c, two modes may propagate. Either mode will contribute to both axial and radial motion at the surface.

The solution to this problem consists of two main elements: First, reducing the amplitude of reflected waves by *pasting a viscous fluid*, which had a dynamic viscosity of about 7 Pas, to the distant end of the specimen, and second, using *piezoelectric excitation of the waves*, which allows selective excitation of different modes and tailoring of the frequency content of the pulse.

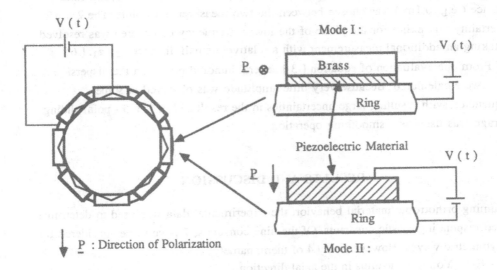

Figure 2. Transducers Built to Selectively Excite Mode 1 and Mode 2 Propagating Waves

In addition, the *signal to noise ratio* can be dramatically *improved by averaging* many experiments. Piezoelectric excitation is very useful for this purpose, because repeating an experiment only involves applying a given voltage another time. In order to eliminate nonaxisymmetric modes of waves, displacements were also *circumferentially averaged* by taking measurements at several angles across the circumference.

The experimental set-up involved a M24 Olivetti PC running under the ASYST Software package. It contained a digital function generator, which provided a transient voltage with a precisely controlled frequency content. Its output was amplified to a peak of 100 V using a Krohn Hite KH7500 amplifier and repetitively applied to one of the transducers as given in Fig. 2. The resulting displacement, which was of the order of 5 nm, was again measured with a heterodyne interferometer [6], that has a noise level in terms of displacement of about 2.5 Angström in a frequency range from 1 kHz to 300 kHz. The demodulator output is bandpass filtered between 2 and 200 kHz with an analog filter (Krohn Hite KH 3550). Its output was captured at a sample frequency of about 3 MHz over 4096 points in a LeCroy 9400 digital oscilloscope which uses 8 bit flash ADC's. It was then averaged over hundreds of experiments to decrease the noise level and fed back to the PC for the data analysis.

Great care had to be taken to minimize aliasing and leakage, when computing the Fourier transform. Phase jumps that occurred as a result of the argument function in Eq. (3) could be easily eliminated due to the extremely low noise levels, if a suitable distance (e.g. 0.1m) was chosen between the two measurement points. The $2 \pi n$ uncertainty that exists for the phase of the lowest frequency considered was resolved by taking an additional measurement with a relatively small distance $z_2 - z_1$. [6]

From one evaluation of equation [3] several hundred points on the dispersion curve were calculated. Because very little amplitude was observed at some frequencies, with resulting large uncertainties in the result of Eq. 3, a 5 - point sliding average was used as a smoothing operation.

RESULTS AND DISCUSSION

Assuming orthotropic material behavior, the experimental data was used to determine the corresponding elastic constants. Of the nine constants, 7 have to be considered for axisymmetric waves. However, only 4 of them, namely

- E_3 Young's modulus in the axial direction
- E_2 Young's modulus in the circumferential direction

- v_{32} Poisson's ratio, which relates strain in the circumferential direction to strain in the z - direction, when a stress in the z - direction is applied
- G_{13} Shear modulus in the r-z-plane (bending range)

will have a major effect on the dispersion for the wavelengths considered in this investigation. For the 0° winding angle, the tubes are transversely isotropic and therefore, the asymptotic theory given by Gasser [3] was applied to obtain approximate values. Shul`ga`s [1,2] method was used for all tubes.

E_3 was computed from the infinite wavelength limit of mode 1: $E_3 = \rho c_0{}^2$. G_{13} dominates the behavior of mode 1 in the upper frequency range considered, while the other 2 constants describe the behavior for the intermediate wavenumbers. The resulting values for the material constants are summarized in Table 1. The uncertainties are estimated at better than 0.5% for E_3, 2% for G_{13}, 5% for E_2 and .02 for v_{32}. For better accuracy, asymptotic theories should be developed for the general orthotropic case or a nonlinear curve fitting algorithm should be combined with the solution of Shul`ga.

As an example, the theoretical and experimental dispersion curves for the 82° winding angle are shown in Fig. 3.

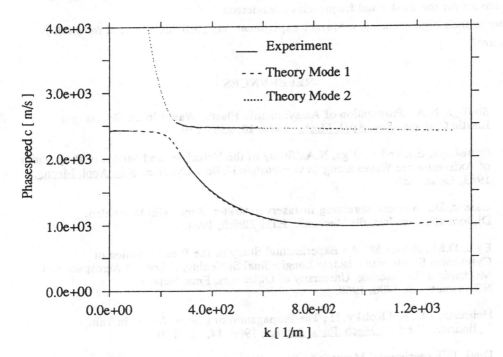

Figure 3. Dispersion Curves for Mode 1 and Mode 2 for One of the 82° Tubes
Theory and Experiment

TABLE 1
Orthotropic Material Constants Evaluated from the Measurements

Winding Angle	E_3 [10^{10} N/m^2]	E_2[10^{10} N/m^2]	G_{13} [10^{10} N/m^2]	v_{32}
0.0°	12.9	0.92	4.9	0.41
22.5°	7.57	1.0	4.7	1.35
45.0°	1.81	2.2	4.5	0.66
67.5°	1.02	8.0	3.5	0.16
82.0°	0.91	11.7	3.3	0.04

CONCLUSIONS

From the excellent agreement between experimental data and theory, it is concluded, that filament - wound shells are *well described by homogeneous orthotropic material* behavior for the modes and frequencies considered.

Due to their high accuracy, resonance experiments are particularly useful for *quality control*.

REFERENCES

1. Shul`ga, N.A., Propagation of Axisymmetric Elastic Waves in an Orthotropic Elastic Cylinder, Sov. Appl. Mech. , 1974, **10**, 936 - 939.

2. Ramskaya, E.I. and Shul`ga, N.A., Study of the Velocities and Modes of Propagation of Axisymmetric Waves along an Orthotropic Hollow Cylinder, Sov. Appl. Mech. , 1983, **19**, 207-211.

3. Gasser, B., Wellenausbreitung in faserverstärkten Kreiszylinderschalen, Diplomarbeit am Inst. für Mechanik, ETH Zürich, 1981.

4. Egle, D.M., Bray,F.M., An Experimental Study of the Free Vibration of Cylindrical Shells with Discrete Longitudinal Stiffening, School of Aerospace and Mechanical Engineering, University of Oklahoma, Final Report, NSF Grant GK-1490, 1968.

5. Heimann, J.H. and Kolsky, H., The Propagation of Elastic Waves in Thin Cylindrical Shells, J. Mech. Phys. Solids, , 1966, **14**, 121-130.

6. Dual, J., Experimental Methods in Wave Propagation in Solids and Dynamic Viscometry, Diss ETH 8659, Zürich, 1988.

TORSION RESPONSE ANALYSIS OF T300 / 914 AND T800 / 914 UNIDIRECTIONAL SPECIMENS

B. Férent, A. Vautrin.
Département Mécanique et Matériaux
Ecole des Mines de Saint-Etienne
158, cours Fauriel,
42023 SAINT-ETIENNE CEDEX 02, FRANCE

ABSTRACT

This paper presents the first analysis of the torsion response of some T300/914 and T800/914 unidirectional specimens.

In a first part we describe the entirely automated experimental set-up we developed to perform the torsion test on square cross-section unidirectional composite specimens.

Then we present results of the monotonic loading of T300/914 and T800/914 specimens and give the main conclusions of our analysis of the characteristics of the torsion response of these specimens pointing out the differences between these different torsion responses obtained in relationship with the nature of the fibres and the fibre content.

INTRODUCTION

The overall behaviour law of a composite material is absolutely required for finite element computation of structures. In particular, both intralaminar and interlaminar shear moduli are necessary. So, one of the most important question that arises is the modelling of the shear response of composite materials based on a well-founded quantitative shear test method.

SHEAR DETERMINATION METHODS

From an industrial point of view, shear performances of composite material are explored using the short-beam flexure test. This test is a standard (ISO 4585) which may lead to shear moduli but cannot lead to reliable shear strength values because of the highly three dimensional stress state. Indeed, this loading does not correspond to a prominent shear stress state.

In a recent review of the in-plane shear determination methods used in laboratory, Lee and Muno [1] finally established that the three most promising in-plane shear test methods are the ±45° tensile test and 10° off-axis tensile test and Iosipescu's test. Besides, these three tests present the best quality/price ratio.

Yet, none of them provides the identification of the whole shear properties as they lead to intralaminar shear properties only. Moreover they fail to induce a pure shear stress state even if the state of stress in Iosipescu's test is actually quite close to pure shear.

Concerning the specimens used, the 10° off-axis tensile test is very sensitive to small misorientations. The Iosipescu's specimen is also difficult to machine, particularly the notches and their positions which are the key of success in introducing quite pure shear in the central area of the specimen.

Whatever the test, the equipment of the specimens (two or three element strain gage rosettes) was revealed quite difficult and too heavy to practise from an industrial point of view.

ADVANTAGES OF THE TORSION TEST

The torsion test of square cross-section bars of unidirectional composite materials was chosen because of the various advantages it presents. This test has been initially compared to other usual shear determination methods by Phang, Vong and Verchery [2]. Chiao, Moore and Chiao [3] have defined the main features an ideal shear test method should respect:
- specimens easily cut-off from industrial products,
- pure shear over a large enough specimen area,
- determination of all the shear properties.

In the case of the torsion test the loading used provides a pure shear stress state on a large enough area of the specimen length even if the stress state is not homogeneous. Both intralaminar and interlaminar shear moduli are provided with one transversely fibre oriented specimen equipped with 2 uniaxial strain-gages [4]. The specimens are easy to obtain as they can be cut from industrial laminates. Futhermore, they are well-representative of the actual industrial material including all the information issued from its elaborating process, post-curing and eventual conditionning.

On a practical field the entirely automated torsion test set-up designed in the Departement Mécanique et Matériaux enables us to perform this test in the best conditions of reproducibility.

EXPERIMENTAL SET-UP

Torsion machine

The torsion machine designed is an horizontal twisting machine (fig.1). It has two rotating grips for holding the specimen in position. One of these two grips is moved by a stepper motor via a planetary reduction gear unit any translation motion of it is forbidden.

The other grip is free in translation on lubricated ball-grooves but it cannot rotate. Thus this grip can move in order to avoid introducing axial force when twisting the specimen.

A cage for measuring torque is placed between the transferring grip itself and the grooves. It is made of three thin plates which transmit the torque to be measured. Two of them are equipped with strain-gages mounted in full-bridge in order to measure their strains. A calibration curve gives us the corresponding value of the applied torque. The design of the thin plates provides the best sensitivity/stiffness ratio. Interferences on the electric signal delivered by the bridge due to the presence of high currents in the stepper motor have to be minimized using a low-pass filter whose bandwith is 4Hz.

Specimens

The specimens are square cross-section bars of unidirectional composite material of dimension of 140 x 7 x 7 mm (fig. 2). The fibres are transversely oriented along the x axis.

Figure 1. Torsion machine.

These specimens are equipped with two strain-gages bonded at 45° with the longitudinal axis of each lateral face. The two measured strains are ε_{yx} and ε_{yz}.

Automation

This torsion machine is entirely automated using the software we developed under ASYST 3.00 on a compatible IBM PC AT3 [5]. It is a user-friendly software which controls the motor motion and drives the acquisition of all the experimental data simultaneously..

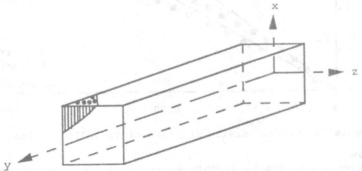

Figure 2. Square cross-section specimens.

The stepper motor can be driven with a standard GPIB IEEE bus via an interface card. Through this connection one can not only control the motor motion but also get information back: errors, position... The motor motion control instruction is a string which specifies the final twist angle and the angular speed. So the natural control of the motor is a twist angle controlled motion. Various loading paths are now available.

As far as data acquisition is concerned, two methods are used in parallel: GPIB bus for the twist angle and analog to digital input for the electric signals from the strain-gages and the torque-meter. ASYST 3.00 features allow to perform a real time plotting on the monitor of the torque versus strains.

The data processing depends on the loading path chosen. It is fundamentally based on the theoretical developpment made by Phang and Vong [4] and Habdelhady and Vautrin [8].

The chosen loading path

To analyse the torsion response of T300/914 and T800/914 unidirectional specimens, we used monotonic loading with constant angular rate.

The measured and acquired values are: the resultant torque, the two strains ε_{yx} and ε_{yz} and the twist angle. Data processing leads to the two tangent moduli and their evolutions during the loading. The documents typed are:
- table of values,
- torque versus strains curves,
- G_{yx} and G_{yz} evolution curves.

ANALYSIS OF T300/914 AND T800/914 TORSION RESULTS

We used the above described experimental set-up to determine the torsion response of T300/914 and T800/914 transversely oriented fibre reinforced unidirectional specimens. A typical evolution of the torque versus strains ε_{yx} and ε_{yz} is shown in figure 3. Such evolutions are currently derived from one single monotonic test at a 10^{-4} s^{-1} shear strain rate.

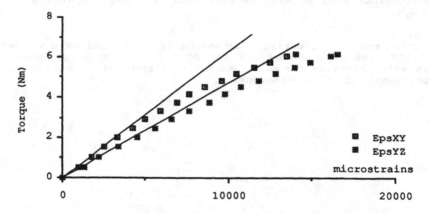

Figure 3. Typical torsion response of T300/914 specimens.

Computation

The above curves can be roughly split in two domains corresponding to apparent linear behaviour and to non-linear behaviour respectively. The software computes the best fitting straight line for the linear range for each evolution. It determines the yield torque beyond which linear approximation is no longer satisfactory [6].

The yield point is the smallest of the two previous yield torques obtained, the yield strains are the two corresponding strains. The slopes of the two lines are needed for the computation of the initial tangent moduli which requires the resolution of the torsion equation using an iterative method.

Concerning the non-linear part of these curves, the software determines the best second degree interpolating polynomials and uses their

145

derivatives to compute the two tangent moduli trought a step by step
numerical process under the assumption the specimen is linear elastic at
each step. This computation leads to the evolutions of the shear moduli
(fig 4 and fig. 5).

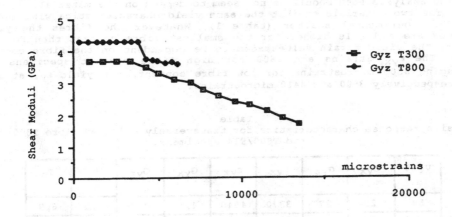

Figure 4. Tangent interlaminar shear modulus for T300/914 and T800/914
specimens versus ε_{yz}.

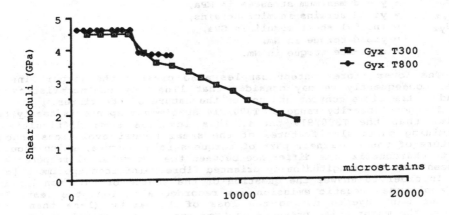

Figure 5. Tangent intralaminar shear modulus for T300/914 and T800/914
specimens versus ε_{yx}.

Analysis
Figures 4 and 5 show that for both T300/914 and T800/914 specimens the intralaminar shear modulus is higher than the interlaminar shear modulus
    Apparently, interlaminar modulus varies little with fibre content and intralaminar modulus increases with fibre content as it can be expected using micromechanics. The small number of specimens do not allow more precise analysis. Both moduli do not seem to depend on the material.
    The two materials exhibit the same yield characteristics with regard to the experimental scatter (table 1). Whatever the fibres the yield torques are a little higher for the smallest fibre content than for the other. The yield strain values seem to be dependent on the fibre content also. The ε_{yx} strains are 2600 for high fibre content specimens and averaging 3300 microstrains for low fibre content, the yield ε_{yz} strains are respectively 3500 and 4410 microstrains.

Table 1
Torsion response characteristics for transversely oriented fibre T300/914 and T800/914 specimens.

	V_f %	σ_{yx}	σ_{yz}	ε_{yx}	ε_{yz}	G_{yx}	G_{yz}	T_l	T_u
T	58	29	27	3370	4540	4,5	3,7	2,0	6,9
300	65	28	21	2600	3540	4,3*	3,8	1,7	5,7
T	57	29	26	3240	4280	4,6	4,3	1,9	4,1
800	63	28	25	2600	3450	5,2	4,2	1,6	3,6

* This values is though to be a little low and is on re-analysis.
σ_{yx} σ_{yz} = yield maximum stresses in MPa,
ε_{yx} ε_{yz} = yield strains in microstrains,
G_{yx} G_{yz} = initial shear moduli in GPa,
T_l = yield torque in Nm,
T_u = ultimate torque in Nm.

    The lower fibre content samples also present the larger linearity range. Consequently we may consider that linearity characteristics only depend on the fibre content and not on the nature of the fibres.
    The non-linearity range of T800/914 specimen response is nearly twice smaller than the T300/914 one's. This involves some difficulties in determining particular features of the shear moduli evolutions, such as curvature of the non-linear part of torque-strains curves, which should be used to characterize the difference between the two material response as it has been done for longitudinally oriented fibre reinforced specimens [6].
    In order to answer the question of the source of this non-linearity high amplitude acoustic emission was recorded all along the test. There were not many events during the loss of linearity (less than 1500). However, the motor noise prevents us from recording eventual low amplitude events under 50 dB. Despite of this we think that this non-linearity is correlated with plastic behaviour of the matrix rather than microcracking of the matrix and the interfaces.
    The most obvious differences between the two types of materials probably concern the ultimate characteristics. T300/914 specimens ultimate torque is nearly twice the ultimate torque of T800/914 specimens for the same fibre content. This is clearly depicted in figure 6 below where the two torque-ε_{yx} curves have been superimposed. Thus the shear performances of T800/914 are lower than the one of T300/914. Futhermore, whatever the fibre content the aspect of the broken section of T800/914 specimens is a

45° oriented plane where as T300/914 specimens are broken into many parts with no particular orientation.

Table 1 shows that ultimate torque decreases when the fibre content increases, such a trend could be related to the intensity of the micro stress field in the matrix.

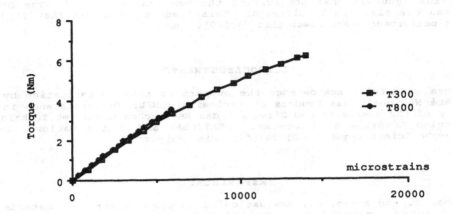

Figure 6. Torque versus ε_{yx} for the two considered materials.

The good agreement between the values of G_{yx} from table 1 and the values of G_{yx} obtained by testing longitudinally oriented fibre specimens in torsion (table 2) is regarded by the authors as a fair indication of the satisfactory working of the test and data treatment performed.

TABLE 2
Comparison between G_{yx} values (GPa) obtained with longitudinally oriented fibre and transversely oriented fibre specimens.

	Vf %	Longitudinal	Transverse
T300	58	4,7	4,5
	65	5,2	4,3
T800	57	4,6	4,6
	63	5,0	5,2

CONCLUSION

A first analysis of the torsion response of transversely oriented fibre unidirectional composite bars was made using an entirely automated torsion test set-up.

The results obtained show that the torsion test of square cross-section bars allows to characterize the shear behaviour of different fibre reinforced composite materials such as T300/914 and T800/914 with various fibre content.

Whatever the material, the monotonic response can be roughly split in an apparent linear range followed by a non-linear range. The characteristics of apparent linearity depend on the fibre content but seem to be independent of the fibre type. Both shear moduli increases with fibre content even if interlaminar shear modulus evolutions is less obvious.

The most striking difference concerns the ultimate behaviour trough the ultimate torque values which are twice higher for T300/914 specimens than for T800/914 ones considering the same fibre content. The broken section are also really different. This leads to consider that T800/914 shear performances are lower than T300/914 one's.

ACKNOWLEDGEMENTS

Authors gratefully acknowledge the support of this investigation by the Société Nationale des Poudres et Explosifs (SNPE) for assistance in the supply of the specimens and Direction des Recherches Etudes et Techniques, Direction Générale de l'Armement (DRET/DGA) and Centre National de la Recherche Scientifique (CNRS) for financial support.

REFERENCES

1. Lee, S. and Munro, M., Evaluation of in-plane shear test methods for advanced composite materials by the decision analysis technique.Composites, 1986, 17, pp 13-22.

2. Phang, C., Verchery, G. and Vong T.S., Etude des tests de cisaillement d'un matériau composite à renfort unidirectionnel. Proceedings of the first Journées Nationales sur les Composites, JNC1, ed. C. Bathias et al., Paris, France, 1978, pp 182-9.

3. Chiao, C.C., Moore R.L. and Chiao T.T., Measurement of shear properties of fibre composites. Composites, 1977, 8, pp 161-174.

4. Phang, C. and Vong T.S., Etude du test de torsion de barreaux composites à section rectangulaire. Research Report n°94, Ecole Nationale des Techniques Avancées, Paris, 1977.

5. Férent, B. and Vautrin, A., Computer aided torsion test. Composite Materials Design and Analysis, ed. J.P. de Wilde and W.R. Blain, Springer-Verlag, Berlin, 1990, pp 323-38.

6. Férent, B. and Vautrin, A., Corrélation entre le comportement mécanique en cisaillement et l'interface fibre-matrice de composites unidirectionnels carbone/époxyde (Correlation between the shear behaviour and the fibre-matrix interface in carbon/epoxy unidirectional composites). Research Report, Ecole des Mines de Saint-Etienne, Société des Poudres et Explosifs, Saint-Etienne (France), Sept. 1990.

7. Surrel, Y., Modélisation du comportement élasto-plastique de composites stratifiés Doctoral Dissertation, Université Claude Bernard, Lyon (France), Jan. 1990.

8. Habdelhady, F. and Vautrin , A., Characterization of the time-dependent shear response of unidirectional carbone/epoxy laminate in torsion. Composites Evaluation, Butterworths, Sevenoaks, 1987, pp. 194-200.

A COMPUTERIZED TEST SETUP FOR THE DETERMINATION OF THE IN-PLANE AND OUT-OF-PLANE SHEAR MODULUS IN ORTHOTROPIC SPECIMENS

D. Van Hemelrijck, L. Schillemans, F. De Roey, I. Daerden, F. Boulpaep, A. Cardon
Composite Systems and Adhesion Research Group of the Free University of Brussels
COSARGUB - VUB - TW - Kb - Pleinlaan 2 - 1050 Brussels - Belgium

ABSTRACT

An automated torsion test is proposed for the determination of the in-plane and out-of-plane shear modulus. The obtained results show that for some composite systems the out-of-plane shear modulus is much different from the in-plane shear modulus, certainly sufficiently to be considered in some design application.

INTRODUCTION

The development of new fibers like boron, carbon and improved glasses has considerably intensified interest in composite materials. The application of these new anisotropic materials in structural engineering requires the production of laminates which have to be designed for a general state of stress and strain in the three dimensions. Unfortunately, to allow such careful and accurate stress and deformation analysis the coefficients of the general hooke's law have to be identified. The determination of the longitudinal and transverse stiffness moduli is quite straightforward. But in the case of composite materials the shear moduli G_{ij} cannot be computed and have to be determined by use of specific shear tests. Conventional shear test methods [1], such as the +- 45° coupon, the rail-shear, the 10 degrees off-axis test are used to obtain the in-plane shear modulus. Following the specifications of Chiao and al. [2] an ideal shear test has to provide all shear properties. Sumsion and Rajapakse [3] described a method to obtain by use of a torsion setup the out-of-plane as well as the in-plane shear modulus. Another interesting point is the increasing concern of the degradation and durability of composite systems. It is our objective to determine the in situ properties of damaged laminates and so to study the torsional fatigue behaviour.

THEORY

The definition of the dimensions used in the following formulas are according to the designation in figure 1.

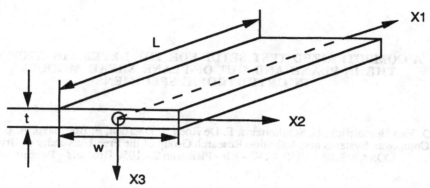

Figure 1. Designation of axis system

The equation relating the applied torque and the angle of twist per unit length for an isotropic rectangular flat beam within the linear elastic range is:

$$M_T = \beta \, w \, t^3 \, G_{12} \frac{\theta}{L} \qquad (1)$$

where

$$\beta = \frac{1}{3}(1 - \frac{192}{\pi^5}\frac{t}{w}) \sum_{n=1.3.5}^{\infty} \frac{1}{n^5} \tanh \frac{n\pi w}{2t}$$

For small values of w/t this equation can be simplified in the following form due to the fact that the tanh(nπw/2t) becomes almost equal to one:

$$M_T = \frac{1}{3} G \, w \, t^3 (1 - 0.63\frac{t}{w}) \frac{\theta}{L} \qquad (2)$$

For an orthotropic rectangular bar the relation between torque and torsion angle becomes:

$$M_T = G_{12} \, \beta(c) \, w \, t^3 \frac{\theta}{L} \tag{3}$$

where
$$\beta(c) = \frac{32 \, c^2}{\pi^4} \sum_{n=1,3,5}^{\infty} \frac{1}{n^4} \left(1 - \frac{2c}{n\pi} \tanh \frac{n\pi}{2c}\right)$$

and
$$c = \frac{w}{t} \sqrt{\frac{G_{13}}{G_{12}}}$$

with G_{12} being the in-plane shearmodulus and G_{13} the out-of-plane shearmodulus.

By combining the measurements of two bars with different width, it is possible to compute the in-plane, as well as the out-of-plane shear modulus:

TEST 1 $\quad w_1, t_1, L_1 \qquad \Rightarrow \quad M_T^1 = G_{12} \beta(c_1) w_1 (t_1)^3 \dfrac{\theta^1}{L_1}$

$\qquad\qquad$ linear regression $\quad \Rightarrow \quad M_T^1 = a_1 \dfrac{\theta^1}{L_1} \qquad\qquad\qquad \Rightarrow a_1 \tag{4}$

TEST 2 $\quad w_2, t_2, L_2 \qquad \Rightarrow \quad M_T^2 = G_{12} \beta(c_2) w_2 (t_2)^3 \dfrac{\theta^2}{L_2}$

$\qquad\qquad$ linear regression $\quad \Rightarrow \quad M_T^2 = a_2 \dfrac{\theta^2}{L_2} \qquad\qquad\qquad \Rightarrow a_2 \tag{5}$

combining (4) and (5) results in

$$\frac{a_1}{a_2} = \frac{G_{12} \beta(c_1) w_1 (t_1)^3}{G_{12} \beta(c_2) w_2 (t_2)^3} \tag{6}$$

when defining
$$\alpha = \frac{c_2}{c_1} = \frac{w_2/t_2}{w_1/t_1} = \frac{w_2 t_1}{w_1 t_2}$$

we have
$$\frac{w_2 (t_2)^3}{w_1 (t_1)^3} a_1 \beta(\alpha \, c_1) - a_2 \beta(c_1) = 0$$

we can now solve this equation for c_1 and use

$$G_{12} = \frac{a_1}{\beta(c_1)w_1(t_1)^3}$$

to obtain the in-plane shear modulus G_{12}

and

$$c = \frac{w}{t}\sqrt{\frac{G_{13}}{G_{12}}}$$

to calculate the out-of-plane shear modulus G_{13}

In order to be able to combine more than two different width's, a Smith least square fit of the function (7) was programmed.

$$\frac{a_i}{wt^3} = G_{13}\beta(\frac{w}{t}\sqrt{\frac{G_{13}}{G_{12}}}) \tag{7}$$

where a_i is the first order coefficient of the linear regression relating torque and torsion angle of each experiment.

EXPERIMENTAL SETUP

The torsion setup, as constructed and used at the Free University of Brussel, is given in figure 2.

Figure 2. experimental setup

It consists of a steppingmotor with a planetary gearbox so that for one step an angular rotation of 0.036 degrees is obtained. The maximum torque which can be applied is 25 Nm. A torque-cell measures the applied torque up to 26 Nm with a minimum resolution of 4 Nmm. The construction is such that one end is completely free to move so that no tensile forces can be introduced. An LVDT with a range of 1.5 mm was added to monitor the axial displacement during testing. A personal computer equiped with an analog-to-digital convertor board is used to control every device and to perform the data-acquisition and processing. The software is written in C and is completely menu-driven.

EXPERIMENTAL RESULTS

Flat aluminium bars were used as testspecimens for the proposed procedure. The results from standard straingage measurements and torsion tests were compared and showed good agreement. The in-plane shear modulus deviated only 1%, the out-of-plane shear modulus 5%. Other materials tested were glassfiber polyester $(0°,90°)_8$ and carbon epoxy (fibredux 914). Results of the tests are summarized in table I. The first column gives the material system and the fiber orientation of the test specimen. The second column lists for each tested specimen the width over thickness ratio. The next column gives the shear modulus calculated with the "isotropic" formula. The last two columns give the in-plane and out-of-plane shear modulus calculated by using the Smith's least squares fit to equation (7) which is valid for orthotropic materials.

	w/t	G_{12} isotropic GPa	G_{12} orthotropic GPa	G_{13} orthotropic GPa
Al	5 10 15	27.5 27.2 27.0	27.0	32.5
GFRP $(0°/90°)_8$	3.0 5.2 8.2	2.7 2.7 2.8	2.8	2.1
Fibredux 914 $(±45°)_{16}$	3.2 5.6 7.4	15.9 20.7 22.1	26.4	3.5
Fibredux 914 $(0°_3,90°_4,0°)_s$	3.3 5.8 8.5	4.7 4.9 5.2	5.1	3.0
Fibredux 914 $(0°)_8$	7.6 11.5 16.8	5.4 5.6 5.5	5.6	3.4

Table I: Shear moduli for different material systems.

The results on aluminium show that there is no increase in shear modulus with increasing w/t ratio. Instead, for cross-ply material there is a consistent increase. The in-plane shear moduli for the +/- 45° composite system is significantly greater than those for the uniaxial composites, the ratio is approximately 4 to 1. The out-of-plane shear modulus (G_{13}), which is generally considered to be the same as the in-plane shear modulus (G_{12}) is in fact appreciably lower: 39% versus 44% in reference []. For the +/- 45° material the difference between the in-plane and the out-of-plane shear modulus is much greater.

ERROR ESTIMATION

Equation (7) was used to calculated a set of 10 simulated experimental data in the form

$$\frac{a^i}{w^i t^3}$$

On this set of simulated data a percentage error ranging from 1% to 10% was superimposed. The resulting in-plane and out-of-plane shear modulus calculated and compared with the expected values. As shown in graph 1 and graph 2 the out-of-plane shear modulus G_{13} is much more sensitive to errors compared to the in-plane shear modulus G_{23}.

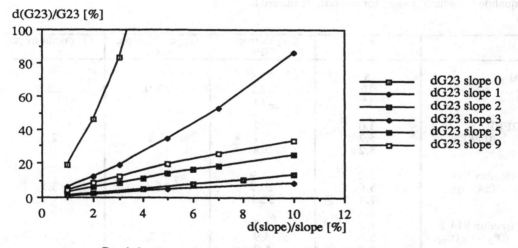

Graph 1: error estimation for the out-of-plane shear modulus.

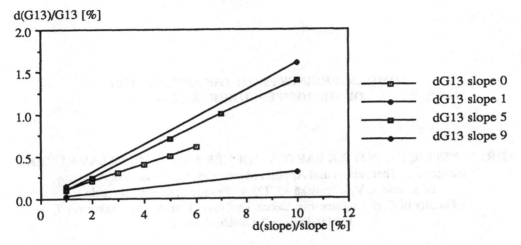

Graph 2: error estimation for the in-plane shear modulus

CONCLUSIONS

Both shear moduli can be determined by using torsion tests of flat specimens with rectangular cross sections. Unfortunately the out-of-plane shear modulus is highly sensitive to errors.

REFERENCES

[1]   Van Hemelrijck D. , 'Materiaalkarakterisatie van vezelversterkte composietmaterialen' Thesis, 1984.

[2]   Chiao C.C., Moore R.L.,'Measurement of shear properties of fibre composites,Composites, Vol. 8,No 3,pp 161-169.

[3]   Sumsion H., Rajapakse Y.,'Simple torsion test for shear moduli determination',Proc ICCM2,1978,pp 994-1002.

[4]   Ferent B., Vautrin A.,'Computer aided torsion test',Proc CADCOMP 90,1990,pp 325-337

SOME EXPERIENCE FROM THE APPLICATION
OF THE IOSIPESCU SHEAR TEST

JIŘÍ MINSTER, FRANTIŠEK BARTOŠ, ZDENĚK FIALA, MILOSLAV RŮŽEK+
Institute of Theoretical and Applied Mechanics, Czechoslovak Academy
of Sciences, Vyšehradská 49, 128 49,Prague 2, Czechoslovakia
+Faculty of Civil Engineering, Czech Technical University, Thakurova 7,
160 00 Prague 2, Czechoslovakia

ABSTRACT

Experience gained using the Iosipescu test for measuring of the static and cyclic shear characteristics of polymer matrix composites are summarized. The stress and strain distributions in the active part of specimen were in the static case determined by means of the reflex photoelasticimetry and time-based photogrammetry respectively. Experimental results were complemented by the theoretical solution applying FEM for two fundamental cases of non mixed boundary conditions. The results of the shear modulus, shear strength and shear fatigue are evaluated on the stress and strain analysis base.

INTRODUCTION

The relatively recent application of the Iosipescu test for the determination of shear characteristics of composite materials resulted in a considerable revival of interest in this experimental method [1]. The contemporary state of development of the method can be characterized by an endeavour to achieve a perfect analysis of stress and strain distribution in the working part of the specimen with regard to its geometry and the loading mode. For this purpose particularly analytical solutions by the finite element method have been used [2]. The scope of application of the method is continuously extended outside the field of measurements of basic mechanical quantities. By way of example it is possible to mention the application of the Iosipescu test to the assessment of fatigue under shear load of quasi-isotropic SMC composites [3], the analysis of crack propagation in unidirectional fibre-reinforced fibrous composites under mixed loading modes using linear fracture mechanics [4] and the measurements of shear photoelastic constants of orthotropic materials [5].

The experimental analysis of displacement field by means of moire interferometry made it possible to explain unexpected results in the case of application of the method concerned to the materials with elastoplastic behaviour of the composite matrix [6].

The simultaneous application of the method without the standardization of experimental fixture and procedure explains some inconsistencies of conclusions obtained in various institutions in recent years. All hitherto achieved results including those attained by the authors of the present paper, however, have confirmed that in the case of anisotropic composites the assumption of the existence of a pure shear field in the specimen area between the notches corresponding with external load is not fulfilled.

This paper presents some experience obtained from the application of the experimental fixture developed for the Iosipescu test.

MODELLING OF STRESS AND STRAIN DISTRIBUTION BY THE FINITE ELEMENT METHOD

The aim of this phase of the solution was to obtain a clearcut picture of mechanical behaviour of specimen in the fixture. The modelling was based on the planar physically linear problem according to programme [7] with triangular elements with six degrees of freedom. The elastic constants of the model orthotropic material for the basic solution were selected identical with the standard values of the carbon-epoxy composite from the Fiberdux 914 C prepreg: E_{11} = 120 GPa, E_{22} = 10 GPa, v_{12} = 0.35, G = 4.5 GPa. The computation was carried out for two basic reinforcement orientations, viz. in the directions parallel to the length of the specimen /0/ and perpendicular to it /90/. The model loading was simulated by two different non-mixed boundary conditions as the loading by forces and displacements.

STRESS DISTRIBUTION

The stress field in the working part of the specimen between the notches was determined by the reflex photoelasticity method. The values of optical birefringence and isocline angles were determined in the vertices of a square mesh shown in Fig.1.

The computation of normal and shear stresses in the measuring points of the optical foil were made numerically by modified Tesař method. The transfer of obtained stresses to the composite material was based on the equality of nominal shear stress beteween the notches and the experimentally ascertained normalized shear stress acting in the foil in the same cross section.

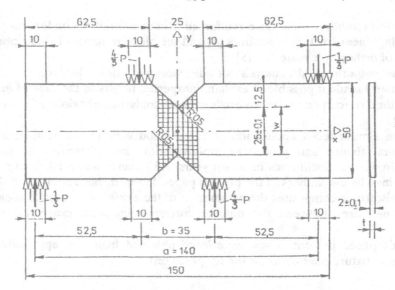

Figure 1. Dimensions of specimen used.

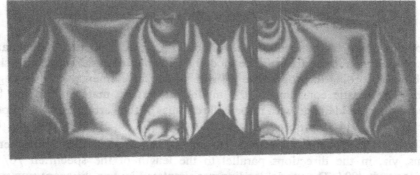

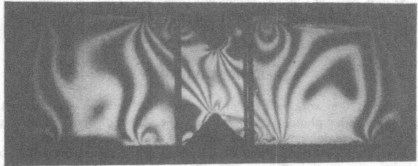

Figure 2. Distribution of isochromatic lines in the specimen loaded only by clamping forces (a) and loaded by external forces in the fixture (b).

The stress distribution is decisive for the values of correlation factors determining the ratio of local shear stress to nominal stress. Normal stresses along the line connecting the notches are much lower for both orientations with the exception of the regions very near the notch roots. Unfavourable stress state combination in the roots makes the /90/ orientation less advantageous for the determination of the shear strength and the /0/ orientation for the measurements of the shear modulus of unidirectionally reinforced composites.

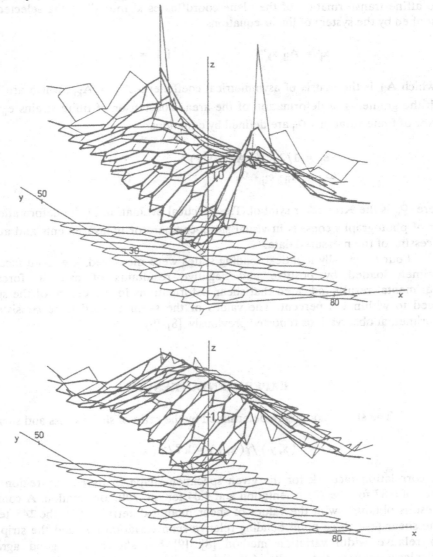

Fig.3. Correlation of theoretical --- and experimental --- normalized shear stress values of orientations /0/ and /90/.

DISPLACEMENT AND DEFORMATION FIELDS

Time-based photogrammetry was used for the determination of changes in the position of objects before and after loading. The photographs must be taken in this case with identical external and internal orientation of the object and the camera. The conditions are simpler, if the objects are as in this case planar. The affine transformation has seemed most suitable for the computation of strains from the ascertained coordinates. The affine transformation of the plane coordfinates x' into x" in the selected area is described by the system of linear equations

$$x_i' = A_{ik} x_k'' \qquad i, k = 1, 2 \qquad (1)$$

in which A_{ik} is the matrix of asymmetrical coefficients $A_{ik} \neq A_{ki}$, which are identical with the gradients of deformation of the area. The tensor of finite strains e_{ij} and the tensor of finite rotations δ_{ij} are defined by equations

$$e_{ij} = 1/2 \, (A_{ki} A_{kj} - \delta_{ij}) \qquad (2)$$
$$\omega_{ij} = A_{ij} - e_{ij} - \delta_{ij} \qquad i, j, k = 1, 2$$

where δ_{ij} is the Kronecker symbol. The practical evaluation of the deformations from pair of photographs consists in visual stereocomparator measurements and automatic processing of the measured data.

Four practically identical loading states were measured, i. e. both faces of the specimen loaded by positive and negative resultants of external forces. The experimental results indicated that shear deformations in the centre of the specimen agreed to within ten percent. The values of the shear moduli were consistent with experimental observations reported previously [8], [9].

EXPERIMENTAL RESULTS

The shear moduli are determined by the ratio of shear stress and shear strain

$$G = \tau(x, y) / \gamma(x, y) = k \bar{\tau} / \gamma (x, y) \qquad (3)$$

The correlation factor k for the given specimen shape in its central region has the values of 0.87 for the /0/ orientation and 1.04 for the /90/ orientation. A comparison of results obtained with the values of shear moduli determined by the ±45 tensile in plane shear test, three-point bending relative span variation test and the strip torsion with relative width variation method [8], [9] has shown very good agreement, particularly between the results of the ±45 tensile and the Iosipescu test of the /90/ orientation. For the tested series of specimens the value of the index determining the ratio of shear moduli determined by the Iosipescu test for the /0/ and the /90/ orientations was 1.16, which corresponds with the experience of other laboratories [5].

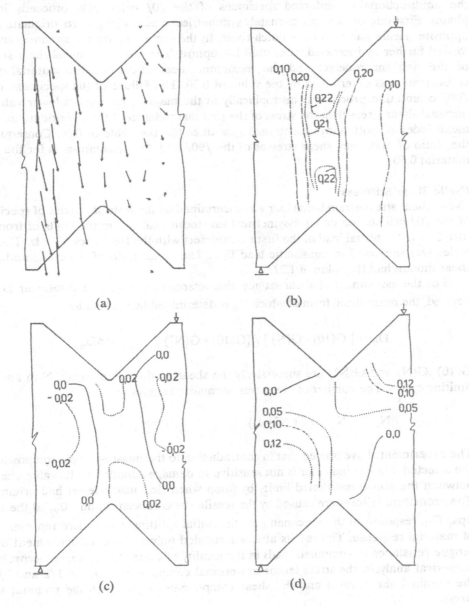

Fig.4. Field of displacements (U + V) (a), contour lines of shear deformations e_{xy} (b) and normal deformations e_{xx} (c) and e_{yy} (d)/ ‰ /.

The shear strength values obtained by the Iosipescu test must be considered as conventional for the given type of the testing fixture and specimen shape. The failure of the unidirectionally reinforced specimens of the /0/ orientation proceeds in two phases. First one or two approximately symmetrical cracks along fibres originate in the opposite tensile parts near the notch roots. In the second phase the specimen may be loaded further by increased loads until its rupture. The ratio of nominal shear stresses of the first macrodefect and the maximum shear stress for the material under consideration is determined by the value of 0.70. The failure of the specimens of the /90/ orientation proceeds catastrophically in the majority of cases at lower values of nominal shear stress than the stress of the first macrodefect of the /0/ orientation. The mean index of both above mentioned quantities has the value of 0.86. Consequently, the ratio of maximum shear stresses of the /90/ and /0/ orientations is for the given material 0.60.

Cyclic Shear Stiffness

Cyclic shear stress-strain behaviour was ascertained by the tests of a series of specimens of the /0/ orientation under assymetrical saw-tooth load within the limits of from 1/6 to 1/3 of the external load of the first macrodefect with the frequency of 1 Hz. The hole series was subjected to quasistatic load first. The mean ratio of the cyclic and static shear moduli had the value of 1.22.

For the assessment of shear-fatigue characteristics the damage parameter D_n was devised, the normalized form of which D_n is determined by the relation

$$D_n = [\, G(10) - G(N)\,] \,/\, [G(10) - G(N^*)\,] \qquad 0 \leq D_n \leq 1 \qquad (4)$$

$G(10)$, $G(N)$ and $G(N^*)$ are successively the shear moduli of the tenth, N-th and life-limiting cycles. The number of cycles was normalized similarly

$$N_n = (\,N - 1\,) \,/\, (\,N^* - 1\,) \qquad 0 \leq N_n \leq 1 \qquad (5)$$

The experiments have proved that in contradiction to the quasi-isotropic composites [3] the selected damage indicator is not sensitive to damage cumulation. Its value changed between the above mentioned limits by jump when first macrodefect had originated. The premature failures are caused by the tensile stress concentration σ_{yy} at the notch tips. The response of the specimen past the initial splitting is not more representative of material response. The results attained enabled only a mediated assessment of the fatigue resistance to combined loads in the locality of macrodefect where, according to numerical analysis, the stress tensor has normal components equal to 1/2 and 3/5 of the nominal shear stress and the shear components equal with the nominal shear stress.

CONCLUSION

The Iosipescu shear test method is a practical technique currently available for measuring of reliable shear properties data of composite materials. The correct interpretation of results afforded by the test on the given fixture must be founded on a thorough knowledge of the stress and strain field in the working part of the test specimen used.

REFERENCES

1.   Adams, P.F. and Walrath, D.E., Current status of the Iosipescu shear test method. J. Comp. Mat., 1987, 21, 494-507.

2.   Adams, D.F. and Walrath, D.E., Further Development of the Iosiplescu shear test method. Exp. Mech., June 1987, 27, 113-19.

3.   Wang, S.S. and Chim, E.S.M., Fatigue damage and degradation in random short-fiber SMC composite. J.Comp. Mat., March 1987, 21, 114-34.

4.   Kumosa, M. and Hul, P., FEM analysis of mixed mode fracture in the Iosipescu shear test. VI ICCM and 2nd ECCM, London, 1987, 3, 243-53.

5.   Sullivan, J.L., The use of Iosipescu specimens, Exp. Mech., Sept. 1988, 28, 326-28.

6.   Pindera, M.J., Ifju, P., Post, D., Iosipescu shear characterization of polymeric and metal matrix composites. Exp. Mech., March 1990, 30, 101-8.

7.   Němec, I., NE 04 Rovinná úloha PDP 11, Dopravoprojekt Brno.

8.   Minster, J., Determination of the shear moduli of carbon-epoxy composites by means of the bending test. Acta Technica CSAV, 1984, 6, 761-74.

9.   Minster, J., Bartoš, F., Náprstek, J., The application of the Iosipescu shear test to carbon-epoxy composites. Proceed. XV. Reinforced Plastics, Karlovy Vary, 1989, 1, 123-29.

10. Lee,S. and Munro,M., Evaluation of testing techniques for the Iosipescu shear test for advanced composite materials. J.Comp.Mat., April 1990, 24, 419-440.

CONSTITUTIVE PREDICTION FOR A NON-LINEAR ORTHOTROPIC MEDIA

Régnier HUCHON *, Jean POUYET * and Jacques SILVY **
* Laboratoire de Mécanique Physique, URA 867 CNRS
Université de Bordeaux I, 351 cours de la Libération 33405 Talence Cedex
** Laboratoire de Génie des Procédés Papetiers, URA 1100 CNRS
Ecole Française de Papeterie, D.U., 38402 Saint Martin d'Hères Cedex

ABSTRACT

Linearly elastic behaviour of paper is described by four independent in - plane constants. Elastic constants depend on probability density functions for length and orientation and also on a coupling network parameter between fibers. It is shown here that the ratio of machine to cross direction Young's moduli determines the fibre orientation parameter. All constants then simply depend on one network parameter and one orientation parameter.

We can describe the behaviour of a thin non-linear orthotropic plate using an asymptotic expansion method. We apply in this paper the theory proposed by Johnson to an assumed form of the elastic strain energy function. Linear or non-linear biaxial behaviour of such a material is then obtained from ordinary tests.

INTRODUCTION

Paper or paper board are multiphase composite materials. Wood fibre itself can be thought of as a natural composite. The ability of self binding of the wood fibres by chemical bonds and the properties of the fibres determine the mechanical properties of the paper. However, in the papermaking machine the fibres are rather oriented in the machine direction (MD) which is the direction of highest stiffness. The machine direction and the cross-direction (CD) are directions of material symmetry.

Thus, paper can be macroscopically modelled as an orthotropic solid with MD(1) and CD(2) axes since out-of-plane anisotropy is not considered in theoretical models regarding the thickness of a paper versus in-plane dimensions. Like composite materials which are both heterogeneous and anisotropic, paper is studied from two points of view : micromechanics and macromechanics.

Linearly elastic behaviour

Linearly elastic behaviour of paper is described by four independent in - plane constants : MD elastic modulus E_1, CD elastic modulus E_2, in-plane shear modulus G_{12} and Poisson's ratio v_{12} or v_{21}. When mechanical sollicitations are applied outside of material symetrical axes, the apparent constants in the new axes reference system are connected to the precedent ones by the relations given in annex (see Jones (1)). The terms U_1, U_4, U_5 are invariant with any axes rotation. Note that in analogy with isotropic case where $U_{1\,iso} = 1/E$ and $U_{5\,iso} = 1/G$, these quantities which are an invariant combination of several intrinsic caracteristics seem to constitute a better reference, for example when studying the influence of any parameter in correlation with paper properties such as machine made elasticity modulus. While E_1 and E_2 are easily measurable, Poissons' coefficients and shear modulus are not. So every forecasting method of these caracteristics is desirable.

Independance of shear modulus
As noted by Schulgasser (2) it seems that paper materials belong to anisotropic material class, previously described by Saint-Venant, for which orientation effects do not influence this modulus. By reference to Campbell (3) and Onogi (4) works then more recently those of Gunderson (5), they produced expressions like :

$$2/E_x = 1/E_1 + 1/E_2 + (1/E_1 - 1/E_2)\cos(2\theta) \tag{1}$$

The study of relations in annex shows that the apparent Young modulus in the direction (Ox) will be identical to the former empirical value if : $U_3 = 0$ or $G_{xy} = G_{12}$. Beyond precedent authors Jones (6) experimentaly showed this independance for five kind of papers by mean of resonant torsional method.

The needs to develop theories explaining correlations between micro and macro properties were first established by Van den Akker (7) : a typical fibre is characterized by its length, its width, its orientation versus a reference direction (θ), and by its mechanical properties. It is crossed by a number of other fibres. The proportion of crossing fibres along the fibre length determines the transfert of loads.

Therefore, in-plane elastic constants depend on probability density functions for length and orientation and also on a coupling network parameter which accounts for the finite fibre length and the coupling efficiency between fibres. Several workers have extended the original model of Cox (8) to take into account many parameters which are often difficult to measure. By an elastic energy method Perkins and Mark (9) relate the elastic constants to four main parameters : fibre modulus, sheet density, fibre orientation and a coupling network parameter.

Density function for orientation f(θ)
The function $f(\theta)$ represents the probability that a fibre is oriented in a sector (θ), $(\theta + d\theta)$:

$$f(\theta) = \frac{1}{\pi} \sum_{n=0}^{\infty} a_n \cos(2n\theta) \quad , \quad a_0 = 1 \tag{2}$$

It is generally an even function to take into account the symmetry in respect with machine and cross directions and its Fourier's development is limited to a few terms. The main representations are the cardioid distribution [(10) $a_2 = 0$], the Von Mises distribution [(11) $a_2 = (a_1)^2/4$], Forgacs's distribution [(12) $a_2 = (a_1)^2/2$], and normal representation [(13) $a_2 = (a_1)^2/8$].

Since fibres are of finite varying lengths, a density function which is the probability that the fibre length is included between (l) and (l + dl) may be the Erlang distribution (12). Assuming the coupling between the two distributions is weak, the functions of length and orientation are independant.

Orientation repartition choice
The choice depends on numerous parameters. When introducing two first terms (a_1 and a_2) limited functions in Cox or in Perkins (9-10) calculations one can get two types of similar expressions :

$$\frac{K}{E_1} = \frac{6 - 4 a_1 + a_2}{2 + a_2 - a_1^2} \qquad \frac{K}{E_2} = \frac{6 + 4 a_1 + a_2}{2 + a_2 - a_1^2} \qquad \frac{K}{G_{12}} = \frac{16}{2 - a_2}$$

$$v_{12} = (2 - a_2) / (6 - 4 a_1 + a_2) \; ; \; v_{12}/E_1 = v_{21}/E_1 \qquad\qquad (3)$$

For Perkin's model : $K = \rho_0 / 16 \, \rho_f \, E_f \, \phi$ where ρ_0 and ρ_f are apparent density of sheet and fibre respectively, ϕ is a complex structural factor which takes into account fibres bonding. Generally this factor is quite close to unity and can be determined by various methods. Calculation of U_3 with its definition and precedent values of mechanical caracteristics gives :

$$U_3 = 2 \, (a_1^2 - 2 \, a_2) / K \, (2 - a_2) \, (2 + a_2 - a_1^2) \qquad\qquad (4)$$

The condition of independance of G_{12} ($U_3 = 0$) leads to the following condition $a_2 = a_1^2 / 2$ which is typical for the above Forgac's distribution :

$$f(\theta)^{-1} = \pi \, (\cos^2 (\theta) + \lambda \sin^2 (\theta)) \qquad a_2 = 2 \, (\lambda - 1)^2 / (\lambda + 1)^2$$

X rays measurements of orientation functions made by Prud'homme (14) or Eusafzai's analyses (12) carried out on high bonded paper gives similar results. So with the different interpretations of K all constants simply depend on a network parameter and an orientation parameter which are both obtainable from a tensile test. The ratio of machine to cross direction Young's modulus determines the fibre orientation parameter ($a_1 = a$) then v_{12} and G_{12} and eventually K are obtained from the following relations :

$$E_1 / K = (2 + a) / (6 - a) = v_{12}$$

$$E_2 / K = (2 - a) / (6 + a) = v_{21} \qquad\qquad (5)$$

$$G_{12} / K = (4 - a^2) / 32$$

This result is in accordance with what J. Silvy (15) deduced by the "pore equivalent" method in which the square of the ellipticity of the " pore " (equal to the anisotropic ratio) controls all the other quantities. Figure 1 shows these quantities in respect with "a" variation.

Experimental confirmation
Schulgasser compared precedent formulations with experiments of Jones (6) which were obtained by pendular method or with those of Craver and Mann by mean of ultrasonic methods (16). Results are in good accordance with E_1, E_2, G_{12}. Rigdahl (17) works upon three values of basis weight (10, 30, 80 g/m^2) and performs in each case five tests with different anisotropic ratios. Distributions are calculated by mean of image analysis and the orientation function is fitted with a Von Mises function. This allowed to perform a_1 and a_2 calculations and let us get anisotropic ratio with relations (3). We can note that using Forgac's function gives the most reliable results between calculation and tests.
If we admit this representation, we can forecast all the mechanical constants. Figure 2 gives results from experiments of J.Silvy (15) , Rigdalh (17) and Perkins (9), Jones (6).
Another way is to calculate K with experimental values of E_1, E_2, then compare the predicted value of E_2 for example with the effective measured value. Figure 3 shows a good

correlation with 27 points taken from four authors' work. For out of basis weights near $80g/m^2$, the comparison is good.

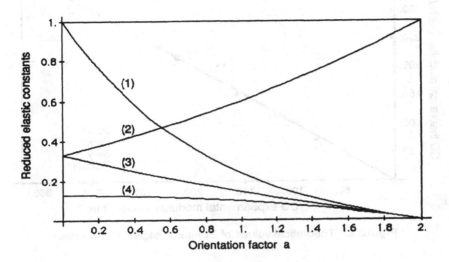

Figure 1. (1) E_2/E_1, (2) E_1/K, (3) E_2/K, (4) G_{12}/K versus a

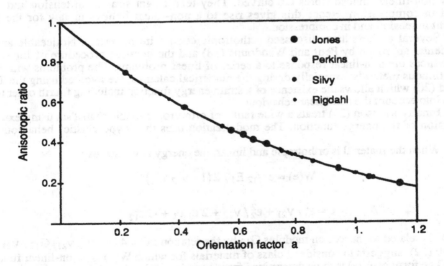

Figure 2. Comparison of theoretical prediction of E_2/E_1 to experimental values

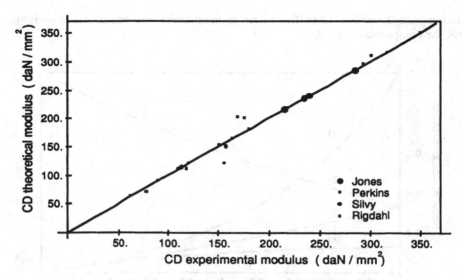

Figure 3. Theoretical values of E_2 versus experimental ones

Non-linear behaviour

The theories mentioned above use linear approaches. In fact the behaviour of this material is mainly non-linear, indeed fibres are curved. They tend to get straight in tension and more curved in compression. Hence this gives rise to a non-linear behaviour law for the fibre different in tension and in compression.

Several authors have discussed orthotropic composite materials. Noticeable are the incremental approach by Petit and Waddoups (18) and the iterative procedure of Jones (19) who reduces the non-linear response to a series of linear problems. The problems with these discontinuous methods, especially during the numerical calculus, are avoided using the Tsai's method (20) which allows the existence of a strain energy function including fourth order terms to take into account the non-linear behaviour.

Finally Johnson (21) treats a wide range of orthotropic media than Tsai using extented expression of the energy function. The modelization uses the "hyperelastic" behaviour law (21).

When the material is orthotropic and linear, the energy is written as :

$$W(e) = e.v_{12} \, E_1 \, / \, 2 \, (1 - v_{12} \, v_{21})$$

$$e = \varepsilon_1^2 \, / \, v_{21} + \varepsilon_2^2 \, / \, v_{12} + 2 \, \varepsilon_1 \, \varepsilon_1 + C \, \varepsilon_{12}^2 \tag{6}$$

where C is related to the torsion modulus G_{12} by the relation : $C = 4 \, (1 - v_{12} \, v_{21}) \, G_{12} \, / \, v_{21} \, E_1.$ Urbanik (22) suggests to consider a class of materials for which W(e) is a non-linear function of "e". The form of W(e) is to be determined from experimental uniaxial data.

The experimental method consists in evaluating the strain energy from uniaxial tensile tests and theoretical expressions formulated in this particular case. The relevant equations are :

$$\sigma_1 = 2\ \varepsilon_1\ (1 - \nu_{12}\ \nu_{21})\ W'(e)\ /\ \nu_{21}$$

$$(7)$$

$$e = \varepsilon_1^2\ (1 - \nu_{12}\ \nu_{21}) / \nu_{21}$$

where σ_1 and ε_1 are tensile stress and strain in the machine direction.
One can see that in this case :

$$\sigma_1\ (\varepsilon_1) = dW(\varepsilon_1)\ /\ d\varepsilon_1$$

This last relation allow us, by modelising the tensile curve, to reach the energy function W(e).

Tensile curve modelization

The expressions suggested to modelise the tensile test are those suggested by El Hosseiny (23), Ramberg (24) and Urbanik (22). A comparison of different models showed that our tested paper samples corresponded best with the three parameters hyperbolic tangent model :

$$\sigma(\varepsilon) = C_1\ .\ \text{th}\ (C_2\ .\ \varepsilon) + (C_3\ .\ \varepsilon)$$

$$(8)$$

This allowed us to calculate the coefficients C_i and then the energy function W(e).

Experimental results

To try and qualify our method, we performed tensile uniaxial tests upon calandered coated sheets of bleached kraft pulp. The following tables show principal papermaking characteristics of this product (table 1) and experimental conditions (table 2) :

Table 1: Papermaking characteristics

Basis weight	92	$g.m^{-2}$
Caliper	76	μm
Taber flexural rigidity	0.28	mN.m
Elmendorf tearing	446	mN
Opacity	83.6	%

Table 2 : Experimental conditions

Strain velocity	0.833	$10^{-2}\ s^{-1}$
Temperature	22	°C
Relative humidity	50	%

We performed a non-linear regression for each (σ - ε) curve according to preceding description. Mean values of C_1, C_2, C_3 deduced from calculations are given for each orientation in Figure 4.

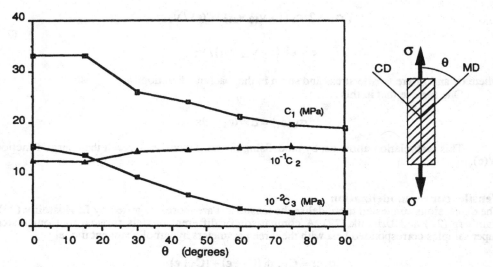

Figure 4. Coefficients C_i of URBANIK's model

The strain energy density function $W(\varepsilon_1)$ is evaluated as mentioned above when integrating the formula (7). The elastic constants related to $W(e)$ calculation are deduced from relations (6). Figure 5 illustrates plots of uniaxial MD and CD deduced from theory. They are correlated with experimental data.

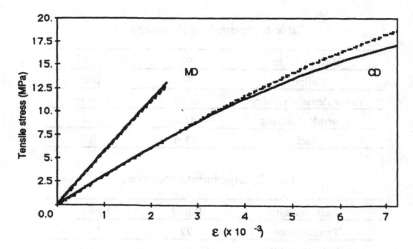

Figure 5. Tensile stress-strain curves
--- hyperelastic model
——— approximated curve

Obviously it is not possible to predict exactly the inelastic behaviour. This is due to the theoretical approximation. However this special theory can be a starting point in predicting multiaxial loadings.

REFERENCES

1. Jones, R . M., Mechanics of Composite Materials , Mc Graw-Hill , New York , 1975 .
2. Schulgasser, K., Fiber Science and technology, 1975, **15**, pp.257-270.
3. Campbell , J.G., J. Appl. Sci., 1961, **12**, p. 356.
4. Horio M., Onogi S., J. Appl. Phys. 22, 7 , p.971, July 1951.
5. Gunderson D., ESPRA Conf., Wisconsin Rapids, Wisconsin, May 1984.
6. Jones A.R., Tappi, vol. 51, n° **5**, May 1968.
7. Van den Akker J.A., The Formation and Structure of Paper, 1962, Clowes and Sons, London, pp.205-209.
8. Cox H.L., Brit. J. of Appl. Phys., **3**, p.72, 1952.
9. Perkins R.W. and Mark R.E., Proc. The Role of Fundamental Research in papermaking. Mech. Eng. Publ. Ltd, London, 1983.
10. Perkins R.W., Proc. of the Conf. of Paper Science and Technology. The cutting edge. Institute of Paper Chemistry, Appleton, p.89, 1980.
11. Perkins R.W., International Paper Physics Conference, p.83, 1983.
12. Eusufzai A., Sheet structure in relation to internal network geometry and fiber orientation distribution, Thesis Univ. of New-York, Syracuse, March 1982.
13. Schulgasser K., J. of Materials Science, **20**, p.859, 1985.
14. Prud'homme R.E., J. Appl. Polym. Sci., **19**, p.2609, 1975.
15. Silvy J., Doc. FCI E.F.P., Grenoble IRFIP, SE 92 46, 1976.
16. Jones R.W., Tappi **63**, p.163, 1980.
17. Rigdhal M., Fibre Science and Technology, **19**, p.927, 1983.
18. Petit, Waddoups, J. of Comp. Mat., vol.**14**, n°1, pp.2-19, 1969.
19. Jones R.W., J. of Comp. Mat., vol.9, n°1, pp.10-27, 1975.
20. Tsai, J. of Comp. Mat., vol.7, n°1, pp.102-118, 1973.
21. Johnson, J. of Appl. Mech., vol. **51**, n°1, pp.146-152, 1984.
22. Urbanik T.J., Effects of paperboard stress-strain characteristics on strength of singlewall corrugated fiberboard : a theorical approach. USDA Forest Service, Research Paper FPL 401, 1981.
23. El Hosseiny, The stress-strain curve of fibrous networks. Tappi **62**, 10, 127, pp.181-187, 1979.
24. Ramberg W., Osgood W.R., Description of stress-strain curves by three parameters. Naca Technical Note n° 902, 1975.

APPENDIX

$$1/E_x = U_1 + U_2 \cos (2\theta) \qquad + U_3 \cos (4\theta)$$

$$1/G_{xy} = U_5 \qquad - 4 U_3 \cos (4\theta) \qquad U_i = U_i (E_i, G_{12}, v_{ij})$$

$$- v_{xy} / E_x = U_4 \qquad - U_3 \cos (4\theta)$$

CREEP AND RELAXATION OF COATED FABRICS UNDER BIAXIAL LOADING

Anne GUENAND, C.E.R.M.A.C. INSA-UCB
Patrice HAMELIN, C.E.R.M.A.C. INSA-UCB

CERMAC   INSA-UCB
Bâtiment 304 - I.N.S.A.
20, Avenue Albert Einstein
69621 VILLEURBANNE CEDEX
Tel : 72.43.82.41
Fax : 72.44.08.00

ABSTRACT

This paper presents a study on the mechanical behaviour of coated fabrics under biaxial loading. The cruciform test specimen investigated is constitued by about thirty percent of UP fibers weaved in two perpendicular direction, protected by a PVC matrix.
The relaxation functions established by curve fitting of experimental points are confronted with those obtained by the modelization based on the Maxwell's model.

INTRODUCTION

The appearence of new synthetic fibers and the development of calculation methods for tensed structures has brought about a renaissance in textile architecture. (TN Mars 1989 "Un matériau composite souple"). Examples can be found in large covered areas, show halls, sports halls and stocking areas. The requirements demanded by the schudle are :

- The structures must be mobile and allow easy assembly and dismanteling.
- Once assembled the structure must hold its initial state without needing re-tension after a few months of use.
- To aim for a durability beyond 15 years.

These requirements may be fullfilled by the use of a soft composite : waterproof membranes with textile reinforcement.

Most of the composite materials structures are subject to multiaxial stresses. It is therefore necessary to define failure criteria adapted to composite materials taking in consideration the multiaxial loading as well as the orthotropic nature of the material.

The criteria thus elaborated like those of Tsaï & Wu (1971) demand multiaxial tests to determine the interaction coefficients.

So, for an example along a plane, the criteria is written :

$$F_1\,\sigma_1 + F_2\,\sigma_2 + F_{11}\,\sigma_1^2 + F_{22}\,\sigma_2^2 + 2F_{12}\,\sigma_1\,\sigma_2 + F_6\,\sigma_6^2 = 1$$

All the coefficients can be determined by traction or simple compression except for F_{12} which requires a biaxial test.

In general, to elaborate new theories of rupture or to calculate coefficients, one needs trial results from multiaxial tensile tests. In the example of a plane the testing equipment must allow us to be able to vary σ_1, σ_2 and σ_6 (the normal and shear stresses respectively).

More over the tests must, as for as possible, fullfil the following conditions :

- The state of strain in the test on a least in its useful part, must be homogeneous.
- The components of the state of strain have to vary independantly.
- The rupture must happen in the area concerned.

The methods of testing proposed all present advantages and disavantages. The tests of strains in a cruciform specimen induce the concentration of stresses at the level of arm (Owen 1983) but has the advantage to permit different types of loading and is a good representation of the unidirectional case or stratified plane.

We have leaved in this study towards the mode of investigation of the state of the limit along a biaxial plane.

PHENOMENOLOGICAL APPROACH OF RHEOLOGICAL BEHAVIOUR

Apparatus and methods

<u>Biaxial apparatus</u> : We study the behavior of coated fabrics UP/PCV, under increasing biaxial loading.

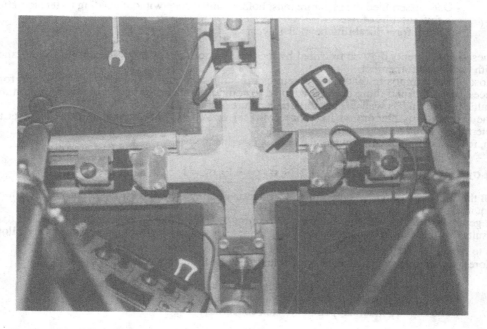

Figure 1.Biaxial apparatus

The biaxial state is obtained by a two-dimensional solicitation in two perpendicular directions on a cruciform specimen.
The displacement of the clamps is obtained thanks to a screw (1,5 mm/t). The maximal strength is about 2000 daN.

Testing principles : One has measure strength put upon the cruciform specimen in both directions. Thus the curve "strength applied as a function of displacement" can be obtained, anables to characterize the biaxial behaviour of the composite material. Gauges are sticked on each axis and connected to an extensometric device.

Displacements in each direction "x" and "y" are measured by photographic negatives or direct measuring in the case of a great displacement..

<u>Results</u> : Major parameters which have an influence on the coated fabrics behaviour are :

- The stress ratio between warp direction and weft direction in the case of biaxial loading.
- The effect of outside conditions (temperature and humidity).

In this study, we work under standard conditions (20°C, 65% relative humidity) and the strength ratio varies from 1 to 4.
The type of curve obtained is as follows   in figure 2

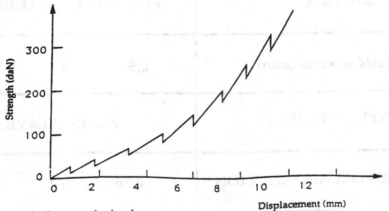

Figure 2. Typical curve obtained.

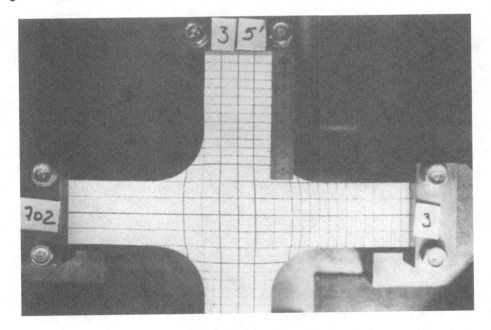

Figure 3.Cruciform test specimen. This shape has been studied by Finit Element Method.

Table 1.Table of industrial characteristics of fabrics

FIBRES	POLYESTER (UP) HIGH TENACITY		
MATRIX	POLYVINYL CHLORURE		
THICKNESS (mm)	0,54	0,7	1
TYPE OF FABRIC	PLAIN WEAVE		
FIBRE VOLUME FRACTION	0,26	0,27	0,29

The strength is a function of displacement. Each vertical step corresponds to a time during which the relaxation is recorded.
Calculing the positive slope allows to measure the evolution of elastic modulus in accordance with displacement.
Relaxation experiment with different strength ratios are as follows.

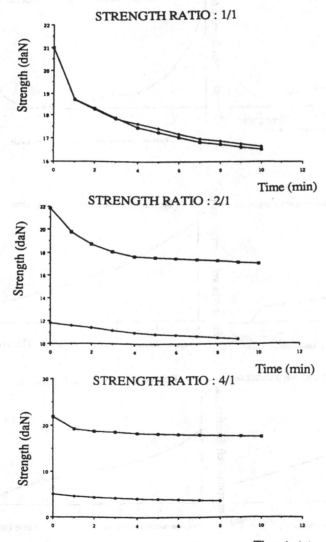

Figure 4. Relaxation curves for different strength ratio.

178

Analysis of rheological behaviour

The relaxation curve fitting allows to obtain the function under polynomial form. (Macintosh II, Software Cricket Graph).

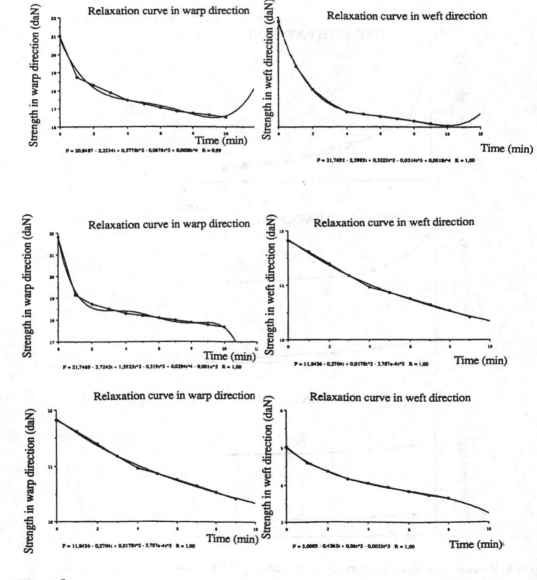

Figure 5.

Modelization of rheological behaviour

After exploiting the experimental results, we choose the Maxwell model to describe the relaxation function of composites.

$$\frac{d\sigma}{dt} = E_m \frac{d\varepsilon}{dt} - \frac{1}{\tau}\sigma$$

$$\sigma = \sigma_0\, e^{-t/\tau}$$

ε_1 E_1, σ_1

ε_2 η_2, σ_2

Taylor's formula of a single variable is as follows:

$$\sigma(t) = \sigma_0\left(1 - \frac{t}{\tau} + \frac{t^2}{2!\,\tau^2} - \frac{t^3}{3!\,\tau^3} + \dots\right)$$

It allows the computing of relaxation time for each relaxation function.

Strength ratio	1/1	2/1	4/1
τ warp direction	9	7,2	3
τ weft direction	3,8	3	6

Table 2. Relaxation times for different strength ratio.

We notice that the increase in strength ratio between warp direction and weft direction implies the decrease in relaxation time in the warp direction.

CONCLUSION

The use of viscoelastic modelization (Maxwell model) enables to have a previsionnal approach of the biaxial behaviour of composite material.
The research aims toward the building of a new testing machine which would allow us to regulate the strength and displacements (optical measuring), which would permit the simultaneousness of the biaxial loading, and eventually the combination of environmental effects and mechanical loads.

REFERENCES

BONNEL P., 1990
"Création d'un logiciel d'aide à la conception des composites stratifiés : Multicouche CERMAC". Mémoire de DEA. Université Lyon 1.

HUET C., 1963
"Etude, par une méthode d'impédance, du comportement viscoélastique des matériaux hydrocarbonés".Thèse de Docteur Ingénieur.Faculté des Sciences à l'Université de Paris.

HUET C., BOURGOIN D., RICHEMOND S., 1984
"Rhéologie des matériaux anisotropes".Compte rendu du 19e colloque national annuel,Paris. Cepadues Editions. 591 p.

JEANNE Ph., 1986
"Contribution à l'étude des zones interfaciales dans les composites polyépoxyde. Fibres de verre".Thèse de Docteur Ingénieur. Ecole Centrale de Lyon. 253 p.

TSU-WEI CHOU et FRANK K.KO, 1989
"Textile structural composites".Volume 3. Elsevier Editions. 387 p.

JOINTS AND ASSEMBLIES OF COMPOSITE MATERIAL STRUCTURAL ELEMENTS: CHARACTERIZATION OF ADHESIVES IN BULK FORM AND IN COMPOSITE STRUCTURES

G. VAN VINCKENROY; W.P. DE WILDE; F. DE ROEY
Composite Systems and Adhesion Research Group of the Free University of Brussels
(COSARGUB)
Free University of Brussels (V.U.B) - Pleinlaan, 2 1050 Brussel - Belgium

ABSTRACT

As far as experiments are concerned, life time prediction of adhesively bonded structures implies an extensive test program. The properties of the joint components, i.e. adhesive and composite adherend are experimentally determined as they serve as data input for finite element model of the bonded joint. The adhesive characterization includes shear tests performed on bulk adhesive material and on thin film form (rail shear and the Arcan ring test), tensile test, torsion test and *resonalyser* test, based on eigenfrequencies measurement of a free plate. An elastoplastic and a viscoelastic model are fitted to the experimental data. The unidirectional carbon-epoxy composite is also characterized, and single lap structures are fabricated and tested. Destructive techniques give information about (the evolution of) the residual strength of the tested joints. The evolution of physical properties are monitored with non destructive techniques like ultrasonic scanning. Experimental results on single lap joints are further used to assess numerical model of bonded structures.

INTRODUCTION

The final objective of the research is to develop life time prediction models together with the investigation of damaged state for adhesively bonded joints. The approach is based on probabilistic methods, as it is well known that generalized loads (i.e. thermomechanical and environmental) applied to an adhesive joint are better described by probabilistic techniques, and advanced composite materials structures have properties which exhibit great variations in space and time, due to intrinsic causes.

The set-up of life time prediction model involves the use of finite element program together with experimental analysis. Indeed, numerical tool is needed to predict mechanical behaviour of bonded joints (stress distribution, failure load, fatigue,...), the input data of this modelization being the properties of the components of the bonded structure. So one of the basic needs in characterizing structural behaviour of bonded structures is the development of an accurate analytical representation of the material properties.

Statistical treatment of experimental data involves an extensive test program in order to obtain the elastic and ultimate properties of materials in such a quantity and quality as to fit an adequate probabilistic model. Due to the large number of tests required, the test method has to fullfill some conditions, as: minimum time consumption, reproducibility as high as possible, small induced experimental errors.

METHODS, RESULTS AND DISCUSSION

The composite material used is a carbon fibre reinforced epoxy, supplied in prepreg form, with a nominal thickness of 0.125mm and 34% resin content. The adhesive under investigation is a modified epoxy adhesive film including a random polyester mat, supplied in prepreg form.

Preparation of Adhesive Specimens

Thick laminates of adhesive film are made by bonding several plies of adhesive together to a thickness of about 2mm for tension and shear tests. Those laminates are then put into an aluminum cadre to avoid resin flow during curing, and cured following the manufacturer recommended cure cycle: heat up in 30 min. to 120°C, hold 70 min at 120°C, with a pressure of 0.28 MPa, cool down in 80 min to 25°C before pressure removal.

Static Tensile Test on Bulk Adhesive

Tensile specimens are cut by means of a water-cooled diamond saw according to the ASTM D 638-M norm. Strain gauges of the Tokyo Sokki Kenkyujo Co. series are attached to the specimens with the P2-adhesive of the same manufacturer. The Young's modulus is determined by means of a linear regression on the linear range of the stress/longitudinal strain curve; the Poisson's coefficient by linear regression on the transversal strain/longitudinal strain curve. The ultimate stress and strain are directly read from the graphs. Various specimens are tested in the same conditions, the data obtained are fitted to a statistical model.

The stress/strain curve obtained can be fit to an analytical model (Figure 1): for high stress levels, the perfectly elastoplastic model with the same ultimate stress and strain has the same energy to failure as the actual stress-strain curve and is adequate to predict the strength. For lower stress levels however, the perfectly elastic model is to be preferred as it describes more accurately the stress at low load level.

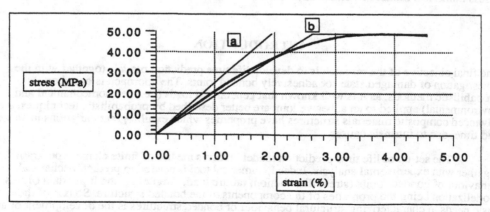

Figure 1. Experimental tensile stress-strain curve and (a) perfectly elastic (b) elastoplastic model.

Creep Tensile Test on Bulk Adhesive

The time dependent effects are taken into account by representation by an viscoelastic model, based on Schapery approach, where the non linear effects are expressed by means of stress-dependent material functions. This model is based on thermodynamics of irreversible processes, and has the advantage of conserving the single-integral formulation even in the non linear range of viscoelastic behaviour.

For uniaxial loading, the constitutive equation is written in the following form:

$$\varepsilon(t) = g_0 D_0 \sigma(t) + g_1 \int_{-\infty}^{t} \Delta D(\psi - \psi') \frac{dg_2 \sigma(\tau)}{d\tau} d\tau \tag{1}$$

where D_0 and $\Delta D(\psi)$ are respectively the initial and transient components of the creep compliance, ψ being the so-called reduced time defined by:

$$\psi = \psi(t) = \int_{t_0}^{t} \frac{d\tau'}{a_\sigma}, \qquad \psi' = \psi(\tau) = \int_{t_0}^{\tau} \frac{d\tau'}{a_\sigma} \qquad a_\sigma > 0 \tag{2}$$

where t_0 is an arbitrary time prior to the first non-zero straining of the material.

g_0, g_1, g_2 and a_σ are the stress-dependent material functions. In the general case these functions are stress, temperature and environmental variables dependent, but in a constant environment, only stress influences these functions. The linear form is recovered when $g_0 = g_1 = g_2 = a_\sigma = 1$, and these values must be approached as strains become sufficiently small.

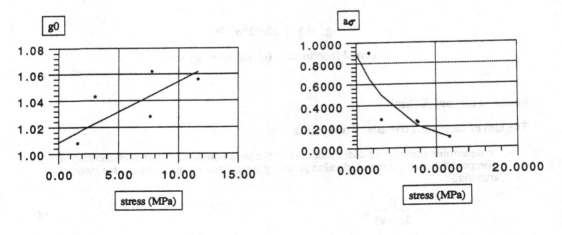

$$g_0 = 1 + 4.57e\text{-}3*\sigma$$

$$a_\sigma = \exp(-0.19*\sigma)$$

Figure 2. (a) values of g_0 vs σ. (b) values of a_σ vs σ.

We follow here the method of Peretz and Weitsman [1], to determine the stress functions from creep-recovery tests at different stress levels.

An additional assumption is made, concerning the form of the transient compliance:

$$\Delta D(\psi) = C\,\psi^n \qquad (3)$$

The power law form is chosen with reference to polymer behaviour and because the parameters can easily be determined. Other forms have been assumed for the transient compliance, each having its own advantages and drawbacks or application range [2].

The specimens used are similar to those used for static tensile tests. The results are illustrated on the figure 2. The scatter on g2 is considerable and not reliable curve fitting could be performed without supplementary data.

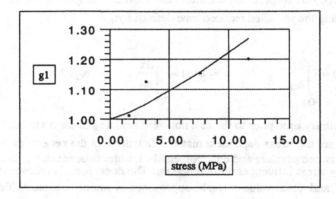

$$g_1 = 1+1.22e\text{-}2*\sigma^{1.26}$$

Figure 2 (continued). (c) values of g_1 vs σ.

Shear Test on Adhesive

The aim of these tests consists of two points:

-   to determine directly the shear modulus of the bulk adhesive. This experimental value is compared with the computed value given by the following relationship, for isotropic materials:

$$G = \frac{E}{2(1+v)}\,, \qquad (4)$$

where   G is the shear modulus,

E the Young modulus and v the Poisson's coefficient, determined by means of tensile tests.

This enables us to control the isotropy of the adhesive material, that is, to quantify, if any, the influence of the carrier used to support the adhesive itself.

- to compare bulk shear properties, i.e. modulus and strength, with the same properties of adhesive in thin film.

Two methods are applied to determine the shear modulus and shear strength of the adhesive material: the rail shear test and the Arcan test. Both methods have the advantage of inducing pure shear stress in the adhesive; the rail shear test is used to determine the properties of the bulk material, and the Arcan test can be performed to characterize the adhesive in thin form or eventually in bulk if the bond line has a thickness comparable to the thickness of a pure adhesive material plate.

Rail shear specimens are cut by means of a water-cooled diamond saw on a rectangular shape (200x150 mm). Holes are drilled and strain gauges are attached to the specimens, at ±45° to the load direction, two at the front side, the two others on the back side at the same location.

Shear stress/strain curves are plotted for various specimens. A regression performed on the linear range of the stress-strain curve gives the in-plane shear modulus (G_{12}), each value being given as an average of a pair of strain gauges placed at the same location. No failure is achieved on the specimen.

It has to be noted that the shear is here applied in a plane perpendicular to the carrier sheet, which is not the situation encountered in bonded joints, where the adhesive is submitted to shear in the carrier sheet plane. To perform shear stress in the carrier sheet plane, Arcan tests are performed, in which the glue line has the same thickness as the bulk material plate.

The **Arcan test** is used to determine the out-of-plane shear modulus (G_{13}) of adhesive in bulk form or thin form. Uniform shear stress is applied to the adhesive layer of an aluminum/steel structure: the thickness of this layer can be thin (0.10mm = 1ply) or thick (2mm = 16 plies) to investigate the thin and thick form properties of the adhesive, making the assumption that the behaviour of the thick adhesive layer is similar to the pure bulk material behaviour.

Torsion and Resonalyser Tests on Bulk Adhesive

Non destructive tests are performed on bulk material to determine the elastic properties. Those methods, namely the torsion [3] and resonalyser tests [4], are less time consuming methods than tensile tests due to the fact that we do not need to attach strain gauges to the specimens. The appropriatness of those tests are similar to the destructive tests, the error being of the same order. Only ultimate properties can not be determined by means of those methods. With both methods, based on a mixed experimental/numerical model, the material under investigation can be considered as isotrope or orthotrope and the properties are determined in consequence.

All results obtained on adhesive in bulk and in thin form are summarized in Table 1.

Composite Characterization

Laminates of unidirectional composite are made by bonding several plies of prepreg together to a thickness of about 1mm (8 plies at 0°). Those laminates are then cured following the manufacturer recommended cure cycle: heat up in 20 min. to 120°C, hold 20 min at 120°C, heat up in 10 min. to 175°C, hold 60 min at 175°C with a pressure of 0.70 MPa, cool down to 60°C before pressure removal.

TABLE 1
Elastic and ultimate properties of adhesive in bulk and thin form

	E_1-mod (GPa)	E_2-mod (GPa)	Poisson's coeff.	in-plane shear mod. (MPa)	out-of-plane shear mod. (MPa)	ultimate stress (MPa)
tension	2.509	2.509	0.3889	903.3 (**)		48.0(*)
rail shear	-	-	-	856.72		-
Arcan	-	-	-	-	751.4	39.8
resonalyser -isotrope -orthotrope	2.503 2.517	2.503 2.513	0.4536 0.4381	861.1 842.8	-	-
torsion -isotrope -orthotrope	-		-	887.33 945.23	887.33 708.79	-

(*)　　Ultimate tensile stress for a test rate of .5mm/min.
(**)　Shear modulus calculated by means of relationship (4) for isotropic material.

Tensile specimens are cut by means of a water-cooled diamond saw according to the ASTM D 3039 norm in the longitudinal and transversal directions and tabs are bonded on the longitudinal specimens. Strain gauges of the Tokyo Sokki Kenkyujo Co. series are attached to the specimens with the P2-adhesive of the same manufacturer. The Young's modulus is determined by means of a linear regression on the linear range of the stress/longitudinal strain curve; the Poisson's coefficient by linear regression on the transversal strain/longitudinal strain curve. The ultimate stress and strain are directly read from the graphs. Various specimens are tested in the same conditions, the data obtained are fitted to a statistical model.

Resonalyser and torsion tests are also applied to the composite material.

Results:

E_1 = 136.4 GPa　　　　E_2 = 12.8 GPa
ν_{12} = 0.344　　　　　ν_{21} = 0.03
G_{12} = 5.1 MPa　　　　G_{13} = 3.0 MPa

Test on Bonded Structures

The aim of those tests is to investigate the influence of variables as joint geometry , surface treatment. Results obtained experimentally will be compared to numerical modelization based on the previously determined characteristics of bond components.

One of the most important factors on which the strength of a joint depends is the degree of adhesion to the bonding surfaces. Adhesion occurs at the adherend/adhesive interface, within a thin layer of molecular dimensions. It is then obvious that great care has to be brought to the preparation of the surface, on which adhesive is to be applied, in order to obtain the optimum state, to induce strong interactions across the interface. Surface contaminants can reduce the bond strength to zero because they are themselves generally weak adherends and prevent a good contact between adhesive and substrate. The aim of all surface treatments is to remove those weak layers and put the surface to be bonded in a suitable state.

As far as composite materials are concerned, degreasing action and mechanical treatments are the more common methods: apart from substrate roughening by abrasion to improve bonding properties,surface treatments of composites are basically concerned with the removal of contaminants as mould release agents, lubricants, oils, water, plasticizers.

A general method of preparing composite substrates prior to bonding consists of three steps:

1. solvent wiping or vapor degreasing to remove oil and solvent-soluble contaminants.
2. water-rinsing or alkaline degreasing to remove water-soluble contaminants.
3. abrasion - by hand or machine - to increase the surface energy of the surface and to improve mechanical interlocking.

Different surface treatment were applied, the most efficient one will be used for further tesing and is described below:

- peel ply removal
- 1,1,1 trichloroethane wiping
- sandpaper (120) abrasion
- distilled water rinsing
- drying at 100°C during 60'
- 1,1,1 trichloroethane wiping
- drying at 100°C during 90'

The cured composite plate is cut following the ASTM standards (D 1002 for metal adherends and D 3164 for plastic adherends) with a water cooled diamant saw. Each pair of plates undergo the previously described surface treatment. They are further on bonded together with the adhesive prepreg, the tabs for the grips are bonded at the same time, with the same adhesive. The tabs are made of the same composite but with a cross-ply layup. The cure cycle used is the one of the adhesive. The specimens for testing are further on cut with a water-cooled diamant saw.

Before being tested to failure, the bonded joints are submitted to an ultrasonic scanning (C-scan). The test coupons are tested in tension at a speed rate of 0.5mm/min on an Instron bench. After failure, scanning electron microscopy (SEM) is used to study the surface morphology and fracture.

Statistical Treatment of Data

The choice of a distribution to represent a physical system is often difficult, depending on operator judgment and prior experience. It is generally motivated by an understanding of the nature of the underlying phenomenon and is verified by the available data. The process is often facilitated by the construction of probability plots on various probability papers so that experimental data will plot approximately as a straight line if the data follow the appropriate underlying distribution [5]. After a model has been chosen, its parameters must be determined and a statistical test should be performed to determine whether the chosen distribution provides an adequate fit to the data.

The statistical distributions used in this research are the two and three parameters Weibull distribution, the normal and the log-normal distribution. Concerning the parameters estimation, general methods do exist, the ones used here are the following: the maximum likelihood method, the method of moments and probability plotting. The statistical test performed to determine whether the chosen distribution provides an adequate fit to the data are the following: the regression analysis and the Chi-square analysis [6].

The properties determined for each test are now used to fit the statistical distributions. The correlation coefficient and the chi-squared test failed to reject any of the distributions for all samples at the 5 percent significance test.

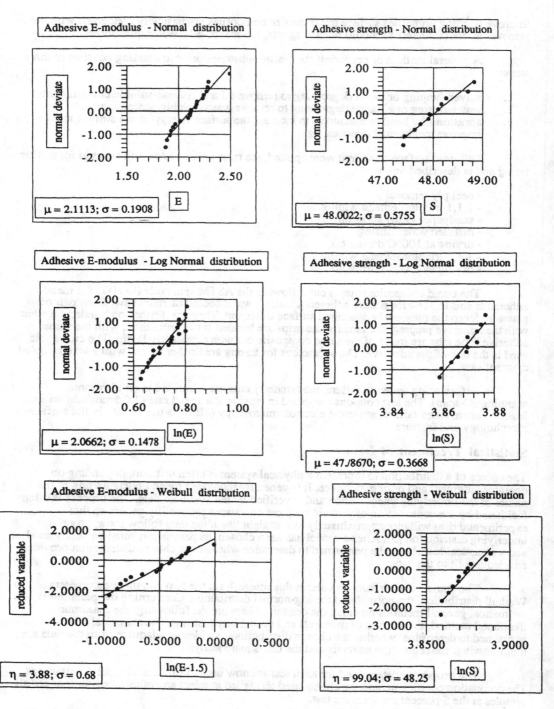

Figure 3. Probability plotting for adhesive E-modulus and strength to fit various probabilistic distributions.

CONCLUSIONS

Complete characterization of materials, more specifically adhesive material, has been carried out, using various mechanical set-ups

- on bulk adhesive under uniaxial tension, under shear and torsion; mixed numerical/experimental technique based on resonant frequencies measurement (resonalyser).

- on thin film adhesive under "pure shear" in bonded joints.

Composite material has been tested in tension, torsion and by means of the resonalyser.

Non destructive testing as the torsion and resonalyser method implies less time consuming techniques but they do not deliver ultimate properties. Due to strain gauges attachment, destructive techniques as tensile tests are time consuming ways to determine elastic as well as ultimate properties on the same test coupons. One could combine those methods, depending on the properties he is interested in, time available and material available.

Material properties are described with probabilistic laws to take the actual variability of data into account.

Destructive tests are performed on single lap composite joints to investigate the effect of surface treatment, and further to assess numerical model of bonded structures.

ACKNOWLEDGEMENTS

The financial support provided by the Instituut voor aanmoediging van Wetenschappelijk Onderzoek in Nijverheid en Landbouw (I.W.O.N.L.) under contract number 5313 is gratefully acknowledged.

REFERENCES

[1] Peretz D., Weitsman Y. Non linear Viscoelastic Characterization of FM73 Adhesive. Journal of Rheology, 1982, 26 (3) 245-261.

[2] Brouwer R., Non linear viscoelastic characterization of transversely isotropic fibrous composites under biaxial loading, PhD. Thesis Vrije Universiteit Brussel, may 1986.

[3] Sol H., De Vissher J., Van Tomme J., : Measurement of complex moduli of composite materials and discussion of some results. Euromech 269, Elsevier Applied Science Publishers,1991.

[4] Van Hemelrijck D., Schillemans L., De Roey F., Daerden I., Boulpaep F., Cardon A., A computerized test setup for the determination of the in-plane and out-of-plane shear modulus in orthotropic specimens. Euromech 269, Elsevier Applied Science Publishers,1991.

[5] Engineering Sciences Data, item n° 68013 to 68017, Enginnering Sciences Data Unit

[6] M.G. Kendall and A. Stuart, The advanced theory of statistics, vol 1, Hafner Publishing Company,New York, 1958.

STATIC AND DYNAMIC CHARACTERIZATION OF COMPOSITES, USING A FOURTH GENERATION PROGRAMMING LANGUAGE ON A MACINTOSH

I. Daerden, D. Van Hemelrijck, L. Schillemans, F. De Roey, F. Boulpaep, A. Cardon.
Composite Systems and Adhesion Research Group of the Free University of Brussels
COSARGUB - VUB - TW - Kb - Pleinlaan 2 - 1050 Brussels - Belgium

ABSTRACT

This paper will present two applications of a fourth generation programming language, called Labview.
Both applications consist of acquisition and -analysis of data obtained from static and dynamic mechanical tests for characterization of composite materials.
Both programs are analysed in detail and results from static and dynamic loading applied on rectangular carbon- and glassfiber reinforced specimens are presented.
An attempt to monitore the evolution of phase-shift between stress and strain during fatigue will be described.

INTRODUCTION

The static tests are performed on a 100 kN Instron mechanical testbench. It is possible to characterize specimens in compression, tension, three and four point bending. A railshear-device is also available.
The dynamic tests are performed on a MTS servo-hydraulic testbench with two degrees of freedom. Specimens can be subjected to alternating load of various waveforms. The frequency is limited up to 20 Hz.
Labview is a fourth generation programming language, which in the matter of reasoning, differs a lot from a conventional programming language. In fact, no program has to be "written" in the real sense of the word. A logical thinking-scheme comes into existence by linking virtual instruments. A virtual instrument is a software program, packaged graphically, to look and act like an instrument. A front panel (with knobs, dials, switches, graphs, meters, etc.) - fully in color - specifies the inputs and outputs and provides the user interface for interactive operation. Behind the front panel is a block diagram, which is the actual executable program. The components of this block diagram are icons of lower-level instruments and program control structures "wired" together. In this way, a program is developed with a logic, comparable to an electronic network-scheme. Labview is an hierarchical system because any virtual instrument can be represented as graphical icon and used in a block diagram to build other virtual instruments. When running the

program, front panels from successively called subvirtual instruments are opened and enable the user to enter complementary input. Drawings made by other software, can be placed into the front panels. Labview provides a lot of time-gaining because standard virtual instruments exist for plotting a graph, inverting a matrix, pattern generation, digital signal processing (calculating FFT, convolution ...), digital filtering, statistical analysis and so on.

STATIC TESTS

Hardware
The data acquisition is performed on a Macintosh II computer with 5 Mbyte memory, equipped with a 16-channel analog to digital conversionboard with a sample frequency of 50 kHz. An external amux multiplexer is provided to allow the sampling of 64 channels.

Software
Acquisition is performed by following steps :

- Input of comment, way of measuring (linear - in this case the step must be specified - or logaritmic for creep-testing), if measuring is stopped by hand or time-limit, speed of displacement of the testbar, number of channels and indexation of each channel.
- Choice of test.
- Input geometry of test set-up.
- For each channel, calibrationvalues are retrieved (for calibration, an other program is written) and conversion-, straingaugefactor (only if straingauges are used) and textnumber (name of channel) must be specified.
- Choice of section.
- Input geometry of section (if arbitrary section is chosen, inertiamoment and distance to outer fiber must be given).
- Real-time monitoring of numeric values of zero-measurement of chosen channels.

At this moment measuring is started by clicking the return button.

- Real-time monitoring of numeric values of acquisition of chosen channels.

When stopped, the datafile is saved under specified name and directory and possibility is given to perform again acquisition for the given test set-up. Then the user is asked wether or not processing is desired.

The data is processed as followed :

- Retreaving of specified datafile.
- Assigning of the axes (figure 1). The parameters are channelnumbers of chosen channels and typical experiment-depending factors, like height of specimen, area of cross-section... For each axe, two channelnumbers are provided to calculate mean values of strains, measured by gauges placed on both sides of the specimen.

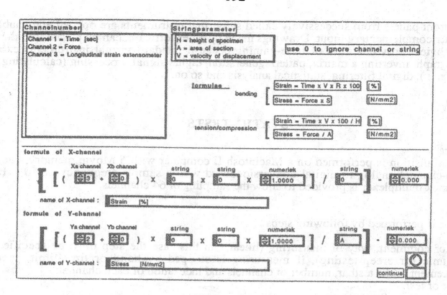

Figure 1.

Figure 2 shows the preview of the plot with possibility to mark out a window wherein the slope and correlationcoefficient will be calculated.

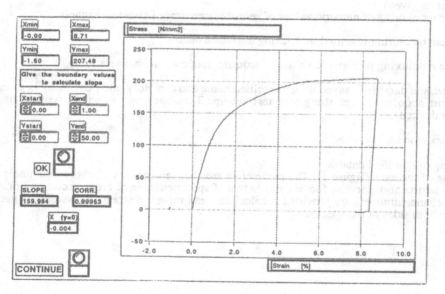

Figure 2.

Then the calculated datafile is saved and possiblity is given to calculate another datafile.
Figure 3 shows the final plot for rapport generation with all the entered remarks, parameters and
results of experiments collected.

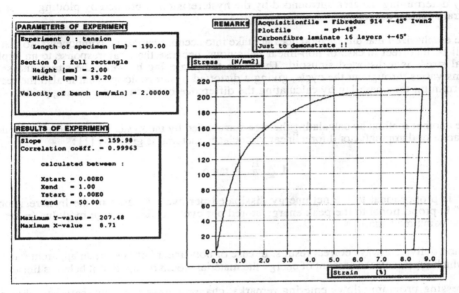

Figure 3.

Acquisition, processing and representation parts of the program can be runned independent from
eachother.
In the example, a tensiontest until failure from a +-45° carbonfiber/epoxy specimen is performed.
One can easily derive following characteristic constants :

Young's modulus $= 16.000 \text{ N/mm}^2$

Failure-stress $\approx 200 \text{ N/mm}^2$

Failure-strain $\approx 8.5 \%$

DYNAMIC TESTS

When a linear viscoelastic material is subjected to a sinusoidal displacement, there exists a phase-
lag δ between the applied stress σ and the strain response ε of the same frequency. The purpose is
to determine the evolution of this phase-shift during fatigue-loading.

Acquisition
Load and displacement or strainsignal are captured via a Harvester (Bergougnan Instruments) data-
acquisitionsystem with a "sample and hold" analog to digital conversionboard, able to sample two
signals simultaneously.

Processing

Determining phase-shift is done in two ways :

1) by calculating Fast Fourier Transform (FFT) of both waveforms of load and strainresponse

2) by determining the area surrounded by the hysteresisloop obtained by plotting stress and strain against eachother.

It must be emphasized that this method does not take into account harmonic components, so that the result will be incorrect in case of non-linear behaviour, because the strainresponse to an applied sinusoidal stress is a distorted sinusoid. Therefore the phase-lag between the stress and strain waveforms varies throughout the cycle, giving a distorted hysteresisloop. That is also the reason why determining the phase-lag by calculating the difference between zero-crossing points is left behind.

Therefore another method by calculating the area surrounded by the hysteresisloop, is also applied. The lossfactor or dampingfactor d for a linear viscoelastic material is given by [1] :

$$ d \triangleq \frac{E'}{E''} = \tan \delta $$

where E' is proportional to the net energy dissipated per cycle (= area of the hysteresisloop). E" is proportional to the peak energy stored per cycle, called storage modulus (= area of enclosing rectangle).

This method delivers always correct results. In case of non-linear behaviour, an apparent δ can be derived which represents the amount of energy the material would dissipate if it behaves linear.

The processing program allows entering remarks, channelnames and -numbers and choice of calculationmethod must be made (figure 4).

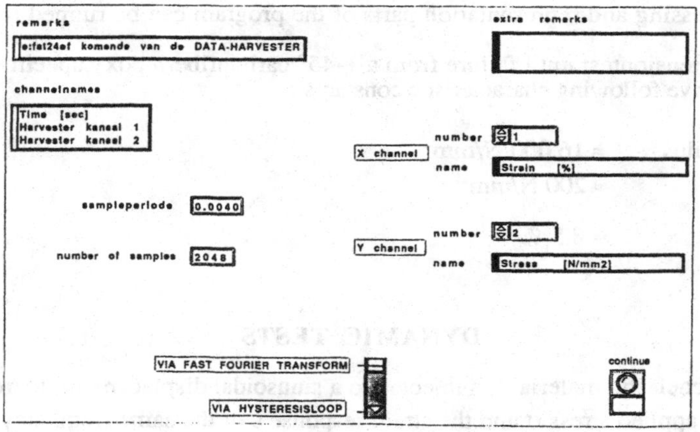

Figure 4.

Figure 5 shows results obtained via FFT. The frequencyspectra of loadsignal and strainresponse are monitored, even so exact frequency of the loadsignal and the phase-shift. Notice that the frequencyspectrum of the amplitude of strainresponse exhibits a number of harmonic components, which suggests non-linear behaviour of the specimen.

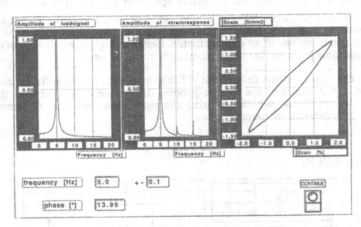

Figure 5.

In figure 6 the panel with results obtained by the hysteresisloop is given. Stress and strain are plotted against eachother for one cycle. For that cycle, number of points, dissipated energy, storage modulus and phaseshift are calculated.

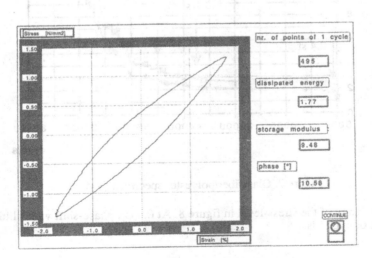

Figure 6.

EXPERIMENTAL RESULTS

Carbonfibre/epoxy and glassfiber/polyester specimens were subjected to a sinusoidal varying load applied in tension with a frequency of 5 Hz. The R ratio was 0.1. The carbon specimen was made of 16 plies of Fibredux 914 prepreg. The glassfibre specimen consisted of 8 plies of bidirectional woven fabric of $300g/mm^2$. Both lay-ups where +-45° oriented in order to get a considerable amount of friction between fiber and matrix.
The strain was measured by use of an extensometer. Because high stress-levels were applied, the strains were considerable high, so that no strain-gauges could be used.

Figure 7 shows that just before failure (at 77.000 cycles), phase-shift is increasing very pronounced. At every moment of the fatigue-proces, difference exists between the phase-shift obtained by the two methods, which confirms the non-linear behaviour. Stress-levels were increased in order to accelerate the damage proces.

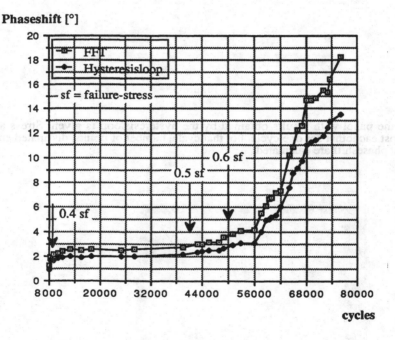

Figure 7. Glassfiber/polyester specimen (+-45°)

Notice the dependency of the stress-level in figure 8. At 0.4 σ_f, phase-shift varies little, no matter what the number of cycles is.

Phase-shift [°]

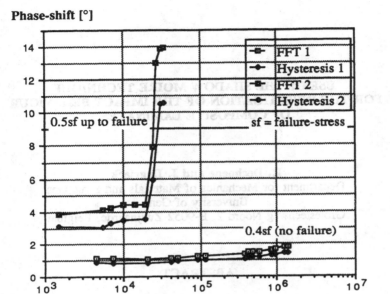

Figure 8. Carbonfiber/epoxy specimen (+- 45°)

CONCLUSIONS

Labview is extremely well adapted to perform data-acquisition and processing.
Evolution of phase-shift during fatigue-loading can be monitored. Unfortunately, measuring errors can be considerably high, especially when phase-shift values are low.

REFERENCES

[1]   Sims G. D. and Bascombe D. "Continuous monitoring of fatigue degradation in composites by dynamic mechanical analysis". ICCM & ECCM volume 4, 161-171.

[2]   De Visscher J. and Bosselaers R. "Experimentele bepaling van de demping van composietmaterialen" (Thesis mechanical engineer 1988).

USE OF THE SHADOW MOIRE TECHNIQUE
FOR THE INVESTIGATION OF THE IMPACT BEHAVIOUR
OF COMPOSITE LAMINATES

R. Dechaene and J. Degrieck
Department for Mechanics of Materials and Structures
University of Gent
Grotesteenweg Noord 2, B-9052 Zwijnaarde, Belgium

ABSTRACT

A method is proposed which allows for the experimental identification of the behaviour of composite laminates during a transverse impact. The method combines the well known shadow moiré technique with high speed streak photography to record the deflections of a laminate during an experiment. Onset and growth of properly oriented matrix cracks and fibre breakage at the back of the laminate also can be followed. From the photographs thus obtained, quantitative information can be extracted for the identification of the impact behaviour and for validation purposes of numerical simulations.

INTRODUCTION

It is well known that certain types of fibre reinforced plastic laminates are prone to damage under low energy impact. Most often impact tests are performed on small plates. But in spite of this simple geometry, the behaviour of a laminate under transverse impact is not fully understood yet. In most current research programmes dealing with impact on fibre reinforced plastics, the attention is focussed on the "post mortem" inspection (nature and extent of the damage) and on the practical problem of the residual strength. The authors of the present paper have set as their main aim the understanding of the behaviour of the laminate during impact: to determine precisely at which moment a particular type of damage occurs and how the conditions leading to damage can be expressed in terms of mechanical variables such as stress, strain and their time derivatives. This understanding may not be necessary if impact testing is used as a means for classifying materials according to standard tests. It is however of paramount importance for the development of laminates with better impact tolerance, and most of all, for the development of material models and the validation of computer simulations of impact on laminates.

Part of the investigation currently conducted consists of numerical simulations of an impact on composite specimens; the other part consists of experimental work using highly instrumented setups, allowing real time recording of the experiments and allowing the experimental identification of the impact behaviour of the materials under investigation. The present paper deals with the basic aspects of the technique used for the recording of the deformation of the specimen.

MEASURING DYNAMIC PLATE DEFLECTION

Known Methods

Few papers are dealing with the measurement of plate deformation or damage development during impact. Chai, Knauss and Babcock [1], for example, used a shadow moiré method and high speed photography to record the growth of delaminations in laminates which were preloaded in static compression, while undergoing a lateral impact. The compressive stress causes the delaminated layers to buckle and the delaminations to grow. The lateral displacements of this buckling mode could be observed by a shadow moiré, and the rate of growth of the delaminations could be measured. Takeda, Sierakowski, Ross and Malvern [2] used high speed photography to record the growth of delaminations in translucent laminates. They found them to propagate at the same velocity as the bending waves. Broutman and Rotem [3] used high speed photography to investigate the deformation of small beams made of fibre reinforced plastic. Gardiner and Pearson [4] used acoustic emission to detect the instant at which damage started, without succeeding fully in ascertaining the location or the nature of the damage.

Use of the Shadow Moiré method to Measure the Plate Deflection

As known, the shadow moiré method is very well suited for the measurement of out of plane displacements of a few millimeters of magnitude. Like most optical methods it has the advantage of requiring no sluggish wires, of being insensitive to electromagnetic effects in such wires and to temperature, and of adding no mass or stiffness to the specimen. The principle is shown in figure 1. A grating on a transparent flat plate is placed at the back of the specimen. It is illuminated by a bundle of light, casting a shadow on the specimen. An observer looking at the scene from a sufficient distance sees dark fringes where the shadow lies between two lines of the grating, and light fringes in between. Let us first assume for simplicity that the light bundle is parallel and that the observer is at an infinite distance.

Let:
c	=	distance between the undeformed specimen and the grating,
p	=	pitch of the grating,
$w(x,y)$	=	deflection of the specimen,
$w'(x,y)$	=	$c - w(x,y)$
	=	distance between the grating and the deformed specimen,
α	=	angle between incident light and normal to the undeformed specimen.

Then it is easily seen that the shadow of a line in the grating is shifted with respect to this line across a distance:

$$w' \, tg\alpha = (c-w) \, tg\alpha$$

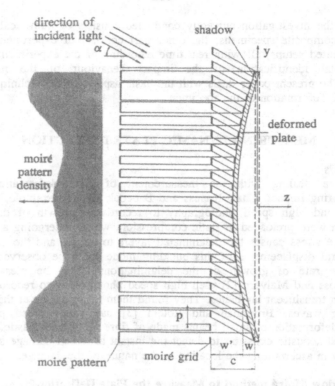

Figure 1. Basic principle of the shadow moiré method for recording plate deflection.

One sees a light fringe when this shadow falls exactly behind another line, i.e. when the shift equals a multiple of the pitch:

$$(c-w)\, tg\alpha = Np \qquad or: \qquad w = c - \frac{Np}{tg\alpha},$$

N being a whole number, the fringe number. The fringes are loci of points on the specimen with equal displacement.

In the current setup the light originates from a diverging light source and the observer (or the camera) is at a finite distance. The moiré pattern of a symmetrically bent plate is no longer symmetric, as shown by the pattern of figure 2. This requires a number of refinements to the equations, which can be found in the work of Buitrago and Durelli [5]. Using the proper equations, the asymmetric moiré pattern yields a symmetric deformation, as expected for a central impact of an orthotropic plate.

Equipment Used to Record the Fringes
An ultra high speed framing camera could be used to record successive fringe images. In a typical test the velocity of the striker is 20 m/s, and the total deflection is 5 mm. In

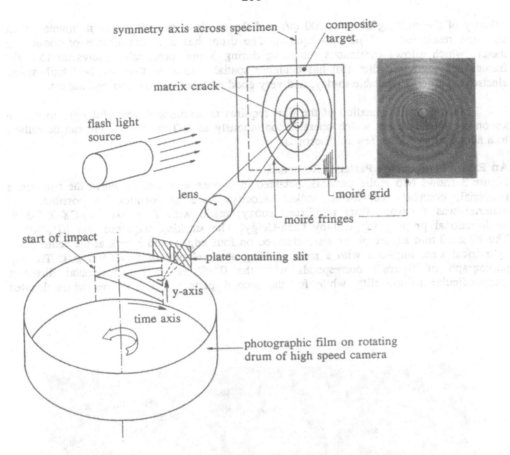

symmetry axis across specimen

composite target

matrix crack

flash light source

lens

moiré grid

moiré fringes

start of impact

plate containing slit

y-axis

time axis

photographic film on rotating drum of high speed camera

Figure 2. Setup used to record the deflection of the specimen along an axis of symmetry, using a rotating drum camera.

order to have a sufficient number of frames to allow for differentiation, one would then require about one frame per 5 μs, with a total number of at least 200 frames. Electronic framing cameras can reach this framing rate, but not the required number of frames.

The solution has been found by limiting the observations to one axis of symmetry of the specimen. As shown in figure 2, a slit is used between the specimen and the image plane, such that only the image of a narrow zone along an axis of symmetry of the specimen is formed in the camera. If the film is placed on a drum rotating about an axis which is parallel to the slit, a continuous picture is formed, giving the successive stages of the moiré fringes on the axis of symmetry of the specimen. If the circumferential velocity of the rotating drum is 200 m/s and the width of the slit is one millimeter, then the time resolution is 5 μs, as required. The drum has a circumference of about one meter, which allows continuous recording during 5 ms, practically equivalent to 1000 frames. Photographic film still has a higher spatial resolution than the best high speed electronic cameras available to-day, and very good pictures can be obtained indeed.

As a 5 μs illumination of the film requires an extremely powerful light source, a xenon arc tube is used, which operates continuously at 200 to 400 W, but can be pulsed to a about 10 kW for a few milliseconds.

An Example of Fringe Patterns Recorded

Figure 3 shows two moiré patterns, obtained in two experiments in which the conditions (material, boundary conditions, striker velocity) were as identical as possible. The material was a carbon fibre reinforced epoxy, made with Fibredux 920-CX-R-7-33% unidirectional prepreg supplied by Ciba-Geigy. The stacking sequence was $[0°_4/90°_4]_s$. The 80 x 80 mm square plates were clamped on four edges, and struck at the centre by a cylindrical steel impactor with a mass of 50 g at a velocity of about 9.5 m/s. The top photograph of figure 3 corresponds with the 0°-fibres in the horizontal direction (perpendicular to the slit), while for the second pattern the 0°-fibres where located

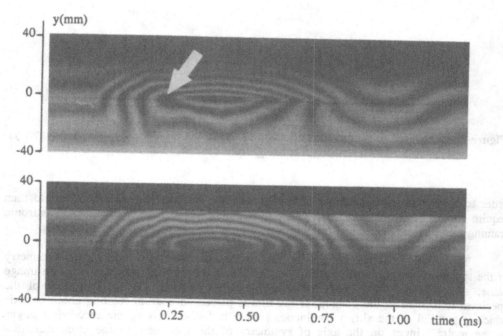

Figure 3. Recorded Moiré patterns along two different axes of symmetry of specimens, which are otherwise nearly identical. The horizontal axis is the time axis.

vertically. The resolution of the moiré patterns (mm deflection/fringe) are slightly different. The top edges of the specimens are not visible because of the shadow cast by the clamping fixture.

The differences between the two images illustrate the anisotropy of the plate deflection. The first fringe to appear, starting from the left, makes angles with the time axis, whose tangents give an estimate of the group velocity of flexural waves. For the top photograph this group velocity is 326 m/s, while for the bottom picture it is 457 m/s. The ratio is 1.40, while it should be 1.47 according to the ratios of the elastic constants involved. The correspondence is quite good. The impactor rebounds after 820 μs, after which the specimen continues its damped vibration.

On the top photograph a thin black line appears at 235 μs, as indicated by the arrow. This is a matrix crack parallel to the fibres, at the back of the specimen. If it is clearly visible at 235 μs, it must already have a sufficient crack opening and it has probably started some time before. As it frequently happens, the instant when such a crack appears coincides with the instant at which the flexural wave, reflected from the clamped boundary, returns to the centre of the specimen. The cracks closes at about 750 μs.

DATA PROCESSING

Numerical Method

The moiré images are scanned and are stored in digital form. The coordinates of the centres of the fringes are extracted by microcomputer from these raw data. Where the quality of the moiré image is not sufficient or the illumination is too uneven, the centres of the fringes are determined manually, using a digitising tablet. In both cases the first result is a set of coordinates of points located on the centre of the fringes.

The second step is the computation of the fractional moiré numbers at the nodes of a grid of equidistant coordinate lines. This is done by least squares adapting of a polynomial to the data, within a certain domain around each grid point, and using the value of the polynomial as the value of the fractional moiré number at this grid point. The degree of the polynomial and the size and shape of the domain are adapted to the local circumstances. Where a crack exists, the domain is always on one side of the crack, allowing for the discontinuity of the displacement across the crack.

The third step uses standard procedures for the computation of derivatives and for the graphical presentation of the results.

Example

Figure 4 shows the deflection in the central part of the axis of symmetry, computed from the top image of figure 3. It is clear that several widely different natural frequencies are operative.

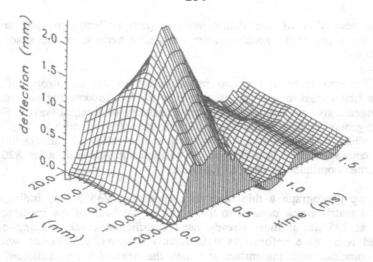

Figure 4. 3-D representation of the deflection of the central part of an axis of symmetry.

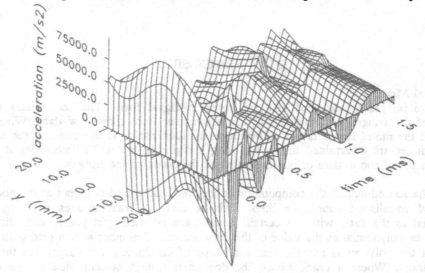

Figure 5. Graphical representation of accelerations.

Taking derivatives of this displacement function w(y,t) amplifies the higher frequencies and the corresponding natural modes. Figure 5 shows the second derivative in the time direction, i.e. the acceleration. The frequency of about 5000 Hz is very strong during the first 800 μs, as long as the impactor remains in contact with the specimen. After this, the first natural frequency of the clamped plate without the mass of the impactor, is dominant. The discontinuity at y = 0 appearing at 235 μs, is due to the matrix crack.

FINAL REMARKS AND CONCLUSION

A method has been developed which allows the measurement of the deflection of a flat plate while it undergoes a normal impact. A continuous record is obtained of the deflection along a straight line on the specimen, for instance along an axis of symmetry. The resolution and the accuracy make possible computation of first and second derivatives with respect to time and space, i.e. velocities, accelerations, slopes and curvatures. At the same time suitably oriented matrix cracks can be detected as they grow. The reproducibility of the results and their correspondence to the results of numerical simulations confirm their reliability.

The results are used to investigate the loading conditions which bring about the nucleation and growth of damage. They are also used to evaluate the results of numerical simulations of impact phenomena, using a suitable model for the development of damage.

ACKNOWLEDGEMENT

The financial support of the Belgian Science Policy Department and the National Foundation for Scientific Research is gratefully acknowledged.

REFERENCES

1. H. Chai, W.G. Knauss, C.D. Babcock
    "Observation of Damage Growth in Compressively Loaded Laminates"
    Experimental Mechanics, Sept. 1983, pp.329-337
2. N. Takeda, R.L. Sierakowski, C.A. Ross, L.E. Malvern
    "Delamination Crack Propagation in Ballistically Impacted Glass-epoxy Composite Laminates"
    Experimental Mechanics, Jan.1982, pp.19-25
3. L.J. Broutman, A. Rotem
    "Impact Strength and Toughness of Fibre Composite Materials"
    Foreign Object damage to Composites, ASTM STP 568, 1975, pp.114-133
4. D.S. Gardiner, L.H. Pearson
    "Acoustic Emission Monitoring of Composite Damage Occurring under Static and Dynamic Loading"
    Experimental Techniques, Nov.1985, pp.22-28
5. J. Buitrago, A.J. Durelli
    "On the Interpretation of Shadow-moiré Fringes"
    Experimental Mechanics, June 1978, pp.221-226

DELAMINATION DETECTION VIA HOLOGRAPHIC INTERFEROMETRY TECHNIQUES

A.C.LUCIA, G.P.SOLOMOS, P.ZANETTA
Commission of the European Communities, Joint Research Centre
Institute for Systems Engineering and Informatics, Ispra (VA) 21020 ITALY

L.MERLETTI, R.PEZZONI
Laboratorio Tecnologie dei Materiali, AGUSTA, Gallarate (VA) 21013 ITALY

ABSTRACT

The method of laser holographic interferometry is utilised for the detection of impact induced damage in graphite-epoxy composites. Damaged specimens, having received three different levels of impact energy, as well as virgin ones are used. Real time holography is applied and the straining of the specimens is achieved by heating. It is observed that as the temperature starts rising ($\Delta T=1$-$3^{\circ}C$), irregular fringe patterns of closed shape, indicative of deformation anomalies, begin to form at the impacted areas. Further, the more damaged the specimen, the sooner the appearance of these patterns. In contrast, the virgin specimens always produce fringes consisting of a system of almost parallel lines. Comparison of the interferometric images with the C-scans, obtained from an ultrasonic examination of the same specimens, verifies these results.

INTRODUCTION

Flaws can be introduced in composite laminates during processing and fabrication, or during in service conditions. In either case, impact constitutes a source of damage, which primarily causes material delamination.

A variety of nondestructive evaluation (NDE) techniques is used for assessing the integrity of these materials [1], and each of them is best suited for detecting some particular type of defect. They fall broadly into the general classes of radiographic, optical, thermographic, acoustic, embedded sensor, ultrasonic and electromagnetic techniques.

Among them, the optical method of laser holographic interferometry has proven to be very efficient, especially in producing qualitative results. Other advantages are that it is non-contact and non-intrusive, it is applied remotely and it yields full-field measurements. Citing some of its drawbacks, its over-sensitivity to environmental disturbances (when a continuous laser is used), and the laboriously derived quantitative results should be mentioned.

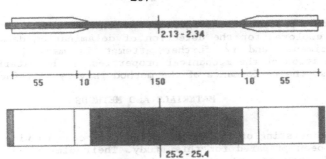

2.13 - 2.34

55   10   150   10   55

25.2 - 25.4

Figure 1. Geometry and dimensions of specimens used.

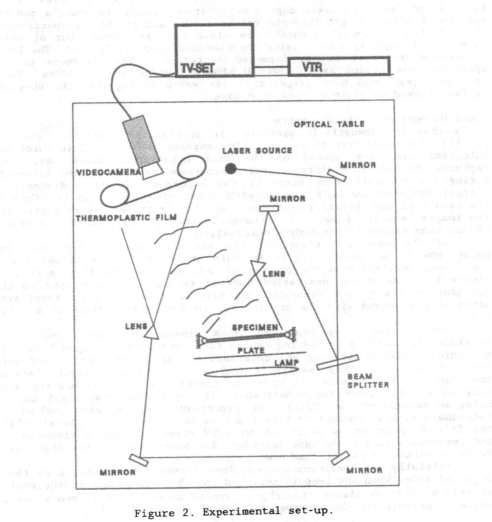

Figure 2. Experimental set-up.

In the present investigation, which is still in progress, this method is employed for the detection of delamination damage in graphite-epoxy specimens, and no further attempt is made for a quantitative characterization of the mechanical properties of the material. As shown by the results, the performance of the method is very satisfactory.

MATERIALS AND METHODS

Specimens

Specimens consisting of 16-ply graphite-epoxy coupons with glass-epoxy end tabs have been prepared for this study. Their dimensions are as shown in Fig.1, and they are of the tensile type as they are intended to be subjected to other relevant tests (acoustic emission). Their stacking sequence is the symmetric $[(45/-45/0/90)_2]_s$ thus producing a quasi iso-tropic material. The fiber volume ratio has a value of $V_f = 42\pm2\%$, while for the unidirectional zero degree orientation lamina the Young's modulus and the tensile strength have the values 115GPa and 1545MPa, respectively.

Twelve specimens in total were planned to be tested, out of which nine were damaged by impact using the standard drop weight test. The level of damage, referred to the impacted energy over the thickness of the specimen, was graded as follows: 0.5J/mm, 1.0J/mm and 1.5J/mm. Three specimens per level were prepared. It is worth noting that the specimen surface showed no visible signs of damage.

Laser Holographic Interferometry

The method is exhaustively presented in specialized textbooks, such as ref.[2]. Its basic version is the double-exposure holography, in which two holograms are superimposed on the same holographic plate; each one captures the body in a different state separated by a fixed time interval. During the reconstruction stage of the hologram, the three-dimensional virtual image of the body appears with a set of fringes on it. This is the result of the optical interference of the simultaneously reconstructed two images, and it is due to the change in the optical path of the second object beam caused by the body's deformation.

Of the several variants of the method, the so-called real-time holography is primarily applied in this study because it allows for a continuous monitoring of the process of deformation. According to this, a single hologram of the unstressed body is recorded and developed in the holographic plate; the reconstructed virtual image is then interfero-metrically compared with the actual body in a different state at a later time.

The specimens are supported in a hinged type manner and their straining is achieved by heating, which is provided by a halogen lamp [3]. Mechanical loading (bending and compression) has also been tried out, but with no satisfactory results. A dark metallic plate is placed between the lamp and the specimen in order to render the specimen heating more uniform and to reduce the undesirable effects of the lamp light on the hologram reconstruction, Fig.2. The experimental set-up also includes a holocamera with thermoplastic film (in lieu of a holographic plate), while the fringe formation is monitored on a TV screen through a videocamera, and recorded with a video tape recorder. The source of coherent light is a He-Ne continuous laser of 35mW power.

Initially the specimens are examined (laser illuminated) from their impacted side. When the lamp is switched on, the temperature at the center of this side rises almost linearly by approximately 2.5°C in two minutes, which is principally the observation time of the experiment.

RESULTS

The results of this investigation consist of interferometric images, which, in a qualitative manner, portray the state of integrity of the material. The principle underlying the method is that, for a stressed zone of a body without geometrical discontinuities, distortions of fringe patterns indicate abnormal surface deformation and, consequently, the presence of some type of material defect.

Figure 3. Fringe pattern of undamaged specimen.

A fringe pattern recorded from a virgin specimen after approximately one minute of heating is illustrated in Fig.3. Clearly, it is made up of almost parallel lines and it corresponds to some flexural mode of deformation. This is exactly as expected, because the specimen heated tends to expand, but constrained at its extremes by the hinge supports, it undergoes compression and is forced to buckle in a mode compatible with some initial shape imperfection. This pattern does not change with

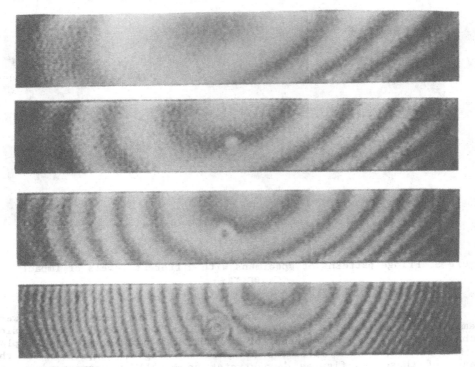

Figure 4. Fringe pattern evolution with temperature.

temperature, apart from an increase of the fringe density, translated into larger bending deflections.

Fig.4 shows the evolution of fringe pattern for a specimen which has received impacted energy of 0.5J/mm. It is observed that as the temperature increases, pattern discontinuities start appearing. They are localized at the zone of the specimen, where the impact took place. Actually, when observing the phenomenon in the TV screen, this zone acts as a source and as a sink of closed fringes for increasing temperature and for the cooling down phase, respectively.

Fig.5 shows a fairly well developed fringe pattern (2 minute heating) of three specimens with different impact energy levels. It is observed that the number of fringes and the extent of the pattern anomaly are proportional to the impacted energy (and, consequently, to the incured delamination). Similar are also the results of the other specimens, which are not reported here due to space limitations.

0.5J/mm

1.0J/mm

1.5J/mm

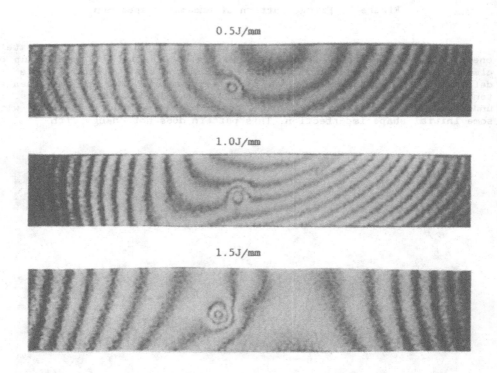

Figure 5. Fringe patterns of specimens with different levels of impact energy

For the verification of the above findings, the same specimens were examined ultrasonically [4], using the double through transmission technique with a signal frequency of 2.25MHz. The amplitude C-scan display of the three previous specimens is shown in Fig.6. Comparison of the images of these two figures demonstrates an excellent correspondence of the results of the two methods employed. The C-scans indicate the position and the extent of delamination (the central white area) exactly as has been revealed by the interferometric images before.

0.5J/mm

1.0J/mm

1.5J/mm

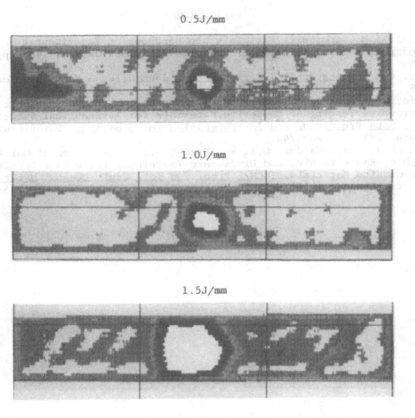

Figure 6. C-scans of specimens with different levels of impact energy.

DISCUSSION-CONCLUSIONS

Laser holographic interferometry has been employed in an efficient manner for detecting impact induced delamination in graphite-epoxy laminates. The method has amply demonstrated all its powerful features. A positive aspect is also its neatness, when contrasted, for example, with the specimen water immersion and the point by point scanning required by the ultrasonics.

It is reminded that an increase of only 2°C was sufficient for making the method work. The potential of this approach as a NDE technique is thus evident, as such a temperature rise can be easily effected (with a hot air gun, or with an IR lamp). The above applied real-time holography can serve for calibrating the double-exposure technique. The approach may, therefore, prove very useful for a qualitative, but reliable, preliminary examination even under field conditions, if a pulsed laser is used.

ACKNOWLEDGMENTS

The technical assistance of Mr. M.Franchi and Mr. M.Facchini is greatly appreciated.

REFERENCES

1. Daniel, I.M., Composite Materials. In Handbook on Experimental Mechanics, ed. A.S. Kobayashi, Prentice-Hall, New Jersey, 1987, pp.814 -890.
2. Vest, C.M., Holographic Interferometry, John Wiley & Sons, New York, 1979.
3. Delvò, P., Ferraro, P., Rizzi, M.L. and Sabatino, C., Holographic Non-Destructive Testing of Composite Materials for Aeronautical Structures, Proc. Data Processing and Interpretation in Holography, Saint-Louis, France, 1988, pp.229-240.
4. Sendeckyj, G.P., Maddux, G.E. and Tracy, N.A., Comparison of Holographic, Radiographic, and Ultrasonic Techniques for Damage Detection in Composite Materials, ICCM/2 Proc. 2nd Int. Conf. Compos. Mater., Toronto, Canada, 1978, Metallurgical Society of AIME, pp.1037-1056.

DYNAMIC FAILURE PROCESSES IN FIBRE-REINFORCED COMPOSITES

Yiren Xia and C.Ruiz

Department of Engineering Science,

University of Oxford, Parks Road, Oxford, OX1 3PJ, U.K.

ABSTRACT

A dynamic failure model is proposed for unidirectionally reinforced composites (UDCs) under impact loading. It comprises two parts: the first part involves the double-edge notched composites with matrix split. The second part is a complete stochastic simulation of sequential failure processes for various coefficients of variation of fibre strength. The predicted final fracture surfaces for both CFRP and GFRP are consistent with observations from experimental work on tensile impact of composites.

INTRODUCTION

The difference between the modes of failure of carbon fibre reinforced plastics (CFRP) and glass fibre reinforced plastics (GFRP) at various rates of strain was described in [1, 2]. The shape of the stress-strain curves remains unchanged for CFRP as the strain rate increases, although there is a slight increase in the tensile strength. Failure is always through fibre fracture and occurs with very little permanent deformation. In contrast, there is a marked change in these curves in the case of GFRP. At very low strain rates, the fracture is again virtually within the elastic regime and occurs as a result of fibre breakage. At higher strain rates, the matrix split, fibres pull out and as they do, serrations appear in the stress-strain curve. The pullout followed by fibre break dissipates energy through frictional effects, and hence results in an increased energy absorption as measured by the area under the curve. This phenomenon is not only reflected by the change in shape of the areas but by the observation of the broken specimens.

The aim of the present paper is to present a method to describe the transient failure process for certain UDCs. First, the dynamic stress field ahead of a notch and a split tip and the failure modes are studied. This is followed by a complete stochastic simulation for UDCs under transient tensile loading. The results to be sought are the various fracture surfaces for both CFRP and GFRP. Fig.1 illustrates the two configurations studied.

EQUATIONS AT THE MICRO-STRUCTURAL LEVEL

The development of the analytical model is based on the shear-lag theory extensively described in [1, 3, 4], therefore, only the final equations need to be repeated.

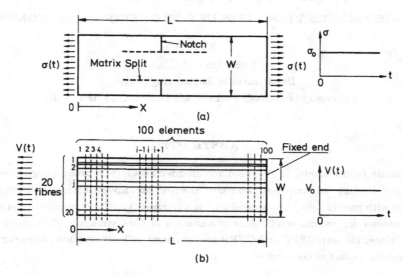

Figure 1: The configurations of the composite models: (a) Double-edge notched strip under symmetrical tensile loading. (b) Stochastic simulation model.

The general governing equations for a monolayer composed of equally spaced fibres are constructed from the dynamic equilibrium of a fibre element i

$$A\frac{d\sigma_i}{dX}dX + (\tau_{i+1} - \tau_i)hdX = \rho A\frac{\partial^2 U_i}{\partial T^2}dX$$

where the shear stress in the undamaged matrix is defined as

$$\tau_i = \frac{Gh(U_i - U_{i-1})}{m}.$$

Define the nondimensional displacement, coordinate and time as,

$$u = U\sqrt{\frac{EAGh}{m}}$$

$$x = X\sqrt{\frac{Gh}{EAm}}$$

$$t = T\sqrt{\frac{Gh}{\rho Am}}$$

where U, X, T are the real physical displacement, coordinate and time respectively. E, A, and ρ are elastic modulus, cross sectional area and density of the fibres, respectively, and G is the shear modulus of the matrix resin. m is the gap between uniformly spaced fibres and h is the height of the monolayer.

This leads to the governing equations:

$$\frac{\partial p_i}{\partial x} + (u_{i+1} - 2u_i + u_{i-1}) = \frac{\partial^2 u_i}{\partial t^2} \quad (i = 2, N - 1), \tag{1}$$

and

$$p_i = \sigma_i A = \frac{\partial u_i}{\partial x}. \quad (i = 1, N) \tag{2}$$

At the edges (i=1 or N), the equations degenerate to:

$$\frac{\partial p_1}{\partial x} + (u_2 - u_1) = \frac{\partial^2 u_1}{\partial t^2},$$

$$\frac{\partial p_N}{\partial x} + (u_{N-1} - u_N) = \frac{\partial^2 u_N}{\partial t^2}.$$

The above N partial differential equations can be transformed into 2N ordinary differential equations along the characteristic lines $\lambda = \frac{dx}{dt} = \pm 1$:

$$\mp dp_1 + d\dot{u}_1 = (u_2 - u_1)dt$$

$$\mp dp_i + d\dot{u}_i = (u_{i+1} - 2u_i + u_{i-1})dt \quad (i = 2, N - 1) \tag{3}$$

$$\mp dp_N + d\dot{u}_N = (u_{N-1} - u_N)dt.$$

The next step is a simple finite difference method discretising Eqn. 3. For any possible fibre breakage, the broken end is split into two nodes with boundary condition $p = 0$ and two independent sets of unknown variables such as u_i, $\dot{u}_i$ and τ_i. The detailed expression of the algorithm can be found in [5]. Matrix split can occur at either upper or lower side of a node regardless of whether it has broken or not. If the matrix at a surface fails ($\tau_i > \tau_c$), then τ_i becomes either nought or a constant frictional stress. Again a lengthy description can be found in [5].

The simulation adopted here for the second part of the work is basically a numerical Monte-Carlo random search routine. The configuration of the composites is a system of N fibres, each fibre being divided into N1 elements in the longitudinal direction. The distribution of strengths of unidirectional composites is expressed in the form of a two parameter Weibull distribution, i.e.

$$F(\sigma) = 1 - exp(-L(\frac{\sigma}{a})^b)$$

where a and b are the two parameters and L is the fibre length. The mean strength of each fibre can be expressed as:

$$\bar{\sigma} = aL^{-\frac{1}{b}}\Gamma(1 + \frac{1}{b})$$

where Γ is the standard gamma function. The coefficient of variation, CV, for this distribution is given by:

$$CV = \frac{\Gamma(1 + \frac{2}{b}) - \Gamma^2(1 + \frac{1}{b})}{\Gamma(1 + \frac{1}{b})}.$$

By giving the mean strength of fibres and the strength scatter CV, the Weibull distribution can be fully modelled.

APPLICATION TO A DOUBLE-EDGE NOTCHED COMPOSITE

Dynamic loading is simulated by applying simultaneously at both ends of the specimen a Heaviside tensile stress σ_0 ($p_0 = \sigma_0 A$). The stress wave velocity in the CFRP is 11335m/s and in the GFRP it is 5324m/s, therefore, the times for the stress waves to reach the notch tip for the CFRP and GFRP are respectively $2.21\mu s$ and $4.70\mu s$. However, the most interesting result is the stress concentration at the notch tip.

It is known that for an intact sample, the stress at $X = L/2$ will be twice the value of σ_0 since the stress waves from both ends meet and superpose there and nowhere in the matrix will there be shear stresses. The presence of stress waves distinguishes dynamic fracture from quasi-static behaviour in that when the wave front reaches the crack plane, parts of it will be reflected to give a stress free condition on the faces whereas in the ligament the wave will be transmitted unchanged. This gives rise to a sharp stress concentration at the notch tip. Fig.2 gives the transient fibre stresses for the fibre just ahead of the notch tip for both CFRP and GFRP cases when the matrix is still intact. The number of broken fibres at either edge is assumed to be 120. The time in the figure is measured from the instant when the external load is applied. The dynamic stress concentration factor is approximately 20% higher for the GFRP than for the CFRP.

The stress distributions at other fibres or across the ligament of the sample at $X = L/2$ are given in Fig.3 in which only the stresses in the first six fibres ahead of the notch are sketched. General profiles are the same for both CFRP and GFRP. However, the sharper drop of stresses in GFRP suggests that the stress concentration is more confined in GFRP than in CFRP. In other word, the order of singularity for GFRP is higher than that of CFRP. The study of shear stress around the notch tip shows a higher shear stress at the interface ahead of the notch in GFRP than in CFRP (by about 30%) [5].

The configuration of initial failure as the combination of a notch and a matrix split is also studied. As addressed in [5], matrix split will greatly reduce the fibre stress concentration at the notch tip, but it will hardly affect the severity of matrix shear stresses ahead of the split.

TRANSIENT FAILURE OF INITIALLY INTACT COMPOSITES WITH FIBRES OF VARYING STRENGTH

For the testing of composites at high rates of strain, it is customary to use a Hopkinson bar apparatus [2]. The specimen is held between an input bar and an inertia (output) bar and a

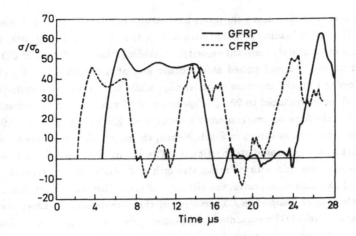

Figure 2: Axial stress in the first unbroken fibre in front of the notch.

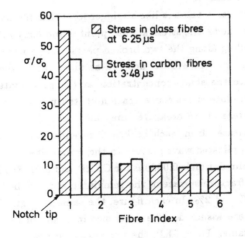

Figure 3: Notch tip stress distributions for six unbroken fibres.

tensile stress wave, travelling along the input bar, results in a tensile pulse in the specimen that resembles the Heaviside loading function assumed in the analysis. In practice, the specimen stretches at an approximately constant velocity. Therefore, the initially intact UDC is assumed to be clamped at one end and pulled at the other end at a constant velocity (10m/s in the following). In order to make the simulation feasible while still keeping the statistical meaning, the number of fibres are reduced to 20. The specimen is 10mm by 2mm. The diameter of fibres is adjusted so that the fibre volume content is kept at 50%. Each fibre consists of 100 elements in the longitudinal direction, as shown in Fig.1. Hence, the probability of failure for all these 2000 (100×20) elements are generated stochastically. The strength of each fibre, however, follows the Weibull distribution rule with its mean strength of 3.5GPa. It is noteworthy that, drawn from the tests [2], at high strain rates, the strength of glass fibre composites is higher than the strength of carbon fibre composites. Therefore, in this simulation, all fibres are assumed to have mean strengths of 3.5GPa regardless of whether they are carbon or glass, just for brevity and for easy comparison. Tests carried out at Oxford [6] on interlaminar shear strengths for carbon reinforced epoxy showed a marked strain-rate effect by making use of the lap shear tests. The average shear strength τ_c at low strain rates is 54MPa while it is 66MPa at high strain rates. For glass fibre composites, the maximum shear stress is about 34MPa [7] statically. In order to do a comparative study, it is assumed that the shear strength of the interface is always 50MPa for both CFRP and GFRP. The value of τ_f (frictional stress after matrix split) is arbitrarily chosen to be 10MPa.

As soon as a fibre breaks, it behaves like a trigger for either the adjacent fibre breakage or the matrix split, because the tensile stress at that point is suddenly released, causing reflected compressive waves travelling along the two broken parts. Large fibre stress concentration as well as shear stress in the matrix will be incurred as discussed in the previous section. The matrix split would reduce the stress concentration at the immediate neighbourhood of the broken ends, but it may induce stress concentration at its tip, leading to fibre breakage at the downstream fibres. Scattered fibre breakages may also occur because of the scatter of their strengths. The waves released from each broken fibre ends would interact with each other, causing an extremely complicated wave pattern in the composites.

The time step in the simulation is 0.0088μs for CFRP and 0.0188μs for GFRP. For a typical sample, Fig.4 shows the fracture surface of the remaining fibres pulled out at the loading end when CV equals 4%, 10% and 2%. In each figure, the same strength distribution of UDC for both CFRP and GFRP are assumed. The difference in shape of the fracture surfaces when CV equals 4% is remarkable. For CFRP, the profiles always tend to be clean cuts of fibres or stepped or multi-stepped surfaces, while for GFRP, extensive fibre debonding and pull-out can be seen, leaving a jagged profile of the surface. In the figures, the coordinate in the fibre direction represents the remaining length of the fibre since initially each fibre is equally divided into 100 elements.

At CV of 10%, even more pronounced pull-out effect can be found for GFRP specimen. For CFRP, the fracture of fibres begins to occur cumulatively with some fibres being pulled out, but the stepped feature of the surface can still be found.

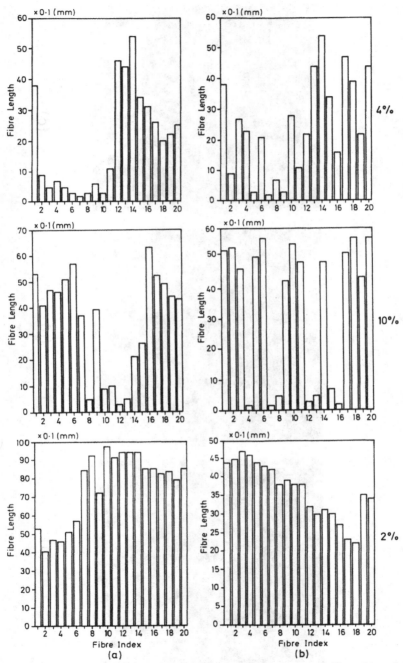

Figure 4: Typical fracture surfaces of composites after impact (c.v.=4%, 10% and 2%). (a) CFRP (b) GFRP.

220

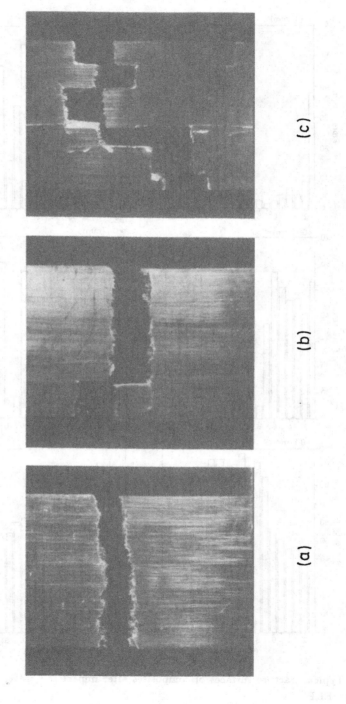

Figure 5: Fracture appearance of CFRP specimens: (a) Single fracture surface. (b) Stepped fracture surface transverse cracking. (c) Multi-stepped fracture surface.

221

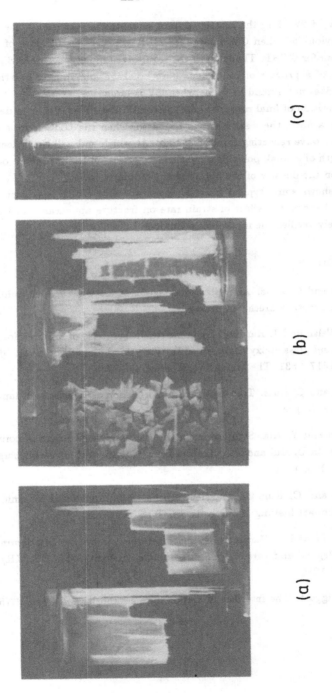

Figure 6: Effect of strain rate on fracture appearance of GFRP specimens. (a) Quasi-static test. (b) Intermediate rate test. (c) Impact test. pull-out of fibre layer in grip section.

At CV of 2%, i.e., the scatter of fibre strengths is very small, the fibre pull-out effect is not as obvious as when CV is large for GFRP. There is still more of clean cut of fibres for CFRP than for GFRP. The locus of the regularly stepped fracture surface for GFRP is clearly indicative of a process of continuous fibre breakage and matrix splitting. The matrix split, however, does not extend as far as when CV is large.

The position of final rupture along the composite is also specimen dependent. Statistically, the longer a fibre, the weaker it will be. Also, when the UDC is intact, tensile stresses only increase on wave reflecting from both ends. At which end will the reflected stress just exceeds the strength of a weak point near that end, depends on the increment of the stress, and hence depends on the density of the fibre and the pulling velocity.

Fig.5 shows some typical fracture appearances of CFRP specimen at all rates of strain, while Fig.6 shows the effect of strain rate on fracture appearance of GFRP specimens. This qualitatively verifies the mathematical model.

References

[1] Y. Xia and C. Ruiz. Analysis of damage in stress wave loaded unidirectional composites. Engng. Fracture Mech. (in press).

[2] L.M. Welsh and J. Harding. Dynamic tensile response of unidirectionally reinforced carbon epoxy and glass epoxy composites. In W.C. Harrigan, editor, *Proc. of ICCM V, San Diego*, pages 1517–1531, The Metal Soc. Warrendale, 1985.

[3] Y. Xia and C. Ruiz. Transient shear deformation at interfaces of laminated beams. Int. J. Fracture (in press).

[4] C. Ruiz and Y. Xia. Significance of the interlaminar strength of composites under impact loading. In D. Hui and T.J. Kozik, editor, *Composite Materials Symposium*, Houston, TX, ASME, Jan. 1991.

[5] Y. Xia and C. Ruiz. A transient stochastic failure model for unidirectional composites under impact loading. Composite Sci. Tech. (in press).

[6] R.K.Y. Li and J. Harding. *Effects of strain-rate on the interlaminar shear strengths of carbon/epoxy and carbon/PEEK laminates*. Report 2057/118 (XR/MAT), University of Oxford, 1989.

[7] M.R. Piggott. The interface in carbon fibre composites. *Carbon*, 27(5):657–662, 1989.

EFFECT OF STRAIN RATE ON THE COMPRESSIVE STRENGTH OF WOVEN GLASS-REINFORCED/EPOXY LAMINATES

JOHN HARDING
Department of Engineering Science
University of Oxford
Parks Road, OXFORD OX1-3PJ, U.K.

ABSTRACT

Two woven glass-reinforced/epoxy laminates are tested in compression at a quasi-static and an impact rate of strain. For the impact tests a split Hopkinson pressure bar is used with two different types of specimen, a solid cylinder as commonly used with the compression version of the Hopkinson bar and a thin strip specimen, waisted in the thickness direction, as recommended for the compression testing of laminated composites. The effect of strain rate and specimen design on the observed mechanical response is discussed.

INTRODUCTION

In recent years several attempts have been made to adapt the split Hopkinson's pressure bar (or Kolsky bar) technique (1) to the study of strain rate effects in composite materials (2,3,4,). This technique, which is well established for the testing of homogeneous isotropic metallic specimens under simple states of stress, allows strain rates of the order of 1000/s to be attained. The specimen is fixed between two long elastic loading bars, an input bar and an output bar, which act as wave guides. In the original version of this apparatus a short cylindrical specimen was subjected to a compressive stress wave resulting from a direct impact on the free end of the input bar. Overall specimen dimensions are required to be small so as to minimise the effects both of radial inertia and of wave propagation within the specimen while care has to be taken that stress concentrations due to frictional end effects do not invalidate the experimental measurements.

When laminated composite materials are to be tested, however, a cylindrical design of specimen is far from convenient while the need for small overall specimen dimensions conflicts with the requirement for a specimen which is large relative to the scale of the reinforcement. In practice, because frictional end effects are likely to be accentuated by the anisotropic nature of the composite, the recommended design of compression specimen for quasi-static testing, in particular for unidirectionally-reinforced material, has the form of a long slowly-tapered strip (5), waisted in the thickness direction, the ends held in massive blocks and constrained to apply a pure compression.

Nevertheless, apart from a very recent investigation (6), all work on composite materials in the compression Hopkinson bar has used short cylindrical specimens. It is necessary, therefore, to consider whether, in such cases, the true compressive impact properties are being determined or whether the results obtained are, at least in part, dependent on the geometry of the specimen. In the present paper, therefore, the results of an earlier, previously unpublished, investigation into the mechanical behaviour of a woven glass-reinforced/epoxy laminate under impact compression, using short cylindrical specimens, are compared with more recent results for a similar, but not identical, woven glass/epoxy laminate, using thin strip specimens.

SPECIMEN MATERIALS AND DESIGN

For the cylindrical specimens a 12.5mm thick plain-weave glass/epoxy laminate of Permaglass 22FE, supplied by Permali Ltd. of Gloucester, was used. The reinforcing fabric was a fine-weave Marglass 116S with 165 ends and 134 picks per 10 cm and the matrix was Ciba-Geigy HY750 epoxy resin (100 parts by weight) with HT976 hardener (36 parts by weight) using a cure cycle of 3 hours at 130°C followed by 4 hours at 180°C. Specimens with a length of 9.525mm and a length/diameter ratio of unity were cut with the cylindrical axis in the plane of the laminate and parallel to a direction of weave. Each reinforcing ply had a thickness of about 0.2mm, giving 48 layers across the full width of the cylinder.

The thin strip specimens were prepared to the dimensions shown in fig. 1 from a plain-weave glass/epoxy laminate supplied by Fothergill and Harvey. The reinforcing fabric was woven from continuous E glass fibres, designation 11 x 2EC5, and had a weight of 96 g/m², an approximate thickness of 0.08mm and a weave geometry of 252 ends and 173 picks per 10 cm. The laminate was hand laid up using Ciba-Geigy XD927 epoxy resin, with 100 parts by weight of resin and 36 parts by weight of hardener and a cure schedule of 24 hours at room temperature followed by 16 hours at

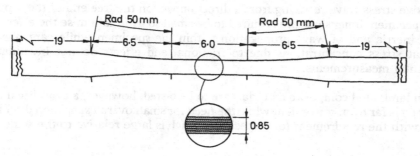

Figure 1. Design of Strip Specimen

100°C. The final laminate contained 14 layers of fabric and had a fibre content of 50% by weight and a thickness of 1.75mm. Thin strips of width 9.525mm were cut from the laminate with the axis parallel to either the warp or the weft fibre tows. The strips were then carefully waisted by hand in the thickness direction to give a total of 7 reinforcing plies in the parallel gauge section. This specimen design is based on that recommended by Ewins (5) for quasi-static compression testing off unidirectionally reinforced CFRP but has been reduced in length so as to ensure a state of quasi-static equilibrium within the specimen at an early stage in the test.

EXPERIMENTAL TECHNIQUES

Quasi-static tests were performed on a standard screw-driven Instron testing machine at a crosshead speed of 0.05 in/min for the cylindrical specimens and 0.2 in/min for the strip specimens. In the latter case a special loading rig was designed to ensure that the loading bars were accurately aligned and to prevent buckling of the loading bar/load cell system between the fixed and moving crossheads. A small strain-gauged load cell and a pair of linear variable differential transformers (LVDT's) in parallel with the specimen were used to determine the load applied to the specimen and the resulting overall displacement.

Impact tests were performed using the split Hopkinson pressure bar technique. For the cylindrical specimens the standard compression Hopkinson-bar was employed with 10.67mm diameter alloy steel loading bars impacted by a falling weight at a velocity of 15m/s. The specimen/loading bar interfaces were carefully lapped to facilitate stress wave transmission and lubricated with molybdenum disulphide to minimise frictional effects. Stress-time records for the input and output faces of the specimen, as determined using the standard Hopkinson-bar wave analysis, were in close agreement, confirming that a state of quasi-static equilibrium may be assumed in the specimen.

Strip specimens were fixed with epoxy adhesive into parallel sided slots in a titanium input bar and a phosphor-bronze output bar, both of which had a diameter of 9.525mm. A small air-gun was used to accelerate a projectile which impacted directly on the free end of the input bar. With this design of specimen the accurate calculation of the stress on its input face is much more difficult and the early attainment of quasi-static equilibrium within the specimen less easy to achieve. However in an earlier investigation (4), a state of quasi-static equilibrium was confirmed in the same design of specimen when loaded in tension.

RESULTS

Cylindrical Specimens
Stress-strain curves: Stress-strain curves for four nominally identical tests at a quasi-static rate of 3.2×10^{-4}/s are shown in fig. 2a. Strain was determined from the Instron crosshead displacement after making a correction for the elastic displacement of the testing machine, load cell and loading bars. This is not an accurate method for determining small strains and great weight should not be placed on the resulting values of compressive elastic modulus. Stress-strain curves for three nominally identical tests

at an impact velocity of 15m/s are shown in fig. 2b. Here the strain was determined using the standard Hopkinson-bar analysis which again does not give very accurate results at low values of strain. The average strain rate in these tests, defined as the failure strain divided by the time to failure, was 860±20/s.

Fracture appearance: Quasi-statically tested specimens were subsequently sectioned and examined under the optical microscope. A section through a specimen after testing is shown in fig. 3a. Failure is by shear on planes inclined at slightly less than 45° to the axis of loading. The width of the shear band is clearly related to the wavelength of the weave. A similar study of the failure process in impact tested specimens was not possible since all specimens disintegrated during the impact.

Effect of Specimen Geometry: In view of the possibility that friction at the specimen interfaces may have a significant effect on the observed mechanical behaviour

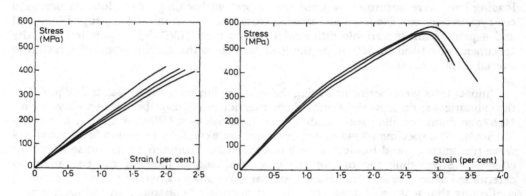

| a) Quasi-static tests | b) Impact tests |

Figure 2. Stress-strain curves for tests on cylindrical specimens

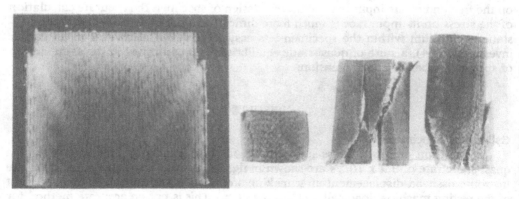

a) Section through standard specimen   b) Effect of length/diameter ratio
Figure 3. Fracture appearance of quasi-statically tested cylindrical specimens

and since both the shear bands apparent in fig. 3a intersected one end of the specimen, some further specimens having an increased length to diameter ratio, 2:1 and 2.5:1, were also tested, in an attempt to obtain a shear failure confined to the central region of the specimen. However, as apparent from fig. 3b, which shows the failure modes in quasi-static tests on the three lengths of specimen, this attempt was not successful, the only effect of increasing length being to introduce longitudinal splitting as an additional failure mode, possibly at a later stage in the test. Nevertheless, the ultimate compressive strength determined in both quasi-static and impact tests, see fig. 4, showed no dependence on specimen length to diameter ratio.

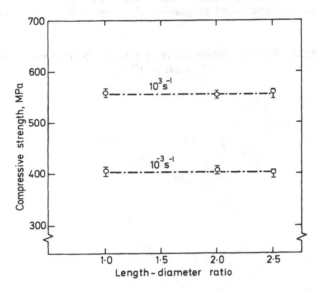

Figure 4. Effect of length/diameter ratio on ultimate compressive strength

Thin Strip Specimens

For the strip specimens strain gauges were also attached directly to the parallel test section, giving a direct measure of the specimen elastic response and hence allowing a more accurate determination of the compressive modulus. Overall specimen strain could also be determined either from the LVDT signals, in the quasi-static tests, or the Hopkinson-bar analysis in the impact tests, as appropriate, so that, in the event of the specimen strain gauges failing, due to surface damage, before final failure of the specimen, the total specimen strain could still be determined.

Stress-strain curves: Tests were performed on specimens loaded in both the warp and the weft directions. Stress-strain curves for four nominally identical tests on specimens loaded quasi-statically in each direction are shown in fig. 5a. The mean strain rate was $\approx 2.3 \times 10^{-3}$/s. In all cases failure of the strain gauges attached to the specimen parallel section followed the attainment of the peak load. This measurement of strain has been used, therefore, in deriving the stress-strain curves shown.

Stress-strain curves for three nominally identical impact tests on specimens loaded in both the warp and the weft directions are shown in fig. 5b. The average strain rate, defined as the strain at failure divided by the time to failure, was 680/s. Strain measurements were taken from the strain gauges attached directly to the specimen parallel section. Only in one test of each set of three did these gauges remain undamaged until the specimen failed. In the other two tests, therefore, it was necessary to use the Hopkinson-bar analysis to determine the strain over the final stages of the test. The procedure used has been described elsewhere (6). Also, as is apparent in fig. 5b, in each set of three tests one specimen showed an initial peak load at an early stage in the test followed by a subsequent further increase in load. This anomalous type of behaviour was not found at quasi-static rates or in the tests on the cylindrical specimens.

Figure 5. Stress-strain curves for tests on strip specimens
a) Quasi-static tests

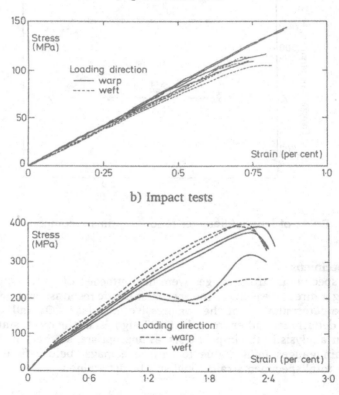

b) Impact tests

Fracture appearance: For the strip specimens, at both rates of loading, failure was by shear across the central parallel region of the specimen on a plane at about 45° to both the loading and the thickness directions and parallel to the transverse direction, as shown in fig. 6a for tests at the quasi-static rate. The damage zone is somewhat larger in the impacted specimens, see fig. 6b, corresponding to the greatly increased strain to failure at this rate of loading. It seems highly unlikely, therefore, that the shear failure in the strip specimens has been initiated by stress concentrations at the specimen ends.

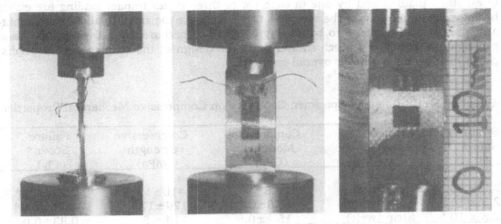

a) Quasi-static tests   b) Impact tests

Figure 6. Failure modes in strip specimens

DISCUSSION

Strain Rate and Specimen Geometry Effects

Mean stress-strain curves for four quasi-static tests and three impact tests are compared in fig. 7a, for cylindrical specimens, and in fig. 7b for thin strip specimens.   For both

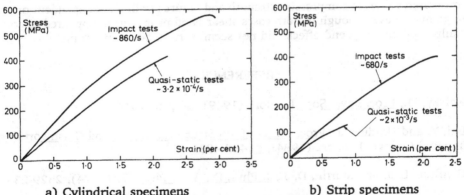

a) Cylindrical specimens   b) Strip specimens

Figure 7. Effect of strain rate on mean stress-strain response

specimen geometries the initial modulus, compressive strength and failure strain all increase with strain rate.   While this increase is greater for the strip specimens, see Table 1, the absolute values obtained are significantly lower.   Thus even under impact loading the compressive strength and the failure strain for the strip specimens remain lower than those for the cylindrical specimens under quasi-static loading.   Since shear band propagation appears to control the failure process for both geometries of specimen it must be concluded that the very low strengths shown by all the strip specimens under quasi-static loading and by one in each set of three under impact loading are either because the dimensions of the specimen test section are too small to allow a representative mechanical response to be reliably obtained or because, by allowing shear band propagation to follow a much shorter path, the design of the strip specimens predisposes them to have a lower overall strength.

TABLE 1

Effect of Strain Rate and Specimen Geometry on Compressive Mechanical Properties

Strain Rate /s	Specimen Geometry	Compressive Modulus (GPa)	Compressive Strength (MPa)	Failure Strain* (%)
3.2×10^{-4}	Cylindrical	20.8±2.0	410±12	2.23±0.22
2.3×10^{-3}	Strip (warp)	16.8±0.7	129±17	0.76±0.04
2.3×10^{-3}	Strip (weft)	15.4±0.6	114± 5	0.83±0.04
860	Cylindrical	29.2±1.5	573±12	2.81±0.06
680	Strip (warp)	27.4±1.8	397±5 (216)**	1.87±0.38
680	Strip (weft)	27.5±1.0	387±8.5 (210)**	2.12±0.33

* i.e. strain at peak load   ** Initial peak load

CONCLUSIONS

Since the laminates used for the two specimen geometries, while similar, were not identical, only general trends in behaviour may be identified. However, it is clear that the measured values of compressive strength and failure strain were dependent on specimen geometry even though in both cases shear band propagation appeared to control the failure process and end effects did not seem to have been important.

REFERENCES

1. Kolsky, H., Proc. Phys. Soc., London, (1949), B62: 676-700

2. Bai, Y. and Harding, J., Proc. Int. Conf. on Structural Impact and Crashworthiness, Imperial College, London, (1984), 2, 482-493.

3. Griffiths, L. J. and Martin, D. J., J. Phys. D., Appl. Phys., 7, (1974), 2329-2341.

4. Harding, J. and Welsh, L. M., J. Mater. Sci., 18, (1983), 1810-1826.

5. Ewins, P. D., RAE Technical Report No. 71217, (1971)

6. Harding, J., Li, Y., Saka, K. and Taylor, M. E. C., Proc. 4th. Oxford Conf. on Mech. Props. of Materials at High Rates of Strain, Institute of Physics Conf. Ser. No. 102 (Inst. of Physics, London and Bristol), 403-410 (1989)

CHARACTERIZATION OF HIGHLY ANISOTROPICALLY REINFORCED SOLIDS IN HIGH VELOCITY TENSION

KOZO KAWATA, MASAAKI ITABASHI AND SHINYA FUJITSUKA
Science University of Tokyo
2641, Yamazaki, Noda, Chiba 278, Japan

ABSTRACT

The characterization method and obtained results for unidirectionally reinforced Kevlar composite in high strain rate tension up to 10^3/s are stated.

INTRODUCTION

It is very important to characterize the uniaxial tensile properties up to final breaking, of highly anisotropic materials under high strain rate tension, not only in fundamental interests but also for the development of advanced technological fields. In spite of the importance, it has remained difficult for a long period.

The difficulty was resolved by the development of KHKK one bar method [1], combined with adopting of very long sheath-adhesion type chucks newly devised. This new type chuck secures the holding of specimen ends through the tensile experiment, preventing for the specimen end to be pulled out by shear fracture in the chuck area, before the tensile fracture in proper gage length occurs. This paper reports the details of the characterization method and obtained results for unidirectionally reinforced Kevlar composite in high strain rate tension up to 10^3/s.

CHARACTERIZATION METHOD

The KHKK one bar method is based upon elastic wave propagation theory and has been used to characterize the uniaxial tensile properties up to final breaking, of many solids such as pure metals, alloys, inorganic brittle materials, polymers and composites, mainly in the strain rate range of 10^3

/s [2]-[11]. The essential point is shown in FIGURE 1. In this method dy-
namic stress $\sigma(t)$, dynamic strain $\varepsilon(t)$ and strain rate $\dot{\varepsilon}(t)$ are derived
from the following two measured values:

$V(t)$: velocity of impact block

$\varepsilon_g(t)$: longitudinal strain at point C of output bar.

The fundamental equations are as follows:

$$\text{Dynamic stress} \quad \sigma(t)=\frac{S_0}{S}E_0\varepsilon_g(t+\frac{a+b}{c})$$

$$\text{Dynamic strain} \quad \varepsilon(t)=\frac{1}{l}\int_0^t\left\{V(\tau)-c\varepsilon_g(\tau+\frac{a+b}{c})\right\}d\tau$$

$$\text{Strain rate} \quad \dot{\varepsilon}(t)=\frac{1}{l}\left\{V(t)-c\varepsilon_g(t+\frac{a+b}{c})\right\}$$

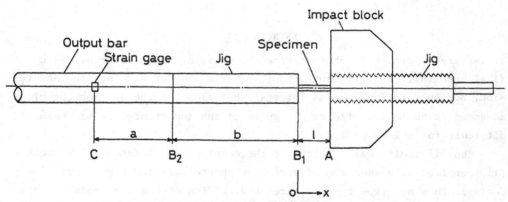

FIGURE 1. The principle of KHKK one bar method

where, S_0, S: cross-sectional areas of output bar and specimen respectively,

E_0: Young's modulus of output bar,

l: specimen gage length,

a: distance from the output bar end B_2 to the strain gage point C,

b: length of the jig B_1B_2,

c: longitudinal elastic wave velocity of output bar.

When the length of 304 stainless steel output bar 2.92m, 1130μs is obtained as the time duration for which data acquisition without the disturbance by reflected wave is secured. The designs of the very long sheath-adhesion type jigs are shown in FIGURE 2. Test material: Kevlar 49/vinyl-polyester is shown in FIGURE 3.

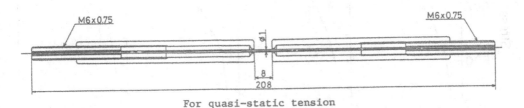

For quasi-static tension

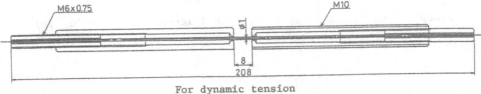

For dynamic tension

FIGURE 2. The designs of very long sheath-adhesion type jigs

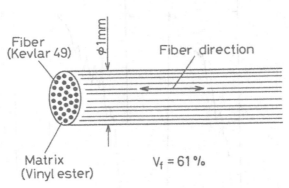

FIGURE 3. Test material

RESULTS ON KEVLAR 49/VINYLESTER COMPOSITE

The obtained results for Kevlar 49/vinylester composite in quasi-static tension ($\dot{\varepsilon}=10^{-3}$/s) and in dynamic tension ($\dot{\varepsilon}=10^{+3}$/s) are shown in FIGURES 4, 5 and 6, and TABLE 1.

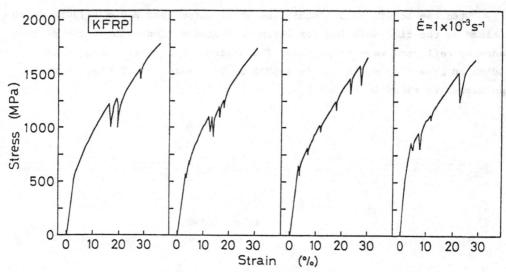

FIGURE 4. Quasi-static stress-strain diagram of KFRP ($\dot{\varepsilon}=10^{-3}$/s)

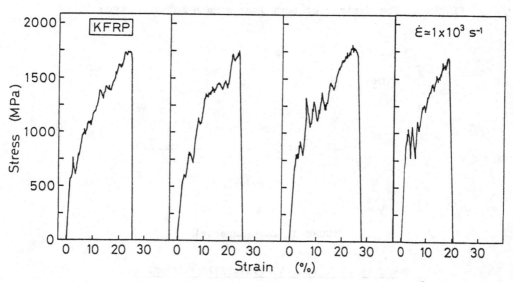

FIGURE 5. Dynamic stress-strain diagram of KFRP ($\dot{\varepsilon}=10^{+3}$/s)

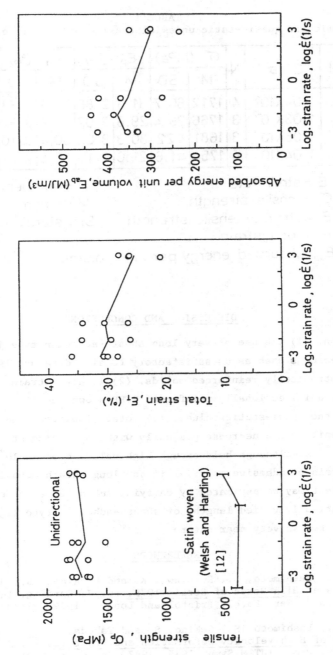

FIGURE 6. Strain rate effects in uniaxial tensile properties of KFRP

TABLE 1
Dynamic and quasi-static uniaxial tensile properties of KFRP

Material	$\dot{\varepsilon}$ (s⁻¹)	N	σ_P (MPa)		ε_P (%)		ε_T (%)		E_{ab} (MJ/m³)	
			M	SD	M	SD	M	SD	M	SD
KFRP	1.04×10^{-3}	4	1712	60.7	31.3	2.88	31.3	2.88	358	37.7
	1.03×10^{-2}	3	1766	79.4	29.6	3.02	29.9	3.37	375	53.4
	1.05×10^{-1}	3	1681	122	30.3	3.02	30.4	3.10	353	60.2
	1.04×10^{3}	4	1754	41.6	21.6	2.19	25.4	2.71	297	40.4

Notes: $\dot{\varepsilon}$ = strain rate
σ_P = tensile strength
ε_P = strain to tensile strength
ε_T = total strain
E_{ab} = absorbed energy per unit volume

N = number of specimen
M = mean
SD = standard deviation

DISCUSSION AND CONCLUSIONS

In conclusion, (1) the use of very long sheath-adhesion type jigs for chucking specimens enables us satisfactory tension data acquisition for highly anisotropically reinforced solids. (2) In high strain rate tension up to 10^3/s, unidirectionally reinforced KFRP shows tensile strength nearly equal to the quasi-static values. (3) Total elongation and absorbed energy per unit volume decrease gradually with increasing strain rate from 10^{-3}/s to 10^{+3}/s, although hold enough high values even in 10^{+3}/s. Degree of shear strain of adhesive material in the long sheath-adhesion type jigs in tension test may be estimated by carrying out a series of experiments with systematically varied lengths of sheath-adhesion type jigs. This point will be reported in very near future.

REFERENCES

1. Kawata, K., Hashimoto, S., Kurokawa, K. and Kanayama, N., Mechanical Properties at High Rates of Strain 1979, ed. J. Harding, Institute of Physics, Conf. Ser. No.47, Bristol and London, 1979, 71-80.

2. Kawata, K., Hashimoto, S., Sekino, S. and Takeda, N., Macro- and Micro-Mechanics of High Velocity Deformation and Fracture, ed. K. Kawata and J. Shioiri, Proc. IUTAM Symp. 1985, 1987, Springer, Berlin, 1-25.

3. Kawata, K., Proc. 5th Japan-US Conf. on Composite Materials 1990, Kokon-Shoin, in printing.

4. Kawata, K., Hondo, A., Hashimoto, S., Takeda, N. and Chung, H.-L., Composite Materials; Mechanics, Mechanical Properties and Fabrication, ed. K. Kawata and T. Akasaka, Proc. Japan-US Conf. on Composite Materials 1981, Japan Society for Composite Materials, 1981, 2-11.

5. Kawata, K., Hashimoto, S. and Takeda, N., Proc. 4th International Conf. on Composite Materials, Japan Society for Composite Materials, 1982, 829-836.

6. Kawata, K., Hashimoto, S. and Takeda, N., Proc. International Cryogenic Materials Conf., Butterworth, Surrey, UK, 1982, 463-466.

7. Kawata, K., Nonlinear Deformation Waves, ed. U. Nigul and J. Engelbrecht, Springer, Berlin, 1983, 237-254.

8. Kawata, K., Hashimoto, S., Takeda, N. and Sekino, S., Recent Advances in Composites in the United States and Japan, ASTM-STP 864, 1985, 700-711.

9. Kawata, K., Hashimoto, S., Miyamoto, I. and Hirayama, T., Composites '86: Recent Advances in Japan and the United States, ed. K. Kawata, S. Umekawa and A. Kobayashi, Japan Society for Composite Materials, 1986, 69-75.

10. Kawata, K., Miyamoto, I., Itabashi, M. and Sekino, S., Impact Loading and Dynamic Behaviour of Materials, ed. C.Y. Chiem, H.-D. Kunze and L.W. Meyer, DGM Informationsgesellschaft Verlag, Oberursel, 1988, 1, 349-356.

11. Kawata, K., Kawagoe, T. and Saino, K., Dynamic Fracture, ed. H. Homma and Y. Kanto, Chuo Technical Drawing Co., 1990, 287-296.

12. Welsh, L.M. and Harding, J., Proc. DYMAT 85, International Conf. on Mechanical and Physical Properties of Materials under Dynamic Loading, Paris, 1985, C5 405-C5 414.

DETERMINATION OF THE EFFECTIVE MECHANICAL RESPONSE OF POLYMER MATRIX COMPOSITES VIA MICROSTUCTURAL DATA

*   **   **   *
A.BOSCOLO BOSCOLETTO, M.FERRARI, I.PITACCO, M.P.VIRGOLINI
(*) Centro Ricerche Montedipe
via della Chimica 5, 30175 Porto Marghera (VE)
(**) Istituto di Meccanica Teorica ed Applicata
Universita' di Udine
viale Ungheria 43, 33100 Udine (UD)

ABSTRACT

Experimental data on the elastic modulus of short-fiberglass-reinforced PBT are compared with the theoretical predictions, obtained using a model which may account for fiber dimensional an orientational distibutions. Experimental techniques for the determination of the microstructural parameters are also presented.

INTRODUCTION

Short fiberglass-reinforced thermoplastic are of considerable relevance in the current technology. Their extensive range of applications continues to generate considerable interest, from the viewpoint of the theoretical modelling of the physical properties. Successful theoretical modelling and performance predictions require the incorporation of significant microstructural data, which are conveniently controlled in the course of the material designing and processing, such as intrinsic properties of each phase and relative concentrations and the morphological and orientational distributions of the embedded phase. The latter aspects are the most troublesome, because the complexity of the theoretical analysis and the difficulty of the experimental determination of the appropriate parameters. In this work, we present some recent results in the theory of homogeneization of short-fiber composites, and some experimental techniques, necessary for its implementation. The material is also characterized from the viewpoint of failure.

For an in-depth presentation of fiberglass-reinforced thermoplastics, reference may be made to [1-5]. The fundamentals of the homogenization theory are given in [6], together with a review of the most widely used effective medium theories. The theoretical developments, leading to the material presented below, are detailed in [7].

MATERIALS USED AND MOLDING PROCEDURE

The material system employed consists of a polybutylene terephtalate (PBT), Montedipe PIBITER TQ 9 T.M., with E-fiberglass reinforcements, of concentrations ranging from 0% to 50% weightwise. The average fiber length and diameter before extrusion are 4.13 ± 0.4 mm and 10 ± 0.4 μm, respectively. The fiber surfaces were coated with a coupling agent, in order to enhance the chemical fiber-matrix bonding [8]. The Young's moduli and Poisson's ratios of the matrix and fiber materials are $(E_m, \nu_m) = (2.5$ GPa, 0.39), and $(E_f, \nu_f) = (72$ GPa, 0.21), respectively.

In order to control the fracture modes of the specimens, and to obtain optimal fiber alignment, particular attention was exercised toward the rheological properties of the system [9,10]. The polymer blends were extruded with a BANDERA TR45 apparatus, equipped with a 45 mm screw. The standard ASTM D638 test specimens were injection molded on a BATTENFIELD BA 750CD. Optimal fiber alignment was obtained by keeping the mold flow in the laminar range, and by an appropriate selection of the injection points.

EXPERIMENTAL PROCEDURE

Standard ASTM D638 tensile tests were performed on an INSTRON 4505, with cross-head speed of 5 mm./min. The axial deformation was measured with an INSTRON 2660-601 strain-gauge. In order to determine the Poisson's ratios of the matrix specimens of unreinforced PBT were also simultaneously tried for longitudinal and transverse deformation by performing a tension test on a ZWICK REL 1852, equipped with a biaxial strain-gauge MTI DSST.

For the micromorphological analysis, an electronic-digital scanning microscope ZEISS DSM 950 was employed. The determination of the fiber length distribution was derived semiautomatically, through a KONTRON-VIDAS image analyser, applied to microphotographs of the ashes of portions of specimens. This method requires the burning off of the polymer, and the spreading of the fibers on a slide. About 1500 such samples were thus tried. The alignment generated by the injection molding was estimated by analyzing same perpendicular and longitudinal polished sections of the tension test specimens from gauge length.

THE MORI-TANAKA EFFECTIVE MEDIUM THEORY

Under the homogeneous displacement boundary conditions

$$u_o = \varepsilon_o x \qquad (1)$$

the effective stiffness tensor C of a composite is defined, in the "direct" approach, as the tensor that maps the applied homogeneous strain ε_o into the average stress $\bar{\tau}$ [6]:

$$\bar{\tau} = C \varepsilon_o \qquad (2)$$

For a biphase composite, the average stress and strain are, respectively,

$$\bar{\tau} = (1-v) \; \bar{\tau}^m + v \; \langle \bar{\tau}^f \rangle \tag{3}$$

and

$$\varepsilon_o = \bar{\varepsilon} = (1-v) \; \bar{\varepsilon}^m + v \; \langle \bar{\varepsilon}^f \rangle \tag{4}$$

where overbars denote volumetric averaging, v is the volume fraction occupied by the fibers, and superscripts m and f denote the matrix and the fiber phase, respectively. For composites with fibers distributed according to an orientation probability density function , ODF,$f(\varphi_1,\varphi_2,\theta)$ the pointed brackets denote f-weighted orientational averaging:

$$\langle . \rangle \equiv 1/8\pi^2 \int_0^\pi \int_0^{2\pi} \int_0^{2\pi} \langle . \rangle \; f(g) \sin\theta d\varphi_1 d\varphi_2 d\theta. \tag{5}$$

Throughout this work, φ_1, φ_2 and θ denote Euler angles, defined according to the convention of [7]: Let $K \equiv \{0, e_i\}$ and $K' \equiv \{0, e_i'\}$ be specimen- and fiber-fixed frames, respectively. φ_1 is taken to denote the angle between the nodal line L and e_1, while φ_2 is the angle between L and e_1', and θ is the angle between e_3 and e_3'. The x_2'-axis is taken to coincide with the fiber axis, for all fibers, while x_3 is the longitudinal specimen direction. For notational ease, the scalar g is used to indicate a triad of Euler angles. Upon introducing the orientation-dependent strain concentration tensor A by

$$\bar{\varepsilon}^f = A \; \bar{\varepsilon} \tag{6}$$

the effective stiffness tensor C may be expressed as

$$C = C^m + v \; \langle (C^f - C^m) A \rangle. \tag{7}$$

Different assumptions on the tensor A correspond to different effective medium theories. In this note, the effective medium theory based on the assumption of Mori and Tanaka [11] is employed. This may be expressed as follows:

$$A = A^{MT} \equiv TH, \tag{8}$$

where

$$H \equiv [(1-v)I + v\langle T \rangle]^{-1} \tag{9}$$

and

$$T \equiv [I + EC^{m^{-1}} (C^f - C^m)]^{-1} \tag{10}$$

In the above, I is the identity element of the set of fourth rank tensors with both minor symmetries and E denotes Eshelby's tensor, as defined in [12]. Since the tensor H may be taken out of the orientational averaging, the e Mori-Tanaka stiffness tensor may be expressed as

$$C^{MT} = C^m + v <(C^f-C^m)T> H . \qquad (11)$$

Eshelby's tensor E depends on the matrix moduli and on the aspect ratios of the ellipsoidal inclusion. Thus, the developed theory accounts for the morphology of the fibers, in the effective moduli's prediction. Through the N-phase extension of the Mori-Tanaka approach, morphological distribution of the fibers may be incorporated [13].

The orientation profile for each morphological and material type is accounted for through the orientation density-weighted orientational averages of type (5). The computational difficulties, arising from the tensorial calculus and algebra, associated with (11) and its n-phase counterparts, have been solved through the joint use of symbolic and numeric codes, in conjunction with an equivalent matrix formalism [13].

DATA ANALYSIS

Figure 1 exhibits the fiber length distribution for specimen with 10%, 20% and 40% fiber reinforcements. Features of interest are:
1) considerable fiber fractioning, with a distributional peak (mode) of about 70 μm, for all fiber concentrations;
2) the number of fibers with lengths contained in the interval 120:500 μm increases with fiber concentration, higher lengths being rarely found at high concentrations.

Table 1 details the distributional data on the length profile. The high degree of fractioning is evident from the fact that the percentage of the fibers, with lengths in the interval 40:500 μm. ranges from 84%, in the 10% concentration case, to the 98% of the 40% concentration case. The average fiber length may also be noted to decrease with the fiber content, as expected [14].

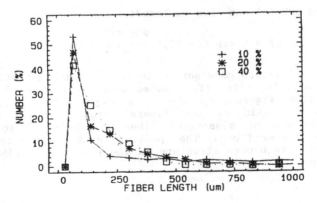

Figure 1. Fiber length distribution in PBT containing 10%, 20% and 40% weigthwise glass fiber

TABLE 1

GLASS FIBER (%)	40 - 500 (um) (%)	500 -1000 (um) (%)	1000 - 2000 (um) (%)	2000 - 3000 (um) (%)	L (um)
10	83.35	12.57	3.60	0.48	267
15	86.75	11.04	2.21	0.00	254
20	92.99	5.01	2.00	0.00	203
25	92.50	7.50	0.00	0.00	198
30	92.78	6.78	0.44	0.00	187
40	98.53	1.47	0.00	0.00	165
50	95.62	3.78	0.60	0.00	165

In order to characterize the fiber orientation profile in the composite, the planar angles α and β are introduced, according to the convention of figure 2, in the method suggested in [15]. The measurement of these angles was performed, in this study, on the polished sections, parallel to the direction of the imposed tensile stress, i.e. x_3. These sections, coinciding with the x_1-x_3 and the x_2-x_3 planes, respectively, are identified by the Euler angle selection $(\varphi_1, \varphi_2) = (\pi, 0)$ and $(\varphi_1, \varphi_2) = (\pi/2, 0)$, in this order. The angle measurement technique consisted of semi-automatically analyzing a backscattering electron image, thus allowing a considerable fiber-matrix contrast. Figure 3 exhibits the fiber orientation distribution in the 30% fiber-reinforced specimen. Figures 3a,b report that the average α- and β-values roughly coincide at 90°. It is believed that these mean values differ from 90° only because of the practical difficulty of perfect alignment of the specimen in the experimental setting. The results of figures 3a,b amount to proving an excellent alignment of the fibers with the direction of the imposed tensile stress, i.e., the vanishing of the mean value of the polar angle ϕ. Since $\theta = \phi + \pi/2$, it was thus shown that $\bar{\theta} = \pi/2$ on the x_1-x_3 and the x_2-x_3 planes. Considering that the α- and β-distributions coincide on these planes, it is now assumed that the θ-distribution is symmetric around the x_3-axis.

The following ODF (5) was thus employed:

$$f(g) = k \exp \left\{(-1/2)[\frac{\phi - \pi/2}{\sigma}]^2\right\} \delta(\varphi_2), \qquad (12)$$

where k is a normalization constant, σ is the standard deviation for α and β, and $\delta(.)$ is Dirac's delta. It is noted that the degree of alignment is also evident from figure 3c, where we report the distribution of minor/major aspect ratio of the fibers' projection on the specimen cross-section. The set of misaligned fibers is believed to be largely composed of the shorter fibers. The specimens with a fiber content larger than 30% exhibit a skin-core structure in the spatial distribution of the fiber orientations.

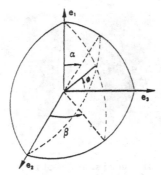

Figure 2. Coordinate system used in determination of fiber spatial orientation

The SEM micrographs of polished sections and of fracture surfaces, shown in figure 4, exhibit the different orientation pattern of the outside and the core regions. These micrographs correspond to the gauge-length of the 50% concentration case. The core region covers about 1.5% of the cross-sectional area, for the specimens at 40-50% concentrations. Thus, the core region does not significantly affect the overall orientation profile. Figure 5 compares the experimentally obtained elastic moduli with the theoretical predictions dededuced via equations (11) and (12). The overall agreement is excellent. In order to underline the importance of employing accurate microstructural data, two other curves are reported, the C1-curve corresponding to correct texture and approximate morphology (the average aspect ratio x = 26 is used), and the C2-curve corresponding to the correct morphological profile under the assumption of perfect fiber alignment. Both of these curves overestimate the actual moduli. The insert in figure 5 exhibits the relative errors for the theoretical, the C1-, and the C2-curve as a function of the fiberglass content.

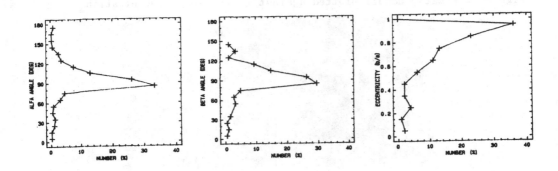

Figure 3. Fiber planar angles distributions (a), (b) and eccentricity distribution (c) in PBT containing 30 % weigthwise glass fiber

244

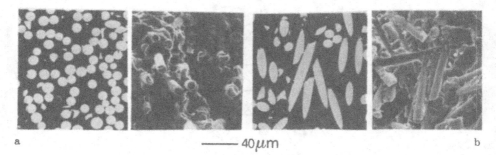

a ——— 40μm b

**Figure 4. Micrographs of polished sections and fracture surface:
(a) outside and (b) core regions**

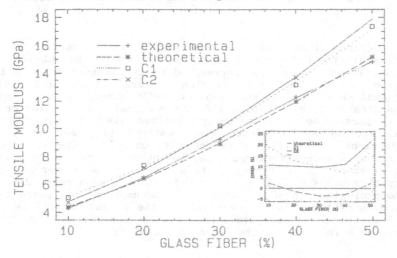

Figure 5. Elastic moduli plotted against glass fiber concentration

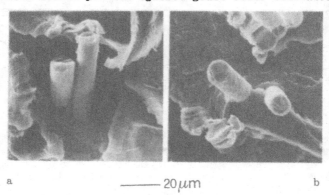

a ——— 20μm b

**Figure 6. Failure modes on the same fracture surface: (a) ductile
behaviour, (b) brittle behaviour**

The failure strain of the composite is a function of the fiberglass concentratio: for concentrations exceeding 25%, the failure strain of the composite falls below the fracture strain of the fiber material, which is of 3%. The unreinforced matrix fails in a ductile way at an ultimate strain of about 70%. Figure 6 reports micrographic evidence of different failure mechanisms contributing to the tensile rupture of the composite: regions of brittle and ductile behavior are there seen to coexist. The ductile behavior is believed to be induced by locally poor fiber-matrix bonding - as shown by figure 6a. Localized weakening of bonding is known to increase with the fiber concentration. Figure 6b shows that brittle failure is associated with higher quality fiber bonding. Within the considered range of orientation profiles, the effect of the misalignment on the failure characteristics proved to be negligible, for the tested materials.

CONCLUSION

The good agreement between the theoretical predictions and the experimental observation of the elastic modulus of short-fiberglass-reinforced PBT has shown:
1) the ability of the Mori-Tanaka theory to model the fiber textural and morphological effects;
2) the adequacy of the experimental procedure, from the viewpoint of the determination of the microstructural parameters;
3) assuming that the fiber length is equal the mean fiber length for all fibers leads to an overestimation of the effective modulus.
The dominant quantity in the composite fracturing was found to be the ultimate deformation, which was approximately constant for several fiber concentration levels.

REFERENCES

[1]   M.J. Folkes, *Short Fiber Reinforced Thermoplastics*, Research Studies Press, Great Britain, 1982.
[2]   R.J. Crowson & M.J. Folkes, *Polym. Eng. Sci.*, 20, 934, 1980
[3]   J.M. Charrier, *Polym. Eng. Sci.*, 15, 731, 1975.
[4]   M.J. Folkes, *Polymer*, 21, 1252, 1980.
[5]   D. Rosato, in *Encyclopedia of Polymer Science and Engineering*, Second Edition, vol. 14,, 327, J. Wiley & Sons, 1988.
[6]   Z. Hashin, *J.Appl. Mech.*, 50, 481, 1983.
[7]   M. Ferrari & G.C. Johnson, *Mechanics of Materials* ,8 ,67 ,1989.
[8]   J.Y. Jadhav & S.W. Kanter, in *Encyclopedia of Polymer Science and Engineering*, Second Edition, vol. 12, 217, J. Wiley & Sons, 1988.
[9]   R. Chung & C. Cohen, *Polym. Eng. Sci.*, 25, 1001, 1985.
[10] V. Gupta et al., *Polym. Composites*, 10, 8, 1989.
[11] T. Mori & K. Tanaka, *Acta Metall.* 21, 571. 1983
[12] T. Mura, *Micromechanis of Defects in Solids*, Martinus Nijhoff, The Hague, 1982.
[13] N. Marzari & M. Ferrari (to appear).
[14] J. Denault et al., *Polym. Composites*, 10, 313, 1989.
[15] L.A. Geattles, in *Mechanical Properties of Reinforced Thermoplastics*, Ed. D.W. Clegg & A.A. Collyer, Elsevier Appl. Sci. Pubbl., 1986.

A MICROSTRUCTURAL METHOD FOR PREDICTING THE DAMPING OF LAMINA

R.Eizenshmits, A.Kruklinsh, A.Paeglitis
Riga Technical University
1, Kalku st. Riga 226355, Latvia, USSR

ABSTRACT

The method for predicting of damping in unidirectionally
reinforced lamina according to the elastic and damping
properties and content of its components is worked out.
The interlayer shear under cyclic loading is taken into
account.

INTRODUCTION

Some authors, for instance [1] have proposed the method for
predicting the damping of laminated fibre reinforced
plastics according to the properties of layers, structure
of lamina and the stress-strain state. In [2] the influence
of interlayer shear is considered too. The layer here is
assumed to be reinforced only in one direction.

MATERIALS AND METHODS

The aim of this paper is to work out the method for predicting of damping in the lamina according to the elastic
and damping properties and content of its components, taking
into account the interlayer shear under cyclic loading.
    As the unit of the damping is used specific damping
capacity (SDC), which is defined as

$$\Psi = \frac{\Delta W}{W} ,$$

(1)

where ΔW - is the energy dissipated in one cycle; W - is the
maximum strain energy during the cycle.
    The degree of anisotropy of deformation and damping
properties of layer depends of the used fibres form. If the

unidirectionally layed fibres are used as reinforcment the
layer is transversal-isotrop. In such a case relative strain
energy can be obtained by the expression:

$$2W = \{\sigma\}^T\{\varepsilon\} = \{\sigma\}^T[s]\{\sigma\} \ , \tag{2}$$

where $\{\sigma\}^T$ - is the stress matrix; $\{\varepsilon\}$ - is the strain matrix;
[s] - is the compliance matrix.
Corresponding relative damping energy can be
expressed by

$$2\Delta W = \{\sigma\}^T[\Delta s]\{\sigma\} \ , \tag{3}$$

where $\{\sigma\}^T = \{\sigma_1, \sigma_2, \sigma_3, \sigma_4, \sigma_5, \sigma_6\}.$

$$[\Delta s] = \begin{bmatrix} \psi_{11}s_{11} & \psi_{12}s_{12} & \psi_{12}s_{12} & 0 & 0 & 0 \\ \psi_{12}s_{12} & \psi_{22}s_{22} & \psi_{23}s_{23} & 0 & 0 & 0 \\ \psi_{12}s_{12} & \psi_{23}s_{23} & \psi_{22}s_{22} & 0 & 0 & 0 \\ 0 & 0 & 0 & 2(\psi_{11}s_{11} - \psi_{12}s_{12}) & 0 & 0 \\ 0 & 0 & 0 & 0 & \psi_{66}s_{66} & 0 \\ 0 & 0 & 0 & 0 & 0 & \psi_{66}s_{66} \end{bmatrix}.$$

$[\Delta s]$ - is the matrix of the short time viskos-elastics
complience: ψ_{ij} - are characteristics of SDC .
    Three characteristics of SDC - $\psi_{11}, \psi_{22}, \psi_{66}$ can be found
directly from experiment, inciting in succession cyclic
longitudional and shear vibrations in the corresponding planes
and directions of the materials elastic symmetry. Terms ψ_{12}
and ψ_{23} can be calculated from SDC coefficients, obtained from
the experimental results of cyclic loading of the material,
for example, under 45° corresponding to the direction of the
elastic symmetry. Experimental coefficients of SDC are $\psi_1, \psi_2,$
$\psi_6, \psi_7 = \psi_{12}(45^O), \psi_8 = \psi_{23}(45^O)$. Taking into account
(1),(2),(3) coefficients of SDC can be calculted from

$$\psi_k = \frac{\{\sigma\}^T[\Delta s]\{\sigma\}}{\{\sigma\}^T[s]\{\sigma\}} \ , \tag{4}$$

where $k = 1,2,6,7,8.$
    Characteristics of SDC can be calculated by the acting of
one (for $\psi_{11}, \psi_{22}, \psi_{66}$) or several (for ψ_{12}, ψ_{23}) conditioned
stresses in equation (4) by following:

$$\psi_{11} = \frac{\Delta s_{11}}{s_{11}} = \psi_1; \quad \psi_{22} = \frac{\Delta s_{22}}{s_{22}} = \psi_2, \quad \psi_{66} = \frac{\Delta s_{66}}{s_{66}} \ , \tag{5}$$

$$\Psi_{12} = \frac{\Psi_7(s_{11} + 2s_{12} + s_{22} + s_{66}) - \Psi_{11}s_{11} - \Psi_{22}s_{22} - \Psi_{66}s_{66}}{2s_{12}}, \quad (6)$$

$$\Psi_{23} = \frac{\Psi_8(s_{11} + 2s_{22} + 2s_{23} + s_{12}) - 2\Psi_{22}s_{22} - \Psi_{11}s_{11} - \Psi_{12}s_{12}}{2s_{23}}. \quad (7)$$

When finding SDC characteristics of lamina in relation to properties of the components and their volume content the following assumptions where used:

1. The reason of damping is inner friction between fibres and binding material. The influence of possible defects and unpredictable porosity has not been taken into account.

2. Amplitude and frequency of vibration do not influence damping of fibres and binding material.

3. All elastic characteristics of binding material and fibres in compression and tension are equal.

Characteristics of SDC in transverse-isotrop lamina are obtained according to the structural element of the lamina, shown in Fig.1.
According to this model, lamina is supposed to be a flaky material, consisting of isotrop layers (A) of pure binding material and ortotropcally mixed layers (AB), consisting of fibres (B) and binder (A). The following assumptions are used:

1. Average deformations in layers (A) and (AB) in their plane are equal to those of a whole lamina

$$\varepsilon^A_i = \varepsilon^{AB}_i = \langle \varepsilon_i \rangle \text{ , where i= 1, 2, 6.}$$

2. The same interlayer shear stresses act in layers (A) and (B) in the plane of isotropy of lamina. They are equal to the average external stresses of lamina

$$\sigma^{AB}_4 = \sigma^A_4 = \langle \sigma_4 \rangle.$$

3. Within the layer (AB) in the direction of fibre orientation components have equal deformations

$$\varepsilon^{AB}_{A1} = \varepsilon^{AB}_{B1} = \varepsilon^{AB}_1.$$

4. Within the layer (AB) components have equal normal stresses perpendicular to the direction of fibres, and equal shear stresses in the plane of the layer (AB)

$$\sigma^{AB}_{Ai} = \sigma^{AB}_{Bi} = \sigma^{AB}_i, \text{ where i= 2,6.}$$

5. Normal stresses acting in components of the layer (AB)
in the direction 3 (Fig.1) are negligible

$$\sigma^{AB}_{A3} = \sigma^{AB}_{B3} = 0.$$

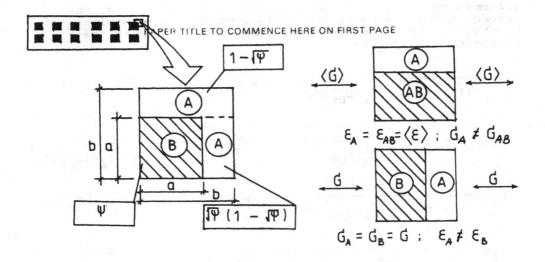

Figure 1. Sheme of stress-strain state of structural element
of the lamina

From the components relation

$$\Delta W = \psi \, \Delta W_B + (1 - \sqrt{\psi}) \, \Delta W'_A + \sqrt{\psi} \, (1 - \sqrt{\psi}) \, \Delta W''_A \qquad (8)$$

where $\psi = a^2 / b^2$ – is relative content of the fibres volume;
the damping in lamina ΔW is a summ of the damping in fibres
ΔW_B and matrix ΔW_A.

The offered method allows to consider the heterogenity of
binders stress-strain state and anisotropy of fibre properties.
The equation (8) can be expressed in matrix form through
elastics deformations as

$$\{\varepsilon\}^T [c][\Delta s][c]\{\varepsilon\} = (1 - \sqrt{\psi})\{\varepsilon\}^T [k_I][c_A][\Delta s_A][c_A] \, x$$

$$x \, [k_I]\{\varepsilon\} + \sqrt{\psi} \, (1 - \sqrt{\psi}) \, \{\varepsilon\}^T [k_{II}][c_A][\Delta s_A][c_A][k_{II}]\{\varepsilon\} + \qquad (9)$$

$$+ \psi \, \{\varepsilon\}^T [k_{III}][c_B][\Delta s_B][c_B][k_{III}]\{\varepsilon\},$$

where $\{\varepsilon\}^T = \{\varepsilon_1, \varepsilon_2, \varepsilon_3, \varepsilon_4, \varepsilon_5, \varepsilon_6\}$ – is deformations matrix;

$[c], [c_A], [c_B]$ - are elasticity matrix of lamina, binder and fibre; $[\Delta s_A], [\Delta s_B]$ - are the matrix of the short time viskoselasics complience of binder and fibre; $[k_I] = [E]$;

$[k_{II}] = \text{diog}[1, k_1, k_2, k_3, k_4, k_5]$; $[k_{III}] = \text{diog}[1, k_6, k_7, k_8, k_9, k_{10}]$.

Matrixes $[k_I]$, $[k_{II}]$ and $[k_{III}]$ show the relation between deformations of lamina components (binder and fibres) and average deformations of the whole-lamina. Those matrixes can be obtained by using assumptions mentioned above in the following way (Fig.1):

$$k_1 = \frac{1}{(1 - \sqrt{\psi}) + (E_A/E_{Br})\sqrt{\psi}} \quad ; \quad k_3 = (1 - \sqrt{\psi}) + (G_{Brz}/G_A)\sqrt{\psi};$$

$$k_2 = \frac{1}{(1 - \sqrt{\psi}) + (E_A/E_{B\theta})\sqrt{\psi}} \quad ; \quad k_4 = \frac{1}{(1 - \sqrt{\psi}) + (G_A/G_{Br\theta})\sqrt{\psi}};$$

$$k_5 = \frac{1}{(1 - \sqrt{\psi}) + (G_A/G_{Brz})\sqrt{\psi}} \quad ; \quad k_8 = \sqrt{\psi} + (1 - \sqrt{\psi})(G_A/G_{Brz});$$

$$k_6 = \frac{1}{\sqrt{\psi} + (1 - \sqrt{\psi})(E_{Br}/E_A)} \quad ; \quad k_9 = \frac{1}{\sqrt{\psi} + (1 - \sqrt{\psi})(G_{Br\theta}/G_A)};$$

$$k_7 = \frac{1}{\sqrt{\psi} + (1 - \sqrt{\psi})(E_{B\theta}/E)} \quad ; \quad k_{10} = \frac{1}{\sqrt{\psi} + (1 - \sqrt{\psi})(G_{Brz}/G_A)};$$

where $E_A, E_{Br}, _{B\theta}$ are modulus of elasticity of binder and fibre; $G_A, G_{Br\theta}, _{rz}$ are modulus of shear of binder and fibre; ψ_{ijA}, ψ_{ijB} are SDC characteristics of binder and fibre.

From one hand equation (9) contains characteristics of SDC of lamina (in matrix $[\Delta s]$) and from the other hand - characteristics of binder ($[\Delta s_A]$) and fibre ($[\Delta s_B]$) - the components of lamina. After some transformations matrix $[\Delta s]$ can be expressed from equation (9) in such form:

$$[\Delta s] = [c]^{-1}\{(1 - \sqrt{\psi})[k_I][c_A][\Delta s_A][c_A][k_I] +$$

$$+ \sqrt{\psi}(1 - \sqrt{\psi})[k_{II}][c_A][\Delta s_A][c_A][k_{II}] + \qquad (10)$$

$$+ \psi[k_{III}][c_B][\Delta s_B][c_B][k_{III}]\}[c]^{-1}.$$

Characteristics of SDC of lamina can be found by dividing the elements of matrix $[\Delta s]$ with corresponding elements of matrix $[s]$.

Elasticity characteristics of lamina in this equation can be obtained according to the sheme showed in Fig.1, as

$$E_1 = E_A (1 - \sqrt{\psi}) + E_{Bz}\,\psi;$$

$$\frac{1}{E^{AB}_2} = \frac{1 - \sqrt{\psi}}{E^*_A} + \frac{\sqrt{\psi}}{E_{Br}} \quad ; \quad E^*_A = \frac{E_A}{1 - \vartheta^2_A} \quad ;$$

$$E_2 = E^*_A (1 - \sqrt{\psi}) + E^{AB}_2\,\sqrt{\psi} \; ;$$

$$\frac{1}{G^{AB}_6} = \frac{1 - \sqrt{\psi}}{G_A} + \frac{\sqrt{\psi}}{G_{Br\theta}} \quad ; \quad G_6 = (1 - \sqrt{\psi})G_A + \sqrt{\psi}\, G^{AB}_6 \; ,$$

where E_A, E_2, G_6 - are modulus of elasticity and shear of lamina; E^{AB}_2, G^{AB}_2 - are modulus of elasticity and shear of mixed layers.

The damping properties of pure fibre and fibre in composite in many ocassions are diferent. For example fibers of organic material after the connection with binder swell out and fulfill almost all cross-section of a layer. According to this characteristicsof SDC of fibre can be obtained from characteristics of SDC of lamina and binder.

A problem of finding SDC of fibres when knowing the characteristics of SDC of the binding material and lamina is very important because of difficulties met in direct experimental determination of all SDC characteristics. Instead, experimental work with patterns from pure binding material and monocomposite and determinating of they SDC characteristics is quite simple.

The matrix $[\Delta s_B]$ for obtaining SDC of fibres can be expressed from equation (9) as:

$$[\Delta s_B] = (1/\psi)[k_{III}]^{-1}[c_B]^{-1}\{[c][\Delta s][c] - (1 - \sqrt{\psi})[k_I][c_A] \times$$

$$\times [\Delta s_A][c_A][k_I] - \sqrt{\psi}(1 - \sqrt{\psi})[k_{II}][c_A][\Delta s_A][c_A][k_{II}]\} \times \quad (11)$$

$$\times [c_B]^{-1}[k_{III}]^{-1}.$$

Characteristics of SDC of fibres can be found by dividing the elements of matrix $[\Delta s_B]$ with corresponding elements of matrix $[s_B]$. In this case characteristics of SDC of lamina can be determined from equations (5) - (7).

If the specimen of pure binder is loaded by cyclic longitudional vibration,characteristic of SDC for binder ψ_{11A} in matrix $[\Delta s_A]$, can be obtained from the equation (5). The second characteristic of SDC - ψ_{12A} then can be defined as:

$$\Psi_{12A} = \frac{\Psi_{11A} c_{11A} - 2\Psi_{A6} c_{66A}}{c_{12A}} ,$$

where $\Psi_{A6} = \Psi_{66A}$ is experimentally obtained coefficient of SDC of binder.

RESULTS

Characteristics of SDC obtained in such a way are constants of the material (binder and fibres) which characterises its damping. Characteristics of SDC – $\Psi_{11}, \Psi_{12}, \Psi_{22}$ and Ψ_{66} in unidirectional glass (GFRP) and carbon (CFRP) fibre reinforced plastic lamina in dependence of relative content of the fibres volume are showed in Fig.2. Curves are designed by using the equation (9)–(10) and data, given in Tables 1 and 2.

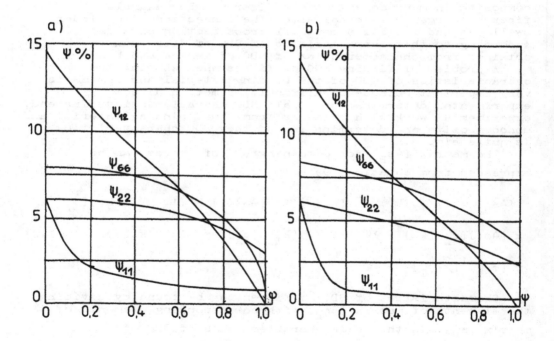

Figure 2. Theoretical dependence of the characteristics of SDC of lamina made from GFRP (a) and CFRP (b) from the relative content of fibre volume

TABLE 1
Elasticity and shear modulus and damping coefficients of lamina

Material of lamina	E_1 (GPa)	E_2 (GPa)	G_6 (GPa)	Ψ_1 (%)	Ψ_2 (%)	Ψ_7 (%)	Ψ_6 (%)	ψ	ϑ
CFRP	172,7	7,2	3,76	0,45	0,45	7,05	6,06	0,5	0.3
GFRP	37,78	10,9	4,91	0,87	5,05	6,91	6,20	0,5	0.3

TABLE 2
Elasticity and shear modulus and characteristics of SDC of fibre and binder

Components of lamina	$E_{1;\theta}$ (GPa)	$E_{2;r}$ (GPa)	$G_{12;r\theta}$ (GPa)	Ψ_{11} (%)	Ψ_{22} (%)	Ψ_{12} (%)	Ψ_{66} (%)	ϑ_{12}
binder	4,06	4,06	2,0	6,0	—	—	8,0	0,36
carbon fibres	341,30	12,60	9,75	0,38	2,43	3,32	4,25	0,20
glass fibres	71,5	71,5	30,15	0,56	2,75	4,47	0,94	0,20

The experimental data from [1] are given in Table 1.
Elastic characteristics of SDC of fibre, obtained by using equations (5),(6),(10),(11) are given in Table 2.
The acquired curves show, that characteristics of SDC of lamina depend of relative content of fibre volume.

CONCLUSIONS

The proposed method makes possible to create lamina of unidirectional reinforced plastics with definite damping properties.

REFERENCES

1. Ni R.G.,Adams R.D.,The damping and dynamic moduli of symmetric laminated composite beams - theoretical and experimental results, J. of composite materials, 1984, 18, pp. 104-121.

2. Крукиньш А.А. ,Паэглитис А.Э., Относительное рассеяние энергии в слоистых армированных пластиках, Механика композитных материалов, 1988, 3, с. 449-456.

THEORETICAL AND EXPERIMENTAL STUDY TECHNIQUE OF THE MECHANICAL PROPERTIES OF FIBER-REINFORCED LAMINATES AND OF THE STRAIN FEATURES OF STRUCTURAL ELEMENTS

V.S.BOGOLYUBOV, A.G.BRATUKHIN, V.N.BAKULIN, G.I.LVOV

Research Institute for Aviation Technology and Production Engineering, USSR, Moscow 103051, Petrovka, 24.

ABSTRACT

Major elements of a technique of the theoretical and experimental study of the mechanical properties of fiber-reinforced laminates are presented. A package of programs is based on macro- and micromechanical models of the composite. Analysis of the composite material strain in the neighbourhood of the loading application and close to the surface of the specimen was carried out on the basis of a finite-elements method.

INTRODUCTION

Complex structure of present-day composites due to strong anisothropy and edge-effect features requires not only broader and more thorough experimental studies but an active application of theoretical study techniques in an experiment preparation and processing of its results. Mathematical models and techniques of the composite mechanics study make it possible to study in detail stress-strain features of specimens during experiments (1). But it is difficult to create a stress-strain state approximating an ideal scheme under experimental conditions (2). In this case techniques using indirect measurments in an experiment deserve attention. Vector $\vec{B}$ of values being measured depends on vector $\vec{P}$ of the required parameters of the material in accordance with the adopted mathematical model of the composite. Equality of vector $\vec{B}$ being calculated and vector $\vec{B_e}$ being measured in the experiment gives an equation of the composite parameters:

$$\vec{B}(\vec{P}) - \vec{B_e} = 0 \qquad\qquad (1)$$

Developed mathematical models of specimens are based on macro- and micromechanics of a composite. The following mathematical models are used:

    (1) bending of a specimen with regard to its laminate structure, additional hypotheses being involved;

    (2) bending and tension of a specimen as a ply-heterogeneous anisothropic body;

(3) bending and tension of a specimen as a microhetero-
geneous body.

Materials and Techniques

Bending of a laminated specimen with considerable differences
in plies elasticity and small ply shear strength is considered on the basis of the theory (3) which does not involve
the assumption that a normal is rectilinear within the
boundaries of each ply. A uniform solution for bending

$$W = a(c_1 e^{ax} + c_2 e^{ax}) + c_3(b - dx^2) + \qquad (2)$$
$$+ c_4(b - 3dx^2) - 6d(c_5 x + c_6)$$

takes into account an edge effect and is used for studying
bending with a three-point loading scheme and other schemes
of loading ($c_1, c_2, ..., c_6$ — integration constants, other
parameters — physical constants of the specimen). It helps, in
particular, to estimate a shear module of shear-pliable
plies by deflection, if other parameters of the laminate are
known. From the point of view of macro-mechanics a specimen
under testing is an anisothropic ply-heterogeneous body.
With regard to the nature of tensile and bending loading a
flat stressed state is considered, the finite-elements method
being used for its estimation. A rectangular element with the
bi-linear approximation of shifts for an orthothropic
material is employed. Solution of algebraic equations is
achieved by an upper relaxation method.
Fig.1 shows a scheme of a laminate specimen tension in which
the clamps create tangential load Q_τ on the edges. The
parabolic law of distribution is assumed to be in action. In
the given case($L = 228$ mm, $h_K = 10$ mm) 800 elements (40 x
20) were sufficient for consideration of one quadrant.

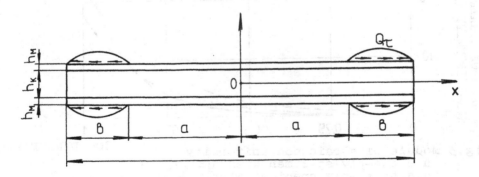

Fig.1 A specimen loading scheme.

Fig.2 Shows the distribution of ε_X strain deviation $\delta_\varepsilon =$ $= (\varepsilon_X - \varepsilon_\infty)/\varepsilon_\infty \cdot 100$ in percentage from uniform ε_∞ strain in an infinitely long specimen. The case of a homogeneous composite has been considered ($E_X = 50$ GPa, $E_y = 10$ GPa, $G_{xy} = 1$ GPa, $b = 36$ мм).

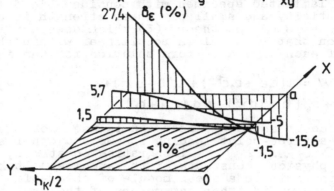

Fig.2 Distribution of strain δ_ε non-uniformity.
Introduction of stiff metal plies ($E_X = E_y = 70$ GPa, $h = 2$ мм) has not changed the mature of the strain non-uniformity distribution. Fig.3 presents $|\delta_\varepsilon|$ in a homogenous composite when y=0 (Curve 1), and in the metal-ply-reinforced composite when y=0,5 h_K (Curve 2) and y=0 (Curve 3). The presence of stiff plies increases non-uniformity ε_X in the area free from clamps, and shifts the maximum absolute value of non-uniformity from the edge y=0,5h_K to the center-line.

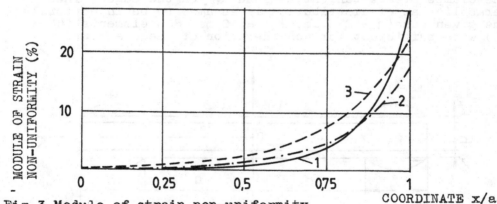

Fig.3 Module of strain non-uniformity
in a one-ply specimen when y=h_K/2 (1)
in a three-ply specimen when y=h_K/2 (2)
and when y=0 (3)
To study micromechanics of composites models and techniques were developed in the form of a package of applied programs "Composite" (4). On its basis, among other things, a study of a reinforced 9-ply composite material strain in the

neighbourhood of a free surface has been conducted.
We consider an orthogonally reinforced plastic material
consisting of alternating plies reinforced at 0° and 90° to
axis X . For simplicity sake let us assume that each
transversely reinforced ply contains one row of fibers.The
strain study has been conducted with the following loading
types:1) tensile load acting along axis X is applied to the
transversely reinforced plies along faces $X=\pm L$;2) ditto,
but the load is applied to the lengthwise reinforced plies;
3)load is distributed among the plies in proportion to their
reduced modules.The summary tensile force is the same as in
the first case.The macroscopic strained state of the stack is
assumed uniform,and the specimen width 2β — big enough to
disregard the influence of boundary plies $z \geqslant -\beta$ and $z \leqslant \beta$
in the circumstances of the above mentioned the boundary
requirements should be only integrally satisfied on faces $z=\pm\beta$

$$\sigma_z = \tau_{xz} = \tau_{yz} = 0 \qquad (3)$$

if we assume that these faces are free of stresses.
Under the given conditions of loading and a constant macroscopic strain this equals the assumption that $\gamma_{xz} = \gamma_{yz} = 0$,
$\varepsilon_x = const$ for the stack as a whole.Value ε_z is related to
macroscopic stresses σ_x and σ_y by equation

$$\varepsilon_z = -\frac{\nu_{xz}}{E_x} \sigma_x - \frac{\nu_{yz}}{E_y} \sigma_y , \qquad (4)$$

where $\nu_{xz}, \nu_{yz}, E_x, E_y$ — are effective elastic constants of
the stack as a whole.Based on the presence of constant
strain ε_z the decisive relations between fiber and matrix
materials take the form

$$\vec{\sigma} = (\sigma_x, \sigma_y, \tau_{xy}, \sigma_z)^T = D\vec{\varepsilon} + \vec{A}\varepsilon_z , \qquad (5)$$

where

$$\vec{\varepsilon} = (\varepsilon_x, \varepsilon_y, \gamma_{xy}, 0)^T; \qquad (6)$$

$$D = \begin{bmatrix} d_{xxxx} & d_{xxyy} & d_{xxxy} & 0 \\ d_{yyxx} & d_{yyyy} & d_{yyxy} & 0 \\ d_{xyxx} & d_{xyyy} & d_{xyxy} & 0 \\ d_{zzxx} & d_{zzyy} & d_{zzxy} & 0 \end{bmatrix} \qquad (7)$$

$$\vec{A} = (d_{xxzz}, d_{yyzz}, d_{xyzz}, d_{zzzz})^{T}, \tag{8}$$

d_{ijke} — component of a matrix elasticity tensor.
Thus, due to the orthothropy and linearity of the material, the problem is split into two two-dimension problems: the problem of the stack strain with a single-axis macroscopic strain ε_z and the problem of a flat strain state. The present paper concentrates on the second problem. The complex field geometry requiring continuity of global approximations on arbitrarily oriented boundaries of finite elements leads to the necessity of using simplex elements; so far as the problem is a flat one, we shall use flat triangular elements with linear approximation of shifts. The finite-element model employed for the solution of this problem consists of 1380 elements and has 1433 degrees of freedom, a quarter of the field under consideration being modelled (which is possible due to the symmetry of the field, loads and structure of the material). The following boundary conditions will be placed on the formed boundaries resulting from symmetry considerations: $U_y=0$ when $y=0$, $U_x=0$ when $x=0$ (9) where U_x and U_y — shifts along axes x and y correspondingly. The boundary plies are the plies with transversely oriented fibers, and thus, the influence of the boundaries on the lengthwise oriented plies will be inhibited. The latter will be modelled as a homogeneous orthothropic material, the effective modules of which are to be calculated in accordance with a specified model of co-axial cylinders (6) for a hexagonal placement. In this case matrices of physical laws will have the following numerical values: for the fiber —

$$D_a = \begin{bmatrix} 7,78 & 1,94 & 0 & 0 \\ 1,94 & 7,78 & 0 & 0 \\ 0,0 & 0 & 2,92 & 0 \\ 1,94 & 1,94 & 0 & 0 \end{bmatrix} \cdot 10^4 \text{ MPa}$$

for the resin —

$$D_s = \begin{vmatrix} 0,598 & 0,369 & 0 & 0 \\ 0,369 & 0,598 & 0 & 0 \\ 0 & 0 & 0,114 & 0 \\ 0,369 & 0,369 & 0 & 0 \end{vmatrix} \cdot 10^4 \text{ MPa}$$

for the orthothropic ply —

$$D = \begin{vmatrix} 4,2 & 0,248 & 0 & 0 \\ 0,248 & 0,528 & 0 & 0 \\ 0 & 0 & 0,342 & 0 \\ 0,248 & 0,158 & 0 & 0 \end{vmatrix} 10^4 \text{ MPa}$$

The most prominent effect of a free surface is exhibited in the fact that the shifts of the points situated on it have a pronounced oscillating character (Fig.4).Maximum absolute values of shifts in the direction of the normal to the free surface are situated among the fibers, and minimum values – close to the centers of fibers (with the given relation of modules and Poisson coefficients of a fiber and a resin). The shifts increase dramatically close to the loaded edge, if the load is applied to the plies with transversely placed fibers, and decrease almost to 0, if the lengthwise reinforced plies are loaded.This fact can be explained by the influence of the loaded edge,if we take into account that at a distance from it the integral tensile forces are distributed among the plies in proportion to their stiffness .The influence of the boundary conditions on the loaded surface is more pronounced at a distance of 3–3,5 ply thicknesses;farther,maximum and minimum shift values remain virtually constant. When the load is applied to the plies in proportion to their stiffness, the shifts oscillate with an almost constant amplitude along the whole length of the free surface $y = h$. Shifts amplitude U_y quickly decreases retreating from the free edge into the stack,and decreases 5,5 times at a distance of a two-ply thickness.

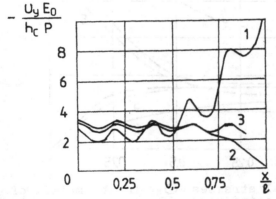

Fig.4 Shifts U_y on free surface $y = h$ under tension along
       axis X :
       1 – the load is applied to transversely reinforced
           plies;
       2 – to lengthwise reinforced plies;
       3 – in proportion to reduced elasticity modules of the
           plies

The strained state of the composite components at a distance from the field edges coordinates well with the solution on an infinite field. Specifically, tangential stresses are equal to zero, with precision compared to approximation errors, on areas $x = const$ and $y = const$ spreading through fibers centers close to the field center, and on areas

x=const – in the middle between fibers.Tensile stresses
σ_x in a matrix reach their maximum in a local zone between
fibers, in the middle of the length connecting the cross-
section centers of adjacent fibers of one row. The calculated value of the stress concentration factor is 1,9,which
is close to the theoretical and experimental results for a
unidirectional glass-fiber-epoxy plastic.So,the paper (5)
presents the value of a stress concentration factor equalling
2 which was found experimentally by photoelastic analysis on
a lexan-epoxy model with a somewhat higher fiber content –
50%.Fig.5 shows the graphs of functions of maximum stresses
σ_x in a matrix among the fibers for the cases when the load
is applied to the lengthwise reinforced plies (solid lines)
and to the transversely reinforced plies (dash lines). As is
clear from the Figure, the maximum stresses in the matrix
change but little at a distrance from the loaded edge exceeding a three-ply thickness.

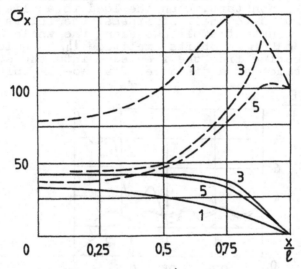

Fig.5 Maximal normal stresses σ_x in the matrix of
transversely reinforced plies (in per cent to the applied stress).Curves numbers correspond to plies numbers.
Solid lines – the load is applied to plies 2,4,6,8;
dash-line – to plies 1,3,5,7,9.

Thus,the influence of the loaded edge is exhibited in a
twofold way:stresses applied to the field extremes,in the
sections remote from the edge,are redistributed both among
the plies with different fiber orientation and through the
thickness of each ply.The redistribution of stresses through
the thickness of the transversely reinforced plies takes
place very quickly and stops at a distance of the double-ply
thickness from the loaded edge, and the redistribution of
stresses among the plies is so slow that the edge influence
holds along the whole length of the finite-element model

and practically stops attenuating at a distance of triple-
ply thickness from the loaded edge, though maximum values of
tensile stresses in the plies vary significantly.So,when
the lengthwise reinforced plies are in tension,maximum
stresses σ_x in ply 1 adjoining the surface are 1,3 times
less than the stresses in middle ply 5, and when the
transversely reinforced plies are in tension, maximum stresses
σ_x in ply 1 are 2,1 times higher than the stresses in ply 5.
The values of stresses in inner plies 3 and 5 with transverse
fiber orientation are close to each other,differing by 10-15%
away from the edge $x = L$. Thus, the developed mathematical
models and techniques make it possible to study the behaviour
of a composite material with regard to the features of its
macro-and microproperties.Comparision of the experimental
and calculation results allows to identify specific features
of the material.

BIBLIOGRAPHY

1. Kural M.H.,Flaggs D.4.A Finite Element Analysis of
   Composite Tension Specimens.-Composites Technology Review,
   Spring,1983, vol.5, No.1,pp. 11-17.

2. Composite Materials (Edited by Vassilyev V.V.,Tarnopolski
   Y.M. Moscow, Mashinostroyenie,1990, 510 p.

3. Rasskazov A.O., Sokolovskaya No. 9.,Shulga N.A. Theory
   and Calculation of Laminated Orthothropic Plates and
   Skins,Kiev,Visha shkola,1986,192 p.

4. Bakulin V.N.,Rassokha A.A.The Finite -Element Method and
   the Holography Interferometry in the Composite Mechanics.
   Moscow,Mashinostroyenie,1987,312 p.

5. Daniel I.M. A Photoelastic Analysis of.Composites//
   Composite Materials.Fr.from Eng. In 8 volumes. Vol. 2,
   Moscow,Mir,1978,pp. 492-552.
6. Sendetsky I. Elastic Prperties of Composites.//Composite
   Materials Fr.from Eng. in 8 volumes. Vol.2,Moscow, Mir,
   1978, pp. 61-101.

DYNAMIC PROPERTIES OF LAYER REINFORCED
BY TWO FAMILIES OF FIBRES

DRAGAN I. MILOSAVLJEVIĆ
University " Svetozar Marković "
Faculty of Mechanical Engineering
Sestre Janjica 6, 34000 Kragujevac, YUGOSLAVIA

ABSTRACT

The object of this theoretical investigation is to determine dynamic properties of such a layer which can provide a standard against which we may compare the results of non-destructive evaluation measurements. It is shown that phase velocity of such waves strongly depends on both direction of propagation and angle between two families of fibres.

Approximate expressions for phase velocities have been given for all three fundamental modes. It is shown that the approximate expressions provide solutions that agree very much with the exact solutions which have been obtained from the model with extensible fibres. The advantage of the approximate solution is that it is easily used to obtain the first information about dispersion of waves in such layers.

INTRODUCTION

This work treats wave propagation in an infinite plate reinforced by two families of straight mechanically equivalent fibres which lie in the planes parallel to stress-free boundaries. This plate is set of layers of fibre reinforced material in which the reinforcements in both preferred directions are families of parallel fibres lying in the plane of the layer. Material is modelled as a homogeneous continuum with two preferred directions having the angle between two families of the fibres, in each point, of 2ϕ .

We consider continuum model for which the stress - strain relation is given by Spencer [1]. This model is a good representation of the solid for consideration of the propagation of long elastic waves. These waves have wavelengths much longer than the inter-atomic spacing and they cannot notice details of the atomic arrangement. Their propagation will therefore depend upon some average properties of an extended region, and thus effect of the atoms may be smeared out into a continuum. Adopting this model for examination of wave propagation is equivalent to assuming that the wavelengths are large compared with the fibre diameter and inter-fibre spacing, so that scattering effects due to individual fibres are negligible. A typical sheet of carbon fibre reinforced prepeg will have thickness of order of

250 μm with the fibre diameter and inter-fibre spacing of order of 6-7 μm .

In order to simplify description of the continuum model of strongly anisotropic solid Spencer [2] introduced a constraint of inextensibility along the fibres. This constraint implies the existence of discontinuous tangential stresses at the upper and the lower surface of the laminate. Corresponding normal stresses become singular in these regions. The physical interpretation of this mathematical phenomenon is in terms of a narrow shear band or boundary layer through which the tangential stress varies rapidly from its value on the boundary to its value very close to the boundary. The inextensibility provides a very simple mathematical tool for description of dynamic behaviour, but, to examine behaviour in regions of rapid changes of stresses, we are obliged to use the model with extensible fibres. Comparison of these two theories is given by Green [3] and Green and Milosavljevic [4] for wave propagation through plate reinforced by one family of the fibres and by Milosavljevic [5-7] for wave propagation through plate reinforced by two families of the fibres.

GOVERNING EQUATIONS

We consider the material with two preferred directions defined by fields of unit vectors $\underset{\sim}{a}$ and $\underset{\sim}{b}$, respectively. All vector and tensor components will be referred to a system of rectangular cartesian coordinates x_i (i=1,2,3). Components of unit vectors $\underset{\sim}{a}$ and $\underset{\sim}{b}$ are then a_i and b_i, respectively. The displacement vector $\underset{\sim}{u}$ has components u_i and infinitesimal strain tensor $\underset{\sim}{e}$ has components e_{ij} so that $e_{ij} = (u_{i,j} + u_{j,i})/2$ where commas denote partial derivatives. The Cauchy stress tensor $\underset{\sim}{\sigma}$ has components σ_{ij} which are derived in [1] and given as

$$
\begin{aligned}
\sigma_{ij} = & \{\lambda e_{rr} + \gamma_3(a_r a_s e_{rs} + b_r b_s e_{rs}) + \gamma_4 a_r b_s e_{rs} \cos2\phi\}\delta_{ij} + 2\mu e_{ij} \\
& + \{\gamma_3 e_{rr} + 2\gamma_1 a_r a_s e_{rs} + \gamma_6 b_r b_s e_{rs} + \gamma_5 a_r b_s e_{rs} \cos2\phi\}a_i a_j \\
& + \{\gamma_3 e_{rr} + \gamma_6 a_r a_s e_{rs} + 2\gamma_1 b_r b_s e_{rs} + \gamma_5 a_r b_s e_{rs} \cos2\phi\}b_i b_j \\
& + \tfrac{1}{2}\{\gamma_4 e_{rr} \cos2\phi + \gamma_5(a_r a_s + b_r b_s)e_{rs} \cos2\phi + 2\gamma_2 a_r b_s e_{rs}\}(a_i b_j \\
& + a_j b_i) + \gamma_7\{a_r(e_{ri} a_j + e_{rj} a_i) + b_r(e_{ri} b_j + e_{rj} b_i)\},
\end{aligned}
$$
(1)

where λ, μ, $\gamma_1, \ldots, \gamma_7$ are even functions of $\cos2\phi$, and 2ϕ is the angle between the two families of fibres. Here and throughout the paper indices take the values 1, 2 and 3 and the summation convention is employed.

The equations of motion, with no body forces, are

$$
\sigma_{ij,j} = \rho \ddot{u}_i,
$$
(2)

where ρ is density and dot denotes differentiation with respect to time.

Choosing the cartesian coordinates x_1 perpendicular to both families of fibres, x_2 and x_3 along bisectors of the two families of fibres, the boundary surfaces of the layer are given as $x_1 = \pm h$ where 2h is the layer thickness. The traction-free boundary conditions are given as follows

$$\sigma_{11} = 0 , \qquad \text{at} \quad x_1 = \pm h . \qquad (3)$$

Suppose that a plane wave propagates with a phase velocity v in the direction which makes an angle α with the x_3- axis in the plane parallel to the stress-free surfaces of the layer. Displacements are assumed in the following form

$$u_1 = U_{(x)} \cos \psi , \qquad u_2 = V_{(x)} \sin \psi , \qquad u_3 = W_{(x)} \sin \psi , \qquad (4)$$

where $\psi = k(sx_2 + cx_3 - vt)$ is a phase, $x = x_1/h$ is a dimensionless coordinate, $s \equiv \sin \alpha$, $c \equiv \cos \alpha$ and k is a wave number. Substituting (4) into (1) and then into (2) the equations of motion become

$$c_{11} U'' - k^2 h^2 (d - \rho v^2) U + khsb_3 V' + khcb_2 W' = 0,$$
$$- khsb_3 U' + c_{66} V'' - k^2 h^2 (g_2 - \rho v^2) V - k^2 h^2 scb_1 W = 0, \qquad (5)$$
$$- khcb_2 U' - k^2 h^2 scb_1 V + c_{55} W'' - k^2 h^2 (g_3 - \rho v^2) W = 0,$$

where prime denotes differentiation with respect to x and

$$d = s^2 c_{66} + c^2 c_{55}, \quad g_2 = s^2 c_{22} + c^2 c_{44}, \quad g_3 = s^2 c_{44} + c^2 c_{33},$$
$$b_1 = c_{23} + c_{44}, \quad b_2 = c_{13} + c_{55}, \quad b_3 = c_{12} + c_{66}, \qquad (6)$$

where c_{ij} are new material constants given in the following way

$$c_{11} = \lambda + 2\mu, \quad c_{12} = \lambda + \sin^2\phi(2\gamma_3 - \gamma_4 \cos2\phi), \quad c_{13} = \lambda + \cos^2\phi(2\gamma_3 + \gamma_4 \cos2\phi),$$
$$c_{22} = \lambda + 2\mu + 2\sin^2\phi[2\gamma_3 - \gamma_4 \cos2\phi + 2\gamma_7 + \sin^2\phi(2\gamma_1 + \gamma_6 + \gamma_2 - 2\gamma_5 \cos2\phi)],$$
$$c_{23} = \lambda + \cos^2\phi[2\sin^2\phi(2\gamma_1 + \gamma_6 - \gamma_2) + 2\gamma_3 + \gamma_4 \cos2\phi] + \sin^2\phi(2\gamma_3 - \gamma_4 \cos2\phi), \qquad (7)$$
$$c_{33} = \lambda + 2\mu + 2\cos^2\phi[2\gamma_3 + \gamma_4 \cos2\phi + 2\gamma_7 + \cos^2\phi(2\gamma_1 + \gamma_6 + \gamma_2 + 2\gamma_5 \cos2\phi)],$$
$$c_{44} = \mu + \gamma_7 + 2\sin^2\phi\cos^2\phi(2\gamma_1 - \gamma_6), \quad c_{55} = \mu + \gamma_7 \cos^2\phi, \quad c_{66} = \mu + \gamma_7 \sin^2\phi.$$

The boundary conditions (3) are then

$$c_{11} U' + khsc_{12} V + khcc_{13} W = 0,$$
$$- khcU + W' = 0, \quad - khsU + V' = 0, \quad \text{at} \quad x = \pm 1. \qquad (8)$$

APPROXIMATE PHASE VELOCITY

The wave number k is related to the wave length Λ by the expression $k = 2\pi/\Lambda$ and the limits $kh \to 0$ and $kh \to \infty$ correspond to waves of infinitely large and vanishingly small wavelengths, respectively. In the case of long waves, when $kh \ll 1$, we may develop a perturbation scheme based on the assumption that kh is a small parameter. Then from equations (5) is obvious that we obtain a regular perturbation expansion. That expansion, however, is valid in very restricted range of wavelengths if we have in mind that material is reinforced with strong fibres. When the fibres become almost inextensible material constant γ_1 become very large and it may be introduced small parameter $\varepsilon \ll 1$ in the following way

$$4\,\gamma_1 = d \,/\, \varepsilon^2 , \tag{9}$$

and both geometric kh and material small parameter ε are to be considered.

Perturbation based on kh $\ll$ 1

Perturbation scheme based on kh $\ll$ 1 may be developed by substitution

$$U = \sum_{\beta=0}^{\infty} U_\beta(kh)^\beta, \quad V = \sum_{\beta=0}^{\infty} V_\beta(kh)^\beta, \quad W = \sum_{\beta=0}^{\infty} W_\beta(kh)^\beta, \tag{10}$$

for displacement and

$$\rho v^2 = \sum_{\beta=0}^{\infty} \rho v_\beta^2 (kh)^\beta, \quad \beta = 1, 2,\ldots,\infty, \tag{11}$$

for phase velocity in equations of motion (5) and boundary conditions (8). The equations of motion and the boundary conditions leads to set of solutions for displacements for each power of kh as well as to the corresponding dispersion equations. It is easy to show that such an analysis leads to expressions for three distinct fundamental modes. A detailed calculation has been given in [6] and here will be reported expressions for phase velocity. First set of solutions leads to squared phase velocity in the form

$$\rho v^2 = \rho v_0^2 + \rho v_1^2 kh + \rho v_2^2 (kh)^2 + O[(kh)^3] = \tfrac{1}{3}\{s^2(g_2 + c^2 b_1)$$
$$+ c^2(g_3 + s^2 b_1) - (s^2 c_{12} + c^2 c_{13})^2/c_{11}\}(kh)^2 + O[(kh)^3] \tag{12}$$

for bending waves which correspond to antisymmetric motion. For extensional waves which correspond to symmetric motion we obtain

$$\rho v^2 = \rho v_0^2 + \rho v_1^2 kh + \rho v_2^2 (kh)^2 + O[(kh)^3] = O[(kh)^3]$$
$$+ \rho v_0^2 \{1 - \tfrac{1}{3}(kh)^2 \frac{[(g_2 - \rho v_0^2)(g_3 - \rho v_0^2) - s^2 c^2 b_1^2]^2}{s^2[c_{12}(g_3 - \rho v_0^2) - c^2 c_{13} b_1]^2 + c^2[c_{13}(g_2 - \rho v_0^2) - s^2 c_{12} b_1]^2}\}, \tag{13}$$

where ρv_0^2 are two solutions of quadratic equation

$$\{\rho v_0^2 - (g_2 - s^2 c_{12}^2/c_{11})\}\{\rho v_0^2 - (g_3 - c^2 c_{13}^2/c_{11})\} - s^2 c^2 (b_1 - c_{12} c_{13}/c_{11})^2 = 0, \tag{14}$$

one of which corresponds to quasi-longitudinal and another to quasi-transverse waves. These waves may be distinguished by considering expressions for displacements.

Perturbation based on two small parameters

When material is strongly anisotropic then the material constant γ_1 becomes large, i.e. $\gamma_1 \gg 1$, and material parameter ε becomes large, i.e. $\varepsilon \ll 1$. It has been shown, in [5] and [6], that in the limiting process which transforms constitutive equation for strongly anisotropic material into inextensible material, given in [1], without loosing any important information significant in dynamic analysis we can set $\gamma_2 = \gamma_3 = \gamma_5 = \gamma_6 = 0$. From equations (6) and (7) it is easy to conclude that g_2, g_3 and b_1 become large constants which may be represented in the following way

$$g_2 = \bar{g}_2/\varepsilon^2 + \hat{g}_2, \qquad g_3 = \bar{g}_3/\varepsilon^2 + \hat{g}_3, \qquad b_1 = \bar{b}_1/\varepsilon^2 + \hat{b}_1, \qquad (15)$$

where bar and hat quantities are of $O(1)$. For $kh = O(1)$ perturbation based on ε becomes singular because the highest derivatives of V and W are multiplied by ε^2. Thus, there is a boundary layer whose solution may be constructed as matched composite solution. When the geometric parameter kh becomes comparable with material parameter ε, then perturbation based on ε becomes again regular. Thus, in the range of wavelengths comparable with γ_1 we have two small parameters involved, and to follow their ratio we introduce yet another parameter $m = kh/\varepsilon$, so that the solution, based on parameter ε, will involve m as a parameter representing inner solution in a composite expansion.

It has been shown in [5] that, for bending waves, displacements and phase velocity may be written as an series in the powers of ε in the form

$$\{ U, V, W, v^2 \} = \sum_{\beta=0}^{\infty} \{ U_{2\beta}, m\varepsilon V_{2\beta}, m\varepsilon W_{2\beta}, v^2_{2\beta} \} \, \varepsilon^{2\beta}, \qquad (16)$$

and their substitution in equations (5) leads to an infinite system of equations which may be solved by successively starting with the lowest order terms. The lowest order solution for squared phase velocity, obtained by imposing the boundary conditions on displacements, take following form

$$\rho v^2_0 = \frac{-1}{\bar{p}_3^2 - \bar{p}_2^2} \sum_{\alpha=2}^{3} (-1)^{\alpha} (d\bar{p}_\alpha^2 - s^2\bar{g}_3 \frac{c_{66}}{c_{55}} - c^2\bar{g}_2 \frac{c_{55}}{c_{66}} + 2s^2 c^2 \bar{b}_1)(1 - \frac{\tanh m\bar{p}_\alpha}{m\bar{p}_\alpha}), \qquad (17)$$

where $\bar{p}_2^2 < \bar{p}_3^2$ are the solutions of equation

$$c_{55}c_{66} \, \bar{p}^4 - (c_{55}\bar{g}_2 + c_{66}\bar{g}_3) \, \bar{p}^2 + \bar{g}_2\bar{g}_3 - s^2 c^2 \bar{b}_1^2 = 0 . \qquad (18)$$

The expression (17), for phase velocity, has remarkable agreement with exact solution, obtained in [6], and, when angle of propagation is perpendicular to one of the families of fibres, it may be reduced to the form

$$\rho v^2_0 = 4 \, s^2 c^2 \frac{c_{55} c_{66}}{d} \left(1 - \frac{\tanh m\bar{p}_3}{m\bar{p}_3} \right) , \qquad (19)$$

but it can be used only as an inner solution in composite expansion.

For the extensional waves, however, the displacements may be written as an series in the powers of ε in the form

$$\{ U, V, W, \} = \sum_{\beta=0}^{\infty} \{ m\varepsilon U_{2\beta}, V_{2\beta}, W_{2\beta}, \} \, \varepsilon^{2\beta}, \qquad (20)$$

and when we assume phase velocity in the form $v^2 = \bar{v}^2/\varepsilon^2 + \hat{v}^2$ substitution in equations (5) leads, again to an infinite system of equations which may be solved by successively starting with the lowest order terms. The lowest order solutions for displacements V and W take the following form

$$V_0 = B_{02} \cosh m\bar{p}_2 x + B_{03} \cosh m\bar{p}_3 x ,$$

$$W_0 = \frac{c_{66}\bar{p}_2^2 - (\bar{g}_2 - \rho\bar{v}^2)}{sc\bar{b}_1} B_{02} \cosh m\bar{p}_2 x + \frac{c_{66}\bar{p}_3^2 - (\bar{g}_2 - \rho\bar{v}^2)}{sc\bar{b}_1} B_{03} \cosh m\bar{p}_3 x ,$$
(21)

where $\bar{p}_2^2$ and $\bar{p}_3^2$ are solutions of equation (18). The boundary conditions lead to the conclusion that equation

$$(\bar{g}_2 - \rho\bar{v}^2)(\bar{g}_3 - \rho\bar{v}^2) - s^2 c^2 \bar{b}_1^2 = 0 , \tag{22}$$

has to be satisfied, leading to two solutions for dominant terms of phase velocity $\rho\bar{v}_\alpha^2$, ($\alpha = 1,2$). Same results follow from equation (14) if we take the dominant terms only. Remaining equations of the lowest order terms lead to the solutions for displacements in the following form

$$U_{0\alpha} = A_{0\alpha} \sin m\bar{q}_{1\alpha} x / m\bar{q}_{1\alpha} ,$$

$$V_{0\alpha} = c_{11}\bar{b}_1 s A_{0\alpha} \cos m\bar{q}_{1\alpha} / [c_{13}(\bar{g}_2 - \rho\bar{v}_\alpha^2) - s^2 c_{12}\bar{b}_1] , \tag{23}$$

$$W_{0\alpha} = - c_{11}(\bar{g}_2 - \rho\bar{v}_\alpha^2) A_{0\alpha} \cos m\bar{q}_{1\alpha} / c[c_{13}(\bar{g}_2 - \rho\bar{v}_\alpha^2) - s^2 c_{12}\bar{b}_1] ,$$

where $\bar{q}_{1\alpha}^2 = \rho\bar{v}_\alpha^2 / c_{11}$, ($\alpha = 1,2$). Thus we obtained two sets of solutions which correspond to quasi-longitudinal and quasi-transverse motion. To obtain the next term of phase velocity we may solve the equations of motion of Order one subjected to the boundary conditions of same Order. This procedure leads to next term of the phase velocity in the following form

$$\rho\hat{v}_\alpha^2 = \frac{\hat{g}_2\bar{g}_3 + \hat{g}_3\bar{g}_2 - \rho\bar{v}_\alpha^2(\hat{g}_2 + \hat{g}_3) - 2s^2 c^2 \bar{b}_1 \hat{b}_1}{\bar{g}_2 + \bar{g}_3 - 2\rho\bar{v}_\alpha^2}$$
(24)

$$- \frac{\text{tg } m\bar{q}_{1\alpha}}{m\bar{q}_{1\alpha}} \frac{c^2 [c_{13}\bar{g}_2 - s^2 c_{12}\bar{b}_1 - c_{13}\rho\bar{v}_\alpha^2]^2}{c_{11}(\bar{g}_2 - \rho\bar{v}_\alpha^2)(\bar{g}_2 + \bar{g}_3 - 2\rho\bar{v}_\alpha^2)} ,$$

where $\alpha = 1,2$. The solution (24) is not valid as $m\bar{q}_{1\alpha} \to \pi/2$ since that would indicate $|\rho\hat{v}_\alpha^2| \to \infty$ what violates the assumption that $\rho\hat{v}_\alpha^2 = O(1)$. Thus, in the case of the symmetric deformations, expressions for the phase velocity $\rho v_\alpha^2 = \rho\bar{v}_\alpha^2/\varepsilon^2 + \rho\hat{v}_\alpha^2$ ($\alpha = 1,2$), are valid for long waves only.

To obtain approximate expressions for the fundamental modes of the phase velocity valid in wide range of wavelengths we recall the result, obtained in [6], for corresponding material with inextensible fibres

$$\rho v^2 = d + c_{11}\pi^2 / 4k^2 h^2, \tag{25}$$

which shows that the cut-off frequency has value $\omega_c = (\pi/2h)\sqrt{c_{11}/\rho}$. The inextensible theory thus predicts that extensional waves with frequency lower than ω_c will not propagate in considered layer. Therefore, we are not able to construct the continuous solution for the phase velocity be- cause at the cut-off frequency, neither expressions for the phase velocity, (24) and (25) are valid. We may obtain, however, the intersection of sol- utions represented by (13) and (25) which is very close to cut-off fre- quency and construct the approximate phase velocity in the form (13) for $kh < kh_c$ and in the form (25) for $kh > kh_c$. Thus although for $kh < kh_c$ ex- pression (24) is more accurate than (13) we construct approximate phase velocity as a combination of expressions (13) and (25). When $m\overline{q}_{1\alpha} \to \pi/2$ from (23) we obtain $V_{0\alpha} \to 0$ and $W_{0\alpha} \to 0$ and displacements switch to the

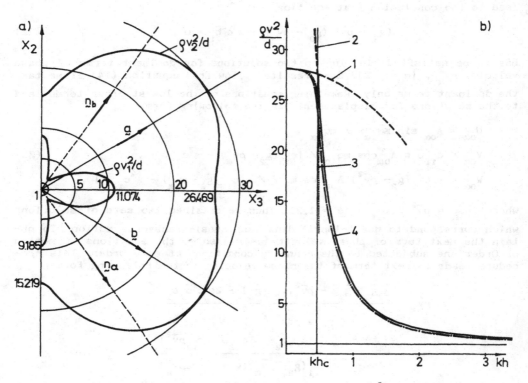

Figure 1. Phase velocity of extensional waves for $\phi=33°12'39''$; a) contours for $kh \to 0$; b) dispersion curves for $\alpha=18°26'6''$ (1- $kh \ll 1$, 2- inextensible, 3- $kh=m\epsilon$, 4- exact)

solutions that inextensible theory predicts in which $V=0$ and $W=0$ for all angles of propagation except when angle of propagation is perpendicular to one of the families of fibres. In that case the equation

$$s^2\sin^2\phi - c^2\cos^2\phi = 0 , \qquad (26)$$

is satisfied which implies

$$\overline{g}_2\overline{g}_3 - s^2c^2\overline{b}_1^2 = 0 , \qquad (27)$$

and (22) leads to the condition that leading term of the smaller phase velocity $\bar{v}_1 = 0$. Thus lower phase velocity is of Order one.

NUMERICAL RESULTS AND DISCUSSION

To illustrate approximate solutions we employ carbon fibres-epoxy resin composite for which nonzero material constants have following values

$$\lambda = 5.65 \text{ GPa} , \quad \mu = 2.46 \text{ GPa} , \quad \gamma_4 = -1.28 \text{ GPa} ,$$
$$\gamma_7 = 3.20 \text{ GPa} , \quad 4\gamma_1 = 220.90 \text{ GPa} . \tag{28}$$

These equations are chosen in such a way that if two families of fibres coincide then material has same elastic constants as those measured in [8] for composite reinforced by one family of the fibres.

Dispersion curves fog bending waves are given in [5] and [6], for material described in (28), and here we are going to illustrate approximate phase velocity for extensional waves.

Figure 1.a. shows contours of the phase velocities when $kh \to 0$ for both quasi-longitudinal and quasi-transverse motion, according to expression (13), when half-angle between two families of fibres has value $\phi = 33^{\circ}12'39''$. When wave propagates perpendicular to one of the families of fibres, i.e. along $\underset{\sim}{n}_a$ or $\underset{\sim}{n}_b$, the lower phase velocity has minimal value.

Dispersion curves, again for $\phi = 33^{\circ}12'39''$, and the angle of propagation $\alpha = 18^{\circ}26'6''$ are plotted on Figure 1.b.. Approximate curves 1, 2 and 3 of higher phase velocity are compared with exact solution 4, obtained in [6]. Intersection of curves 1 and 2 is assigned as kh_c and it indicates wavelength at which cut-off frequency occurs. Approximate phase velocity as a combination of expressions (13) and (25) is plotted as a solid line.

REFERENCES

1. Spencer, A.J.M., Constitutive Theory for Strongly Anisotropic Solids, in Continuum Theory of the Mechanics of Fibre-Reinforced Composites, ed. by A.J.M. Spencer, Springer-Verlag, Wien-New York, 1984, pp. 1-32.
2. Spencer, A.J.M., Deformation of Fibre-Reinforced Materials, Clarendon Press, Oxford, 1972.
3. Green, W.A., Bending Waves in Strongly Anisotropic Elastic Plates, Quart. J. Mech. Appl. Math., Vol. 35, 1982, pp. 485-507.
4. Green, W.A., Milosavljević, D., Extensional Waves in Strongly Anisotropic Elastic Plates, Int. J. Solids & Structures. Vol. 21., No. 4., 1985, pp. 343-353.
5. Milosavljević, D.I., Approximate Phase Velocity for Bending Waves in Elastic Plate Reinforced by two Families of Strong Fibres, in Mechanical Behaviour of Composites and Laminates, ed. by W.A. Green and M.V. Micunovic, Elsevier Appl. Sc., London and New York, 1987, pp. 110-119.
6. Milosavljević, D.I., Wave Propagation in Elastic Plates Reinforced by two Families of Fibres, Ph.D. thesis, Belgrade, 1986.
7. Milosavljević, D.I., Rayleigh Waves in Materials Reinforced by Two Families of Fibres, in Recent Developments in Surface Acoustic Waves, ed. by D.F. Parker and G.A. Maugin, Springer-Verlag, 1988, pp. 251-259.
8. Markham, M.F., Measurement of the Elastic Constants of Fibre Composites by Ultrasonics, Composites. 1, 1970, pp. 145-149.

MICROMECHANICAL MODELLING OF UNIDIRECTIONAL GLASSFIBER REINFORCED POLYESTER: EFFECT OF MATRIX SHRINKAGE

A. TEN BUSSCHEN
Laboratory of Engineering Mechanics
Faculty of Mechanical Engineering and Marine Technology
Delft, University of Technology, The Netherlands

ABSTRACT

In literature micromechanical predictions are given for the transverse strength of Unidirectional Glassfiber Reinforced Polyester. However, these predictions overestimate measured strengths. For this reason the properties on microscale are investigated experimentally and three effects appear to be important for the transverse strength that are not found in literature. One of these is shrinkage of the polyester matrix which is discussed in this article. Based on observations in practise this effect is thought to be responsible for the major reduction of the transverse strength. With Finite Element calculations the transverse strength is predicted by incorporating matrix shrinkage. For different fiber volume fractions these predictions can give a good agreement with experimental measurements.

INTRODUCTION

General.

Fiber reinforcement of plastics comprises potential for high strength and high stiffness at a relatively low specific weight. However, extensive use of these materials for structural applications is obstructed by the low transverse strength. This circumstance causes premature microcracking and delamination of fiber reinforced structures, especially in the neighbourhood of edges and joints [4, 8].

It is tried to explain the phenomenon of low transverse strength based on properties on microscale (properties of the matrix, of the interface, of the fiberarrangement, etc.). This micromechanical modelling is carried out by various authors to predict the low transverse strength e.g. [3, 6, 9]. However, these predictions overestimate measurements in an unacceptable manner, as shown in this article and in [9].

Properties on Microscale.
Obviously the assumptions for properties on microscale are too limited. For this reason for Unidirectional Glassfiber Reinforced Polyester (UD-GRP) these properties are investigated experimentally. From this the following phenomena appear to be important for the assumptions of micromechanical modelling:

* Polyester is a brittle material but the critical mode-I stress intensity factor (K_{Ic}) is too high to justify a fracture mechanics approach on microscale (K_{Ic}=0.6 MPa$\sqrt{m}$, [11]).
* Polyester has a non-linear stress-strain response. However, the material shows no plasticity and no creep deformation in the period tested [2]. The behaviour can be characterized with non-linear elasticity and hysteresis.
* During solidifying polyester shows a volumetric shrinkage of about 8.0% which corresponds to a linear shrinkage of 2.7%. Part of this shrinkage occurs in the socalled gel-phase when relaxation times are still very short [1, 12].
* The mechanical behaviour of the interface (between fiber and matrix) is complex but when a special coupling agent is present on the fiber the matrix is weaker than the interface [1].
* In transverse direction the fiber arrangement is random. Thus, the fiber volume fraction, v_f, will not be the theoretically highest possible (v_f=0.907, hexagonal). The highest fraction attainable is: $v_f \approx 0.75$.

Effects to investigate.
The effects of non-linearity of polyester, of shrinkage of polyester and of random fiber arrangement on the transverse

strength of UD-GRP are investigated. In this article the effect of matrix shrinkage is incorporated in the micro-mechanical modelling and is verified with experimental measurements that are carried out. Based on observations in practise it is supposed that it covers the major discrepancy between predictions of various authors and measurements. Micromechanical modelling is carried out by means of Finite Element calculations because of accurate results that can be obtained and also because of the flexibility to incorporate different effects [3, 6]. The FEM-predictions are experimentally verified by means of bending of a thick, curved beam of UD-GRP made by filament winding.

Unidirectional GRP.

Unidirectional Glassfiber Reinforced Polyester (UD-GRP) consists of parallel glassfibers (E-glass, M084, Silenka bv. [13]) embedded in a polyester matrix (isoftalic based, Synolite 593-A-2, DSM-Resins [13]). As an illustration, (approximate) values of tensile (t) strength and stiffness of UD-GRP (v_f=0.60), of glassfibers (f) and of polyester (m) are given in Table 1 [9, 13]. The longitudinal direction is denoted by 'L', the transverse direction by 'T'. Composite strength and stiffness are homogenized quantities and are therefore macroscopic properties. In order to distinguish, macrostresses are indicated by 'Σ' and microstresses by 'σ'.

TABLE 1

Measured strength and stiffness of UD-GRP and its components

	UD-GRP	glassfiber	polyester
strength	Σ_{Lt} = 1500. MPa Σ_{Tt} = 15. MPa	σ_{ft} = 2500. MPa	σ_{mt} = 87. MPa
stiffness	E_L = 45. GPa E_T = 10. GPa	E_f = 70. GPa	E_m = 3.5 GPa

MICROMECHANICAL CALCULATIONS

Basic Assumptions.

For Finite Element calculations in this article a regular square fiber arrangement is chosen. From microscopic observations of cross sections this appears to be a better approximation than a hexagonal arrangement. In the present calculations both components of the composite are regarded to be linear elastic until fracture. The fiber is assumed not to fail which is likely because of high fiber strength and is confirmed by observing fracture surfaces. Fibers and matrix are assumed to be bonded perfectly. Though the behaviour of the interface is complex [1], the coupling agent on the fibers can be chosen to make matrix failure occur earlier than interface failure [1]. From biaxial experiments on polyester in our laboratory it appears that fracture of the matrix can be described by a criterium for polymers expressed in principal stresses ($\sigma_1 \geq \sigma_2 \geq \sigma_3$) as suggested in [12]. Variables that are necessary to formulate this criterium are the (uniaxial) tensile strength (87 MPa) and the compressive strength (140 MPa) [13] which yield the following result:

$$(\sigma_1 - \sigma_3)/2 + 0.12 \ (\sigma_1 + \sigma_2 + \sigma_3) = 53.7$$

Strength of the composite is defined to be the lowest macroscopic transverse stress (Σ_{Tt}) at which the matrix fails according to this criterium.

Incorporating Matrix Shrinkage.

As already indicated, polyester shows a linear shrinkage of about 2.7% during solidifying. However, stresses occurring due to shrinkage will partly vanish in the gel-phase due to relaxation. Although this process is extremely complex, it is assumed that only a part of the linear matrix shrinkage is effective to yield shrinkage stresses. Four different values of linear matrix shrinkage are incorporated in Finite Element calculations (0%, 1.0%, 1.2% and 1.4%), and are carried out by means of initial strains (ε_{0m}).

Numerical Results.

Based on symmetry conditions the calculation is restricted to a unit cell containing a quarter of a fiber cylinder (the calculations could even be restricted to a smaller unit cell). The normal displacements of the six faces of the cell have a constant value on every face. The Finite Element calculations are carried out with the standard program DIANA [7], using 8-node and 6-node linear interpolation volume elements, see Figure 1.

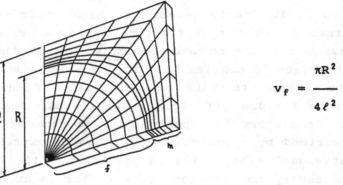

$$v_f = \frac{\pi R^2}{4\ell^2}$$

Figure 1. Finite Element mesh of the unit cell.

Values of the transverse tensile strength (Σ_{Tt}) for different fiber volume fractions (v_f) and for different values of linear matrix shrinkage (expressed in ε_{om}) are given in Figure 2.

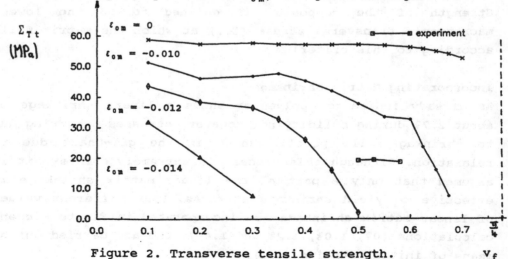

Figure 2. Transverse tensile strength.

RESULTS FROM EXPERIMENTS.

Testing Principle.

In the Laboratory of Engineering Mechanics thick curved beams of UD-GRP are made by filament winding as drawn in Figure 3. By pulling apart this specimen as indicated in Figure 3 a maximum bending moment occurs at the opposite side of the specimen (indicated by 'part in consideration').

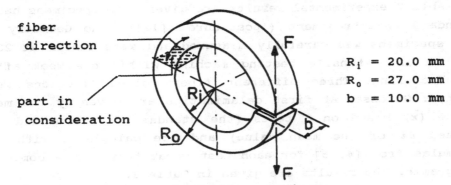

R_i = 20.0 mm
R_0 = 27.0 mm
b = 10.0 mm

Figure 3. Testing specimen.

The test was recently proposed by Kedward [4] and is based on the fact that when a thick, curved beam is loaded by a bending moment, tensile stresses occur both in tangential (θ) and in radial (r) direction, see Figure 4.

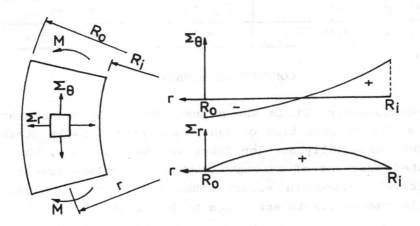

Figure 4. Stresses in a thick, curved beam.

In the part in consideration the highest bending moment occurs and due to relatively high longitudinal strength and stiffness of UD-GRP (see Table 1), transverse fracture is controlled by the maximum radial tensile stress, Σ_r. From Leknitskii [5] the analytic expressions for the radial and tangential stresses in a curved, unidirectional material are known.

Experimental Data.

In Table 2 experimental results are given. The specimens had a standard heat-treatment ('post-cure' [12]). The Geometry of the specimens was carefully measured and were tested at 23C, 50% rh in a tensile testing machine within one week after manufacturing. Three different fiber volume fractions were tested. Stresses at first delamination are given by the mean value ($\bar{x}$) based on 8 tests (the standard deviation did not exceed 8% of the mean value) and are calculated with the formulas from [4, 5] for each test separately with a computer programme. The results are given in Table 2.

TABLE 2

Test Results at first delamination

v_f	$\Sigma_{Tt} = \Sigma_r(\max)$ (MPa)	$\Sigma_\theta(r=R_i)$ (MPa)	$\Sigma_\theta(r=R_o)$ (MPa)
0.500	$\bar{x} = 19.19$	$\bar{x} = 300.$	$\bar{x} = -250.$
0.529	$\bar{x} = 19.35$	$\bar{x} = 300$	$\bar{x} = -250.$
0.580	$\bar{x} = 18.63$	$\bar{x} = 300$	$\bar{x} = -250.$

CONCLUDING REMARKS

By FEM-calculations it is shown that shrinkage of the matrix yields a strong reduction of the transverse tensile strength of UD-GRP, especially at high fiber volume fractions. The calculated decrease of strength with fiber volume fraction is qualitatively found in experiments. With this the initial strain in the matrix is estimated to be -1.1%.

REFERENCES

1. Berg, R.M. van der
   Fragmentatieproeven met Glasvezel/Polyester en Glasvezel/
   Epoxy. Report LTM-918, July 1990, Engineering Mechanics,
   TU-Delft, The Netherlands.

2. Busschen, A. ten
   Creep Properties of Synolite 593-A-2. Report LTM-906,
   February 1990, Engineering Mechanics, TU-Delft.

3. Guild, F.J., Hogg, F.J., Davy, P.J.
   A predictive Model for the Mechanical Behaviour of Transverse Fibre Composites. IMechE, April 1990, UK.

4. Kedward, K.T., Wilson, R.S., McLean, S.K.
   Flexure of Simply Curved Composite Shapes.
   Composites, Vol. 20 No. 6, November 1989, pp. 527-536.

5. Leknitskii, S.G.
   Anisotropic Plates.
   Gordon and Breach Science Publishers, 1968.

6. Moran, B., Gosz, M., Achenbach, J.D.
   Effect of a Viscoelastic Interfacial Zone on the
   Mechanical Behaviour and Failure of Fiber-Reinforced
   Composites. IUTAM-Symposium on Inelastic Behaviour of
   Composites, June 1990, Troy, New York, USA.

7. Standard programme DIANA. Release 3.2.
   TNO-IBBC, Rijswijk, The Netherlands.

8. Pagano, N.J., Pipes, R.B.
   The Influence of Stacking Sequence on Laminate Strength.
   J. of Comp. Mat. 5, 1971, pp. 50-71.

9. Schneider, W.
   Mikromechanische Betrachtungen von Bruchkriterien uni-
   direktional verstarkten Schichten aus Glassfaser/Kunst-
   stoff. Thesis, D17, 1974, Darmstadt, Germany.

10. Whitney, J.M., Daniel, I.M., Pipes, R.B.
    Experimental Mechanics of Fiber Reinforced Composite
    Materials. SEM, 1984, USA (ISBN 0-912053-01-1).

11. Williams, J.G.
    Fracture Mechanics of Polymers.
    Ellis Horwood Limited, 1984.

12. Young, R.J.
    Introduction to Polymers. Chapman and Hall, 1981.

13. Brochure E-Glass. Silenka bv., 1990, The Netherlands.
    Brochure Synolite. DSM-Resins, 1990, The Netherlands.

PLASTIC FAILURE OF UNIDIRECTIONAL FIBROUS COMPOSITE MATERIALS WITH METAL MATRIX IN COMPRESSION

GUZ A.N.
Institute of Mechanics, Nesterov str. 3, Kiev,
252057, USSR

ABSTRACT

The systematic presentation is given of foundations of continual theory in case of plastic failure of unidirectional composite materials with metal matrix in compression, comparative evaluation is presented of theoretical and experimental results for two different composite materials.

INTRODUCTION

It is known that in compression of unidirectional composite materials the stability loss in material structure in many cases is the major failure mechanism which determines completely the initiation of failure. Continual theory of compressive failure of composite materials with elastic-plastic matrix was originally proposed in [3]. This theory is the generalization and development of continual theory of brittle fracture [1] with application to composite materials with metal matrix - since it is necessary to account for effects of plastic deformations. Various aspects of this theory were subsequenty discussed in [4-6,10,12,13] with application to the case of uniaxial compression. The development of this theory with application to the case of combined stressed state is presented in [8]. In the report the continual theory of compressive failure of unidirectional fibrous composite materials with metal matrix is presented.

MAIN RELATIONS AND STATEMENTS

We consider fibrous unidirectional composite material with metal matrix, its fibers are directed along the OX_3-axis (Figure 1), under uniaxial compression along the OX_3-axis.

In continual approximation in the plane $x_3 =$ const (Figure 1)
material may be assumed isotropic. A characteristic feature
of composite materials with metal matrix is deformation of
matrix in plastic region. In accurate modelling, in continual
approximation the composite material with metal matrix is
reasonable to assume as compressible, since the property of
compressibility is evident in the filler or in the matrix. In
view of foregoing considerations we discuss main statements
of continual theory [3] with application to unidirectional
fibrous composite materials.

1. In continual approximation, the fibrous unidirectional composite material (Figure 1) is modelled by transversely-isotropic (ox_3 -axis is the axis of isotropy), elastic-
plastic compressible (in accurate modelling) and incompressible (in approximate modelling) body.

2. Compressive failure occurs as a result of loss of
stability in composite material structure (interior instability [11]). In this process levels of failure loads (theoretical strength limit) are determined from condition of transition of main system of equations in system of hyperbolic
type.

3. The analysis of interior instability is made within
the framework of three-dimensional linearized theory of deformable bodies stability (see, for example, [7]), with application of the second variant of the small precritical deformations theory (precritical states are small and precritical state is determined with the use of geometrically linear
theory).

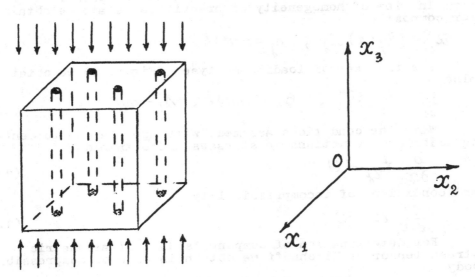

Figure 1. Loading of composite material

4. In the analysis of stability of elastic-plastic body
(within the framework of three-dimensional linearized theory

of deformable bodies stability) the generalized conception of continuing loading is used [2,7] (of Shenley type). Consequently, the change of unloading zones in the process of loss of stability is not accounted for.

5. In the analysis of stability static method is used of analysis of problems of three-dimensional linearized theory of deformable bodies stability, since it is proved strictly [7] that in the model considered under the action of "dead" loads - sufficient conditions of applicability of this method are satisfied.

6. The loading is assumed to be applied in such a manner that in deformation along OX_3-axis (Figure 1) in precritical state following conditions are satisfied

$$\varepsilon^{o}_{33}{}^{a} = \varepsilon^{o}_{33}{}^{m} \tag{1.1}$$

Here and in the following by index "0" values of precritical states are denoted, by index "a" all values are denoted referred to the filler (to reinforcing elements), by index "m" all values are denoted referred to the binder (to matrix). The equality (1.1) follows from the condition that along the OX_3-axis (Figure 1) in precritical state equal shortenings for reinforcing elements and matrix are prescribed.

7. For the case considered (Figure 1) it is assumed that the precritical state is a homogeneous one. This condition is satisfied strictly, when filler and binder are modelled by incompressible bodies, in all other cases this condition is only approximately satisfied. Taking into consideration the foregoing statements we now analyse main relations.

In view of homogeneity of precritical state we obtain for composite

$$u^{o}_{j} = (\lambda_j - 1)\, x_j \; ; \quad \lambda_j = const \tag{1.2}$$

For the case of loading analysed (Figure 1) we obtain also

$$\sigma^{o}_{jj} = \delta_{j3}\, \sigma^{o}_{33} \; ; \quad \sigma^{o}_{33} = const \; ; \quad \lambda_1 = \lambda_2 \tag{1.3}$$

With the conditions assumed, with application to central equilibrium - equations in stresses are obtained

$$\frac{\partial}{\partial x_i}\, t_{ij} = 0 \tag{1.4}$$

and conditions of incompressibility

$$\frac{\partial}{\partial x_i}\, u_i = 0 \tag{1.5}$$

For determination of components t_{ij} of asymmetric stress tensor of Kirchhoff we obtain in case of compressible body

$$t_{ij} = \omega_{ij\alpha\beta}\, \frac{\partial u_\alpha}{\partial x_\beta} \; ; \quad \omega_{ij\alpha\beta} = \delta_{ij}\, \delta_{\alpha\beta}\, A_{i\beta} + (1 - \delta_{ij})(\delta_{i\alpha}\delta_{j\beta} +$$

$$+ \delta_{i\beta}\delta_{j\alpha})\mathcal{M}_{ij} + \delta_{i\beta}\delta_{j\alpha}\sigma_{\beta\beta}^{0} \; ; \quad \omega_{ij\alpha\beta} = const \qquad (1.6)$$

and in case of incompressible body

$$t_{ij} = \mathcal{X}_{ij\alpha\beta}\frac{\partial u_{\alpha}}{\partial x_{\beta}} + \delta_{ij}P \; ; \quad \mathcal{X}_{ij\alpha\beta} = \delta_{ij}\delta_{\alpha\beta}A_{i\beta} + (1-\delta_{ij})\times \qquad (1.7)$$

$$\times(\delta_{i\alpha}\delta_{j\beta} + \delta_{i\beta}\delta_{j\alpha})\mathcal{M}_{ij} + \delta_{i\beta}\delta_{j\alpha}\sigma_{\beta\beta}^{0} \; ; \quad \mathcal{X}_{ij\alpha\beta} = const$$

Values $A_{i\beta}$ and $\mathcal{M}_{ij}$ for various compressible and incompressible bodies are determined from relations [7]. In accordance to [7] solutions of equations (1.4)-(1.7) are expressed by solutions of following equations

$$\left(\Delta_{1} + \xi_{1}^{2}\frac{\partial^{2}}{\partial x_{3}^{2}}\right)\psi = 0 \; ; \quad \Delta_{1} = \frac{\partial^{2}}{\partial x_{1}^{2}} + \frac{\partial^{2}}{\partial x_{2}^{2}} \; ; \qquad (1.8)$$

$$\left(\Delta_{1} + \xi_{2}^{2}\frac{\partial^{2}}{\partial x_{3}^{2}}\right)\left(\Delta_{1} + \xi_{3}^{2}\frac{\partial^{2}}{\partial x_{3}^{2}}\right)\mathcal{H} = 0$$

In view of foregoing considerations the values of failure loads are determined from following condition: in solutions of equations (1.8) should appear solutions of equations of hyperbolic type. It should be remarked that values ξ_{j}^{2} ($j = 1,2,3$) (1.8) are determined from components of tensors ω and $\mathcal{X}$ with the use of known expressions [7]. In the following analysis we assume that the conditions are satisfied

$$\mathcal{I}m\,\xi_{\xi_{j}}^{2} = 0 \; ; \quad j = 1,2,3 \qquad (1.9)$$

for specific models of compressible bodies [7]. When conditions (1.9) are satisfied, in solutions of system (1.8) solution of hyperbolic type equations will appear, if at least for one value of the condition is satisfied $Re\,\xi_{j}^{2} < 0$. For this reason values of failure loads (corresponding to boundary of the region where $Re\,\xi_{j}^{2} < 0$ are satisfied) are determined from the condition that at least for one value $j = 1,2,3$ the relation holds

$$\xi_{\xi_{j}}^{2} = 0 \qquad (1.10)$$

Equations (1.10) taking into account expressions for determination of $\xi_{\xi_{j}}^{2}$ [7] lead to equation

$$(A_{33} + \sigma_{33}^{0})(\mathcal{M}_{13} + \sigma_{33}^{0}) = 0 \qquad (1.11)$$

For the case represented in Figure 1, taking into account notations (1.6) and (1.7) the inequality holds

$$A_{33} > \mathcal{M}_{13} \qquad (1.12)$$

From (1.11) taking into account (1.12) following equation is obtained

$$\mathcal{M}_{13} + 6_{33}^{o} = 0 \tag{1.13}$$

It should be noted, that for non-linear models (in the case considered - for elastic-plastic body) the relation holds

$$\mathcal{M}_{13} = \mathcal{M}_{13}(6_{33}^{o}) \tag{1.14}$$

Consequently (1.13) together with (1.14) is the equation for determination of magnitudes of failure loads, the relation (1.14) is computed for every specific model.

For these reasons it is required to determine the value $\mathcal{M}_{13}$ (1.14) for composite material in elastic-plastic deformation model discussed here. In accordance with (1.6) and (1.7) the value $\mathcal{M}_{13}$ is used in the sense of shear modulus in the plane $x_3 0 x_1$ or $x_3 0 x_2$ (transversely-isotropic body in view of statement 1), computed at the moment of stability loss. As a result of statement 7, precritical state in the filler and the binder is homogeneous; consequently, separately for the filler and the binder relations (1.6) and (1.7) hold, it is only necessary in all values to write the index "a" or "m". Thus the values $\mathcal{M}_{13}$, $\mathcal{M}_{13}^{a}$ and $\mathcal{M}_{13}^{m}$ are used in the sense of shear moduli in the plane $x_3 0 x_1$ or $x_3 0 x_2$, they are computed at the moment of stability loss. For this reason, for computing $\mathcal{M}_{13}$ using $\mathcal{M}_{13}^{a}$ and $\mathcal{M}_{13}^{m}$, expressions for elastic bodies may be used. However computations of $\mathcal{M}_{13}$, $\mathcal{M}_{13}^{a}$ and $\mathcal{M}_{13}^{m}$ should be made at the moment of stability loss. In the following, we introduce notations S_a and S_m - concentrations of reinforcing elements and matrix.

DETERMINATION OF THEORETICAL STRENGTH LIMIT FOR SPECIFIC MODEL

On the basis of the results of previous paragraph we consider the determination of theoretical strength limit in compression along the $0x_3$-axis. We determine the value (Π_3^{-})$_T$ and the theoretical value of ultimate shortening $\mathcal{E}_T = \mathcal{E}_{33}^{o}$ with application to the case when filler and binder are modelled by incompressible isotropic elastic-plastic body (deformational theory with power relation between stresses and deformations intensities). In this case following relations hold (for the model considered)

$$6_{ij}^{a,m} = \delta_{ij}\,\hat{6}^{a,m} + \frac{2}{3}\,\frac{6_u^{a,m}}{\mathcal{E}_u^{a,m}}\,\mathcal{E}_{ij}^{a,m} \; ; \quad \mathcal{E}_{ii} = 0 \; ;$$

$$6_u^{a,m} = A_{a,m}\,\mathcal{E}_u^{a,m\,k_{a,m}} \; ; \quad A_{a,m}, k_{a,m} = const \tag{2.1}$$

It should be pointed out that the model considered includes cases when filler is modelled by incompressible isotropic elastic body; in this case in (2.1) we assume

$$A_a = E_a \; ; \quad k_a = 1, \tag{2.2}$$

where E_a – Young modulus.

It should be remarked that in the case of contitutive relations of the form (2.1) in [9] it is shown (in our case – separately for filler and binder) that conditions (1.9) are not satisfied and the analysis of the problem of internal instability should be made by procedure, different from that used in previous paragraph. For this reason doubts may arise on the possibility of application of results of previous paragraph to (2.1). These doubts are unfounded for following two reasons. First – in the present paragraph relations are used for isotropic filler and binder in the form of (2.1). In this case the composite becomes transversally-isotropic and for such material relations will differ from (2.1); consequently, results of [9] are not applicable to composite. Second – in the present paragraph not the method of internal instability analysis is used, considered in previous paragraph-but the exact solution in the form of (1.13), which has the closed form for compressible material and from which as limit results – relations for an incompressible material should follow.

For the case considered (Figure 1) and for relations (2.1) taking into accountthe main statements of the theory discussed – following relations hold

$$\mathcal{E}_{33}^{o} \equiv \mathcal{E}_{33}^{o\,a} \equiv \mathcal{E}_{33}^{o\,m} = -\mathcal{E} \; ; \quad \mathcal{E}_{11}^{o\,a} = \mathcal{E}_{22}^{o\,a} \; ; \quad \mathcal{E}_{11}^{o\,m} = \mathcal{E}_{22}^{o\,m} \; ;$$

$$6_{11}^{o\,a} \equiv 6_{22}^{o\,a} \equiv 6_{11}^{o\,m} \equiv 6_{22}^{o\,m} = 0 \; ; \quad \mathcal{E} = const \; ; \tag{2.3}$$

$$6_{33}^{o\,a} = const \; ; \quad 6_{33}^{o\,m} = const$$

Taking into account the incompressibility condition for the precritical state from (2.1) and (2.3) we obtain

$$\mathcal{E}_{11}^{o\,a} \equiv \mathcal{E}_{22}^{o\,a} \equiv \mathcal{E}_{11}^{o\,m} \equiv \mathcal{E}_{22}^{o\,m} = -\frac{1}{2}\mathcal{E}_{33}^{o\,a} = -\frac{1}{2}\mathcal{E}_{33}^{o\,m} =$$

$$= -\frac{1}{2}\mathcal{E}_{33}^{o} = \frac{1}{2}\mathcal{E} \; ; \quad \mathcal{E}_{u}^{o\,a} \equiv \mathcal{E}_{u}^{o\,m} = \mathcal{E} \tag{2.4}$$

Taking into account (2.1), (2.3) and (2.4) for the case considered we obtain

$$6_{33}^{o\,a,m} = -A_{a,m}\,\mathcal{E}^{k_{a,m}} \; ; \quad \mathcal{M}_{13}^{a,m} = \frac{1}{3}A_{a,m}\,\mathcal{E}^{k_{a,m}-1} \tag{2.5}$$

In accordance with (2.5) for theoretical ultimate strength the expression is obtained

$$(\Pi_3^-)_T = \mathcal{E}_T\,S_a\,A_a\left(\mathcal{E}_T^{k_a-1} + \frac{A_m\,S_m}{A_a\,S_a}\mathcal{E}_T^{k_m-1}\right) \tag{2.6}$$

where ε_T – theoretical value of maximum shortening along Ox_3 –axis (failure deformation) is determined, in view of (2.3) and (2.4), by following expression

$$\varepsilon_T = \varepsilon = -\varepsilon_{33}^o = -\varepsilon_{33}^{o\,a} = -\varepsilon_{33}^{o\,m} \tag{2.7}$$

Upon substitution of (2.3)-(2.5) into (1.13) and taking account of (1.5), after simplifications we obtain the equation for determination of ε_T in the form

$$\varepsilon_T^{1+k_a-k_m} + \varepsilon_T\, B - B_2 = 0 \tag{2.8}$$

COMPARISON OF THEORETICAL AND EXPERIMENTAL RESULTS

We determine the value $(\Pi_3^-)_T$ – theoretical ultimate strength and the theoretical value of ultimate shortening $\varepsilon_T = -\varepsilon_{33}^o$ with application to the case when binders are modelled by incompressible isotropic elastic-plastic body (deformational theory with power relation between stresses and deformations intensities). A comparison will be presented of theoretical results for determination of theoretical value of longitudinal shortening and of theoretical ultimate strength (in first approximation) for two different composite materials with metal matrix.

First material [14]. Filler – steel wire, binder – aluminium. In [14] results of experimental studies are presented for various concentrations of the filler (value S_a); we present the comparison only for the case $S_a \geqslant$ 10% since at very low concentration of the filler the continual theory is evidently incapable to describe satisfactorily the processes considered. In computations values were assumed (taking into account the information presented)

$$E_a = 2\cdot10^5 \text{ MPa}; \quad A_m = 100 \text{ MPa}; \quad k_m = 0,25 \tag{3.1}$$

Results of comparative evaluation at $10\% \leq S_a \leq 35\%$ are presented in Figure 2, where the solid line represents values of theoretical ultimate strength, the dashed line shows the theoretical values of maximal shortening. In Figure 2 results of experimental studies [14] are also presented: "circles" denote the maximal shortening, "stars" – the ultimate strength.

Second material [10]. Filler – boron fibers, binder – aluminium. In [10] the results of experimental studies of two cases are presented: case A – annealed aluminium, case B – non-annealed aluminium. Following values were used in computations (detailed presentation in [10])

$$E_a = 400 \text{ GPa}; \quad \text{A:} \quad A_m = 130 \text{ MPa}, \quad k_m = 0,43;$$
$$\text{B:} \quad A_m = 70 \text{ MPa}, \quad k_m = 0,25. \tag{3.2}$$

In [10] experimental results were obtained for large

number of specimens, for this reason in comparison of theoretical and experimental results maximal and minimal experimental values are considered as well as average values for the total series of specimens. Comparison of experimental and theoretical results is shown [10] in the Table.

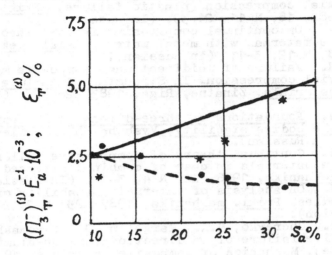

Figure 2. Dependence of theoretical ultimate strength and of theoretical value of maximum shortening – on filler concentration

TABLE

Material	(Π_3^-) experim , MPa			$(\Pi_3^-)_T$ MPa
	max.	min.	average	
annealed	965	501	665	736
non-annealed	1716	1049	1282	1467

It follows from the Table, that a good agreement was found between average experimental values and theoretical values. Here some inaccuracies are eliminated which were found in previously published articles.

REFERENCES

1. Guz, A.N., On determination of theoretical compressive ultimate strength of reinforced materials. Doklady AN Ukr. SSR, 1969, series A, N 3, 236-38. (In Ukrainian).
2. Guz, A.N., Stability of three-dimensional deformable bo-

dies, Naukova Dumka, Kiev, 1973, 272 p. (In Russian).

3. Guz, A.N., On continual theory of compressive failure of composite materials with elastic-plastic matrix. Doklady AN USSR, 1982, 262, N 3, 556-60. (In Russian).

4. Guz, A.N., On failure mechanics of composite materials under axial compression (plastic failure). Prikl. mekhanika, 1982, 18, N 11, 21-39. (In Russian).

5. Guz, A.N., On continual compressive failure theory of composite material with metal matrix. Prikl. mekhanika, 1982, 18, N 12, 3-11. (In Russian).

6. Guz, A.N., Failure of unidirectional composite materials under axial compression. In Strength and failure of composite materials, Zinatne, Riga, 1983, 284-92. (In Russian).

7. Guz, A.N., Foundations of three-dimensional theory of deformable bodies stability, Vyshcha Shkola, Kiev, 1986, 512 p. (In Russian).

8. Guz, A.N., Continual theory of compressive failure of composite materials in case of combined stressed state. Prikl. mekhanika, 1986, 22, N 4, 3-19. (In Russian).

9. Guz, A.N., On analysis of internal instability of deformable bodies. Prikl. mekhanika, 1987, 23, N 2, 24-38. (In Russian).

10. Guz, A.N., Cherevko, M.A., Margolin, G.G., Romashko,J.M., Compressive failure of unidirectional boron-aluminium composites. Mechanics of composite materials, 1986, N 2, 226-30. (In Russian).

11. Biot, M.A., Mechanics of incremental deformation, Wiley, N.Y., 1965, 506 p.

12. Guz, A.N., Fracture of unidirectional composite materials under the axial compression. In Fracture of composite materials, Proc. of Second USA-USSR Symp., Bethlehem, 1981, Martinus Nijhoff Publ., 1982, 173-82.

13. Guz, A.N., Cherevko, M.A., Fracture mechanics of unidirectional fibrous composites with metal matrix under compression. Theor. and Appl. Fracture Mechanics, 1985, 3, N 2, 151-55.

14. Pinnel, M.R., Lawley, A., Correlations of uniaxial yielding and substructure in aluminium-stainless steel composites. Metallurg. Frans, 1970, 1, N 5, 1337-48.

ANALYTICAL MODEL OF FIBRE PULL-OUT MECHANISM

JANUSZ POTRZEBOWSKI
Institute of Fundamental Technological Research
Polish Academy of Sciences, 00-049 Warsaw, Świętokrzyska 21, Poland

ABSTRACT

The main assumption for the proposed model is that the interfacial stress τ is related to the fibre-matrix slip and can be treated as one combined stress field regardless of the nature of that stress: bond or friction. To determine this relation a series of tests was performed in which fibre-matrix slips were measured and interfacial stress distributions were derived. The searched bond-slip relation has been found out as a function of two variables: slip s and distance x along the embedment zone. Agreement of test and model simulation results confirms the assumptions of the model.

INTRODUCTION

The most important among the fracture mechanisms which can be distinguished in fibre reinforced concrete elements seem to be the debonding and pulling out the fibres from the matrix. These two mechanisms are mainly responsible for the reinforcing of the matrix by the fibres in the FRC. Up to now analytical approaches of debonding and pull-out mechanisms were analysed separately and then combined into an overall behaviour of the fibre matrix interface.

To analyse the former mechanism of fracture, namely the interfacial debonding, two principal approaches can be applied. The first one is based on fracture mechanics according to Griffith criterion, i.e. an interfacial crack between fibre and matrix propagates when the rate of energy release is just equal to the critical fracture energy consumption rate. In this approach no assumptions regarding shear stress distribution at the interface are necessary, however the compliance of the fibre-matrix system has to be predicted, [1],[2],[3].

The second approach assumes that debonding is initiated when the shear stress in the interface between fibre and matrix (bond stress) exceeds the interfacial shear strength. To model the process of debonding using this approach the real distribution of bond stresses along the embedded fibre should be known. Up to now no experiments have been performed in which the real bond stresses inside the fibre-matrix interface were measured. The measurement methods applied successfully to conventional reinforced concrete, i.e. methods based on strain gauges are not quite appropriate for

FRC because of small dimensions of reinforcement. In several published models, taking into account the lack of suitable test results, the distribution of bond stress was merely theoretically predicted or the relation between the bond stress and fibre-matrix slip was intuitively assumed, usually as a simple linear function $\tau = ks$, where: τ - interfacial stress, k - const. and s - slip, [4],[5],[6].

The latter fracture mechanism, namely fibre pull-out, is dominated by frictional shear stress acting over the debonded interface, both during and after debonding. In most models published by several authors this frictional stress was assumed as uniformly distributed over the debonded length. Such assumption was not confirmed by any reliable tests up to now.

The comparison of these two approaches to analyse the fibre debonding indicates that the second one based on the shear strength criterion seems to be more suitable for analytical modelling. This approach allows to analyse both debonding and pull-out against friction as one combined field of interfacial stresses without any precise determination which of these mechanisms actually occurs.

The paper presents an analytical model based on the combined stress field. To create that model the real interfacial stress distribution and relation $\tau = f(s)$ between interfacial stress and fibre-matrix slip are required. Therefore a special measurement method has been employed.

TEST METHOD AND RESULTS

To obtain the interfacial stress distribution and the relation $\tau = f(s)$ the speckle photography with double exposure was used. This method assuring high accuracy allows to measure under increasing loading the increment of fibre displacements with respect to the matrix. More details on the test method and preparation of the specimens are given in [7]. It is enough here to mention that the loading has been carried out stepwise and the measurements were performed after each successive loading step. Each specimen was cast of cement paste with a single plain fibre exposed on its external face. The discrete values of the fibre-matrix slip at successive steps of loading have been measured with precision of fractions of one micrometer. All data and diagrams given below are examples and concern one of the tested specimens. Then, for each level of loading these discrete values were approximated by continuous curves which represented slip distributions along the fibre, Fig. 1. It was observed that the entire debonding of fibres understood as non-zero slips of embedded fibre ends occurred before the pull-out load reached 30÷50% of its maximum value. After entire debonding the pull-out process was assured only by friction. One of the conclusion of the test is that the friction plays the crucial role in load transfer between fibre and matrix.

Then, by double differentiation of the slip distributions with respect to the embedment length the interfacial stress distributions were determined, Fig. 2.

Appropriate slip and interfacial stress distributions at the same levels of loading combined together produced the reliable relation between interfacial stress and fibre-matrix slip, Fig. 3. That is represented by a set of curves, one for each loading step, and not by a single curve, as it might be expected.

The main result of the tests was that a general relation between interfacial stress and fibre matrix slip cannot be represented as a function of one variable. Therefore, it is proposed to define it as a function of two variables: $\tau = f(s, x)$, where x is the distance along the embedment zone.

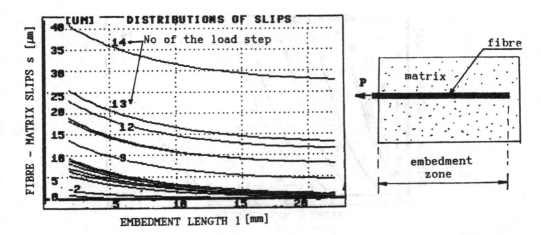

Figure 1. Distribution of slips along the embedment zone at successive steps of loading.

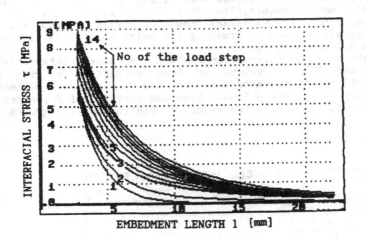

Figure 2. Distribution of interfacial stress along the embedment zone at successive steps of loading.

DETERMINATION OF FINAL MODEL FORMULA $\tau = f(s, x)$

The analysis of interfacial stress – slip relation plots enables to understand that for one set of curves (see Fig. 3) an arbitrary point at one curve and its counterparts at all other curves of that set represent the discrete local relation between interfacial stress and slip for one point of fibre–matrix interface at successive steps of loading. Fig. 4 shows such local interfacial stress – slip relations for four arbitrary points A, B, C, D which are situated as indicated in Fig. 4. Points in Fig. 4

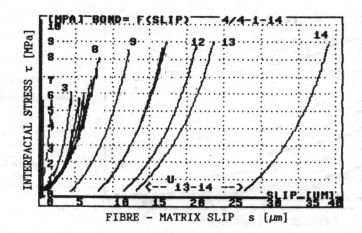

Figure 3. Relation interfacial stress versus slip.

representing the discrete values were linked by sections of line only to make the plot more readable. Therefore it is clear that the interfacial stress depends upon two variables: namely the slip s and the distance x along the embedment zone determining the point where the local stress is defined, i.e.: $\tau = f(s, x)$.

The shape of the curves representing the local interfacial stress - slip relations (Fig. 4) depends on bond properties of the fibre-matrix interface which is characterised by two parameters: ultimate slip s_u corresponding to maximum pull-out load and coefficient β proportional to the rate of interfacial stress increment in relation to slip growth. The curves

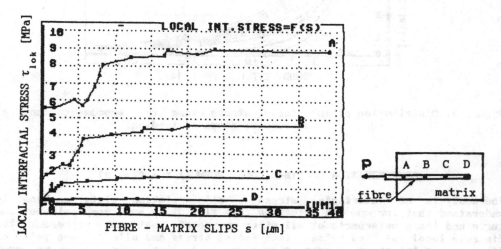

Figure 4. Local interfacial stress - slip relations for 4 arbitrary points A,B,C,D selected along the embedment zone.

can be defined by formula: $\tau = tgh(3.5 \ (s/s_u)^\beta) \ \tau_{maxloc}$, where τ_{maxloc} is maximum local interfacial stress.

The relation between the maximum local interfacial stress and distance x determining where the local stress is defined can be described by the function: $\tau_{maxloc} = \tau_{max} \ exp(-3.38 \ x/l)$, where l is the length of embedment zone and τ_{max} is treated as a parameter which is the maximum value of the interfacial stress, (Fig. 4).

Combining two above determined formula global formula as a function of two variables: s and x is proposed in the following form:

$$\tau = f(s,x) = \tau_{max} \ tgh(3.5 \ (s/s_u)^\beta) \ exp(-3.38 \ x/l) \qquad (1)$$

This formula can be understood as the constitutive equation of the interface in the proposed model valid up to the maximum pull-out load.

DESCRIPTION OF THE MODEL

The embedment zone is composed of two material phases, (Fig. 5):
- ductile fibre which is treated as a linear elastic bar with the cross-sectional area A and perimeter S embedded along the length l,
- brittle matrix - treated as a rigid halfspace.
The problem is considered as one-dimensional, Poisson ratio is assumed equal to 0.

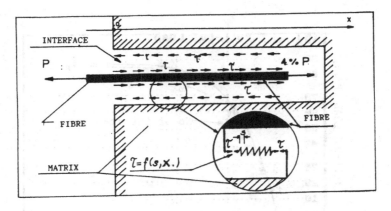

Figure 5. Model of the embedment zone.

The fibre is subjected to the pull-out load P, stress tangent to the lateral surface of fibre equal to the bond or frictional stress, and small force equal to about 4% of pull-out load P at the end of the fibre. The debonding and pull-out mechanisms can be considered as a common unstable state of interfacial stress field regardless of the nature of that stress i.e. as a homogeneous and continuous process in which the adhesive and cohesive bond is replaced continuously by friction. Therefore, it is not necessary to divide the embedment zone into two regions, i.e.: bonded region where interaction is assured by bond and debonded region in which only friction exists.

The interaction between fibre and matrix is assured by constraints imitating the interface properties. The behaviour of the constraints is determined by the stress τ – slip s relation which is given by formula (1). From the equilibrium condition for an infinitesimal section of the fibre it is obtained:

$$A \ E_f \ \frac{d^2 \ s(x)}{dx^2} = f(s(x),x) \ S \ dx \qquad (2)$$

where: A,S – area and perimeter of the fibre cross-section,
$\quad$ E_f – Young modulus of the fibre,

The function s(x) determining slip distribution along fibre is the solution of proposed model. The determination of this function allows to obtain the force distribution along the fibre and the interfacial stress distribution along the embedment zone.

VERIFICATION OF THE MODEL

The verification of the model was performed by numerical solution of the equation (1) to find out the function s(x). Pull-out force or displacement of free end of a fibre were considered as input data. Fig. 6 shows a plot typical for pull-out test, i.e. the relation between pull-out load and displacement of fibre free end. Discrete test results are plotted as points, whereas continuous curve represents the model solution. Good agreement of

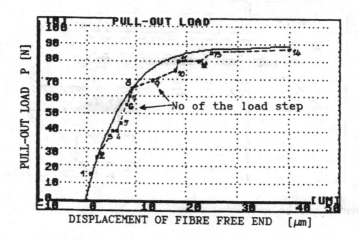

Figure 6. Relation pull-out load versus displacement of fibre free end, points – test result, curve – model simulation.

test and model results is found out. Also good agreement was obtained by comparison of diagrams of distribution of slips, internal forces and interfacial stresses along the embedment. Comparison of test slip distributions with model simulation results for several arbitrary selected steps of loading is shown as an example in Fig. 7.

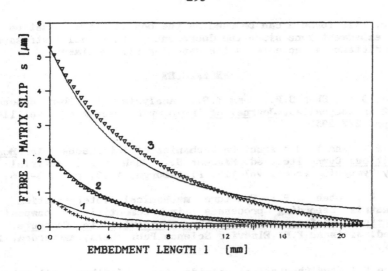

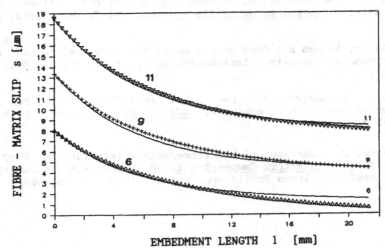

Figure 7. Distribution of slips along embedment zone in successive steps of loading, symbols - test results, curves - model simulation.

CONCLUSIONS

1. Measurements of fibre-matrix displacements confirm the results of model calculations for the pull-out process.
2. In proposed model based on nonlinear interfacial stress-slip relation defined by the function of two-variables only three parameters are required: τ_{max}, s_u, and β to describe bond and frictional properties of the interface, i.e. only by one parameter more than in the models in which the simple linear bond-slip relation was applied.
3. The formula (1) $\tau = f(s,x)$ cannot be considered as a constitutive equation of interface material because of its dependence on distance x.

However, this formula can be used as the constitutive equation for the entire embedment zone since the decrement of interfacial stresses related to the distance x increase is the same for all specimens.

REFERENCES

1. Morrison J.K., Shah S.P., Jenq Y.S.: Analysis of fiber debonding and pullout in composites, Journal of Engineering Mechanics, vol.114, No.2, 1988, pp. 277-295.

2. Shah S.P., Jenq Y.-S.: Fracture mechanics of interfaces. In Bonding in Cementitious Composites, ed. Mindess S., Shah S.P., Material Research Society Symposium Proc., vol.114, Pittsburgh, 1988, pp. 205-216.

3. Stang H., Shah S.P.: Fracture mechanical interpretation of the fibre/matrix debonding processes in cementitious composites, In International Conference on Fracture Mechanics in Concrete, Losanne, 1985, ed. Wittmann F.H.,Elsevier Science Publishers, Amsterdam, 1985, pp. 339-349.

4. Lawrence P.: Some theoretical considerations of fibre pull-out from an elastic matrix, Journal of Materials Science, vol.7, No.1, 1972, pp. 1-6.

5. Nammur G. Jr, Naaman E.: Bond stress model for fiber reinforced concrete based on bond stress-slip relationship, ACI Journal, vol.86, No.1, 1989, pp. 45-57.

6. Lim T.Y., Paramasivam P., Lee S.L.: Analytical model for tensile behaviour of steel-fiber concrete, ACI Journal, vol.84, No.4, 1987, pp. 286-298.

7. Potrzebowski J.: Behaviour of the fibre/matrix interface in SFRC during loading, In Brittle Matrix Composites 1, ed. Brandt A.M., Marshall I.H., Elsevier Applied Science Publishers, London, 1986, pp. 455-469.

NONLINEAR VISCOELASTIC BEHAVIOR OF THE FIBER-MATRIX INTERPHASE: THEORY AND EXPERIMENT

E. Sancaktar
Mechanical and Aeronautical Engineering Department
Clarkson University
Potsdam, NY 13699 USA

INTRODUCTION AND THEORETICAL CONSIDERATIONS

In fiber-matrix bonding, which is the topic of this paper, a common method of measuring the quality of adhesion is the single fiber fragmentation test procedure [1]. Various elastic and elastic-plastic approaches have been used in the literature to study the micro-mechanics of stress transfer between fiber and matrix. The most commonly used relation assumes uniform fiber strength and expresses the adhesion strength, τ_c, in the form

$$\tau_c = (\sigma_f \ d) / (2 \ l_c) \tag{1}$$

where, τ_c is the shear strength at the fiber resin interface, d is the fiber diameter, l_c is the critical fiber fragment size and, σ_f is the fiber tensile strength. For the present work single fiber tension test specimens were prepared by embedding single carbon fibers (7μm Celion G30 500 sized and unsized) in Epon 815 epoxy (Shell) cured with DETA and Armocure 100 (Akzo Chemie America) into 2 X 3 X 140 mm dog-bone tensile coupons with 50 mm gage length.

I - The Effects of Test Rate-Temperature and Interphase Thickness

In order to describe the rate dependence of limit stresses and elastic limit strain for adhesive materials in the bonded form Sancaktar et al. [2] utilized a semiempirical approach proposed by Ludwik (reported by Thorkildsen) in the form:

$$\theta = \theta' + \theta'' \log(\gamma/\gamma') \tag{2}$$

where, θ represents elastic limit stress or strain, γ is initial elastic strain rate and, θ', θ'', and γ' are material constants. Sancaktar et al. also proposed superposition of temperature effects on these equations [1].

Theoretical basis of equation (2) can be found in Eyring Theorem according to which the mechanical response of an adhesive is a process that has to overcome a potential barrier, (ΔE). The author thinks that this barrier decreases not only with increasing stress but also with increasing temperature. Based on the Eyring Theorem the relationship between a limit stress, say θ, and the strain rate can be written as:

$$\gamma = A \exp[(\theta V - \Delta E)/RT] \tag{3}$$

where, A is a pre-exponential factor, R is the gas constant, and V is the activation volume. Note that equation (3) can be expressed in logarithmic form similar to equation (2). If the energy barrier term $\Delta E/V$ of equation (3) in logarithmic form is also affected by temperature then one needs two shift factors in equation (2) multiplying θ' and θ'' terms to describe rate-temperature effects on limit stress-strain equations.

In order to analyze the effects of rate and temperature on the interfacial strength (as measured by fiber fragment length) one can initially use a simple energy approach in the following manner: A critical energy level, W_c, is used to represent interfacial failure. In the presence of matrix and interphase (which transmits the matrix energy to the rigid fiber) the combined elastic energy of the matrix and the interphase can be written as:

$$(1/2)\gamma^2_{matrix} \, G_{matrix} + (1/2)\gamma^2_{interphase} \, G_{interphase} = W_c. \tag{4}$$

Now, if one considers equation (4) in conjunction with equation (2) to include rate-temperature effects, then it can be easily deduced that higher W_c levels are obtained at high rate and/or low temperature levels.

Nonlinear Viscoelastic Stress Analysis of the Interphase. Recent nonlinear viscoelastic stress analysis of the fiber/matrix interphase by Sancaktar et al. concurs with this simplified finding [3]. In this analysis the nonlinear viscoelastic material behavior of the interphase zone and the matrix is represented using a stress enhanced creep [3]. In order to determine the time dependent variation of the interfacial shear stresses different nonlinear creep compliances are assigned to the fiber-matrix interphase and to the matrix.

Results of this analysis revealed that the magnitude of rate/temperature effects on interfacial strength should depend on the thickness of the interphase and relative magnitudes of interphase and matrix viscoelastic material properties. The fiber maximum stress which exists at the fiber ends increases with decreasing interphase thickness while it is shown to decrease with decreasing time exposure to external loading (Fig. 1). Existence of such high stress concentrations at fiber ends usually result in interphase/matrix plastic deformation and/or cracking at those locations which result in redistribution of stresses and, possibly, in lower values of average shear stresses along the fiber. Such cases are not likely to make optimum use of fiber support since major portion of the stress distribution is concentrated right at the fiber ends and the rest of the interphase carries little shear load. Consequently, shear stresses more uniformly distributed along the fiber with smaller peaks at the fiber ends make more efficient use of the fiber support and are thought to result in shorter fragment lengths during a single fiber tension test. Results of this analysis also reveal that depending on the relative magnitudes of interphase and matrix viscoelastic material properties it is possible to have stress reductions or increases along the fiber [3] (Fig. 2), and hence, the critical fiber length obtained will also vary accordingly. This result reveals that not only the quality of adhesion but also the material properties of the bulk matrix and the interphase, all of which can be affected by cure conditions, can affect the critical fiber length which is widely used as the gage by which fiber-matrix interfacial strength is perceived.

Evidence from our current experiments indicate that energy controlled fracture processes may dominate the fiber fragmentation process. This comment is made on the basis of our observations of shorter fragment lengths resulting from higher cross-head rate experiments. In such cases total elastic energy available from the composite system may indeed be the controlling factor for the fragment lengths obtained as indicated by equation (4). Therefore, (slower) processes involving viscoelastic dissipation over longer periods of time would be expected to result in longer fragments.

II - The Effects of Cure Temperature-Time and Curing Agent Content

Epoxy resins are rigid, amorphous, glasslike solids. Theories and evidence of electron microscope suggest that cured epoxy resins have many relatively small three-dimensional braches network or agglomerates of moderate molecular weight material. Due to steric hindrance that minimizes intermolecular reactions, the final size of agglomerates during cure is limited. When the molecules polymerize, their rotational and translational freedom is reduced, which reduces the chances for primary bonds to be set up with adjacent molecules. When the epoxy resin is subjected to cure at elevated temperatures or postcure, the increased thermal agitation permits an additional number of sterically hindered epoxy groups to react. Thus, the size of the agglomerates is increased at the expense of the surrounding lower molecular weight materials, which has lower strength, lower hardness and lower solvent resistance than the highly cross-linked agglomerates. As mentioned earlier, however, that mechanical properties of thermosetting polymers can only be improved up to a limit with increasing cure temperature and time values.

Above the glass temperature the oxygenolysis and degradation increases quickly thus deteriorating the matrix material properties. For example, the tensile strength when plotted against the cure temperature behaves as a bell shaped function: first increases up to a limit and then decreases at elevated temperature cure.

Experimental results [4] reveal that increases in the cure temperature results in the reductions in cure time requirements along with increases in the bulk strength until the maximum strength is obtained. Further increases in the cure temperature may increase the kinetic process but does not result in an increase of the bulk strength but infact may induce molecular changes which would trigger different failure mechanisms and result in strength reduction. It should be noted that the diffusion process which may govern the adhesion mechanism is also controlled by similar constraints.

Fiber-Matrix Interfacial Shear Stress. Pritykin et al. [5] offer a general theory for the strength of the adhesive-adherend interaction, F_{ad}, on the basis of "strength of interaction between the repeating unit of the adhesive and adherend, F_{if}," and "the ratio of the real interfacial contact area to its maximum value" i.e extent of molecular contact. The process of molecular contact formation is presented as adsorption of consecutive repeating units on the adherend. It is argued to depend not only on the temperature, internal stresses and molecular mass, but also on the energy and length of interfacial bond at the molecular level [5]. Accordingly, they express the F_{ad} function in the form

$$F_{ad}(t) = \int_0^t A(\tau) \, [\partial N(t - \tau, x)/ \partial \tau] \, F_{if}(x) dx \, d\tau \qquad (5)$$

where, $A(\tau)$ represents the time dependence of the interfacial contact area and, $N(t - \tau, x)$ represents the number of repeating units of the adhesive at the distance x from the adherend surface at the moment $(t - \tau)$.

Polymer diffusion across the interface between the adhesive and adherend which, in our case, is carbon fibers has also been referred to as the "reptation theory". The diffusion is governed by time and temperature. Based on this theory, the following relation has been used for the adhesion strength σ_{ad}, in thermoplastic/carbon composites [6]

$$\sigma_{ad} = m(a_t t_{c,ref})^n \qquad (6)$$

where, m and n are constants, a_t is the WLF shift factor and, $t_{c,ref}$ is the contact time. The shift factor a_t is defined by the relation

$$\log a_t = [-a_1(T - T_{ref})] / [a_2 + (T - T_{ref})] \qquad (7)$$

where, a_1 and a_2 are constants, T is the temperature and, T_{ref} is the reference temperature. Consequently, equation (6) dictates that the equivalent contact time at temperatures other than a reference temperature (such as the glass transition temperature T_g) is defined by equation

$$t_c = a_t t_{c,ref} \tag{8}$$

Note that equation (6) constitutes a specific form of equation (5).

Elevated cure temperature increases the mobility of polymer molecules, therefore promoting diffusion, and, hence, adhesion but only up to a level of cure temperature. We note that the two important aspects of diffusion mechanism are the following: i) The interdiffusion coefficient is inversely proportional to the polymer viscosity; ii) The activation energies of diffusion and viscous flow are equal.

Based on these premises we can conclude that the intensity of the molecular motion in a polymer can be estimated by its viscosity under given thermal conditions. Increases in the contact temperature, plasticization, and dissolution of the polymer are known to increase the rate of diffusion. Increases in the molecular chain stiffness, the presence of strong polar groups and crosslinking, however, are known to reduce the rate of diffusion while increasing viscosity. For example, Sancaktar et al. [7] showed that the bulk solid viscosity coefficient values calculated on the basis of a viscoelastic model increase with increasing molecular weight. Such viscosity values have been shown to be affected by the cure temperature and time values. Furthermore, increases in molecular weight and consequently the bulk solid viscosity of structural adhesives have been shown to increase ultimate shear stress and maximum creep (rupture) stress levels in bonded single lap joints [7]. There is evidence, however, that such increases in strength values level off and may even drop beyond a certain molecular weight value [7].

Diffusion in polymers is governed by an activation-kinetic process which can be described by an Arrhenius type equation:

$$t = A_o e^{Q/RT} \tag{9}$$

where, t is the time required for completion of a given stage of the diffusion process, A_o is a constant which depends on the thickness of the material and has the dimensions of time, Q is the activation energy of diffusion and, T is the cure temperature in degrees Kelvin. Note that equation (9) provides a general form for equations (7) and (8) combined for the cases when the cure temperature is substantially different than the reference temperature T_{ref}, which is usually assumed to be the glass transition temperature. This constraint arises from the fact that mathematically, for the form of equations (8) and (9) to be similar, the term $\exp(a_2^2 a_1^{-4} \Delta T^{-2})$, where $\Delta T = T - T_{ref}$ and similar terms with higher orders of $1/\Delta T$ in the series expansion of equation (7) need to be approximately equal to one.

An equation similar in form to equation (9) has been used in the literature to represent the polymer viscosity, η, as a function of temperature and the degree of cure, α.

The molecular weight of a polymer, which is expected to increase with increasing crosslinking for thermosetting materials, has been related to its viscosity with the equation: $\eta = A(M)^B$ where, A, B are constants which depend on the temperature and the nature of the polymer and, M is polymers molecular weight. For polymers having molecular weights above the critical level, M_c, which corresponds to the level at which polymer chains become long enough for chain entanglements to occur, Nielsen [8] reports the values $A = 4.5 \times 10^{-14}$ Mpa-sec and $B = 3.4$. Bulk solid viscosity values for thermoplastic polyimidesulfone calculated by Sancaktar et al. [7] on the basis of a viscoelastic model resulted in the values $A = 2.0 \times 10^{-55}$ and, $B = 13.1$. In either event, it is clear that increasing crosslinking which results in increased molecular weight also results in increased polymer viscosity as solidification proceeds in the polymer.

Viscosity values of polymers measured by rheometers have been used to assess their wetting and adhesion properties [1].

III - Electric Resistive Heat Curing of the Interphase

This investigation deals with creating a new interphase between the carbon fiber and the epoxy matrix by producing high temperature gradient around the fiber. Electric resistive heating is used for this purpose. The interphase near the fiber develops higher strength due to the postcure effects derived from electric heating. Consequently, stronger adhesion and better resistance to crack propagation are expected.

Thermal Considerations and the Heat Transfer Problem. The thermal fields produced by electric resistive heating of carbon fibers are different from that produced by convective thermal heating. The temperature distribution around the fiber and in the interphase is needed for considering the effects of electric heating on carbon fiber/epoxy resin interphase. For this purpose a purely heat conduction problem can be assumed if the radiation heat can be neglected. For this problem the general governing equation for heat conduction can be reduced to Laplace equation which can be solved using the methods of numerical analysis. A difference analysis method is used to obtain numerical solution. Energy balance principle requires that the amount of electric heat produced in the carbon fiber should be equal to the amount of heat transferred from the specimen surface to the ambient air under steady conditions. Furthermore, it is assumed that heat transfers perpendicularly from the carbon fiber surface through the specimen surface to air. Thus an additional equation $Q = \lambda u \Sigma \Delta T_{av,s}$ is established where u is a chosen length along fiber axis, Q is the electric energy per second produced in fiber having length u, λ is the thermal conductivity of epoxy resin and $\Sigma \Delta T_{av,s}$ is the sum of the average temperature differences perpendicular to the surface of specimen for each grid contacting air [9]. In our calculation: $\lambda = 2.1 \times 10^{-3}$ J/cm $\cdot$ °C $\cdot$ s; $Q_{I=3.5mA} = 0.0573$J/s; $Q_{I=5.0mA} = 0.117$J/s; $Q_{I=6.5mA} = 0.198$J/s; u = 10mm.

Inverse Matrix Method is used to solve the difference equations with the aid of a computer program was prepared for this purpose [9]. The results reveal that for carbon current I = 3.5mA, the carbon surface temperature is $T_c = 38.67$°C, for I = 5.0mA, $T_c = 52.34$°C and, for I = 6.5mA, $T_c = 74.44$°C. The results reveal that the temperature gradients around the carbon fiber increase as the current is increased.

Fracture Considerations. The Irwin criterion for fracture is based on equating the strain energy with the crack surface energy at the instant of crack propagation. Such energy balance results in the equation $G_c = \sigma_c^2 \pi a / E$ where G_c, being an index of fracture toughness, is the strain energy release rate (related to fracture surface energy), E is the elastic modulus, σ_c is the stress at crack initiation and, a is one half of initial crack length.

Obviously, if "a" is constant, then σ_c^2/E can be used as an index of G_c. Higher G_c means higher toughness. Failure and cracks are unlikely to propagate easily in high G_c regions.

RESULTS AND DISCUSSION

I - The Effects of Test Rate-Temperature and Interphase Thickness

Experimental results revealed that the level of global strain above 4% does not affect the average fiber fragment length significantly. We should also note that the ultimate elongation for the carbon fiber is 1.62% (1.39% minimum) as reported by the manufacturer.

For both unsized and sized fibers experimentation proved that fiber fragment length is smaller at higher cross-head rates. Consequently, the interfacial strength is interpreted to be higher at higher cross-head rates. For unsized fibers a tenfold reduction in cross-head rate from 254 mm/min results in 21% increase in the average fiber fragment length and, and other eightfold reduction in cross-head rate from 25.4 mm/min results in another 20% increase in the average fiber fragment length. Therefore, a 46% decrease in interfacial strength is demonstrated for unsized fibers over an eighty fold reduction in cross-head rate (Fig 3).

When sized fibers are used an eightfold reduction in cross-head rate from 25.4 mm/min results in 32% increase in the average value of fragment lengths. This 32% increase in the fragment average value is much higher than the 20% increase observed for unsized fibers at the same range of change in the cross-head rate. It should also be noted that the fragment lengths obtained are shorter for sized fibers in comparison to unsized fibers at comparable cross-head rates. This difference, however is not constant but gets larger at higher rates, i.e. the average length of sized fiber fragments being 10% shorter at 3.18 mm/min and, 22% shorter at 25.4 mm/min.

In the introduction section we stated that for polymeric materials the increase of strain rate is equivalent to a lowering of temperature. So far we showed that increases in cross-head rate result in shorter fiber fragment lengths. Our experimental results also revealed the applicability of rate-temperature superposition, as stated, on the fragment length and, consequently, on interfacial strength since reductions in the environmental temperature result in shorter fragment lengths.

The analytical and experimental results presented so far reveals the possibility for mutual reduction between rate, temperature, and interphase thickness. If we assume that application of fiber sizing results in thicker interphases and that the cross-head rate, v, can be related to the shear strain rate, γ, and the interphase thickness, δ, with the relation: $v = f(\gamma) \, g(\delta)$ based on classic shear lag analysis (also based on definition of shear strain), we can then argue that for a fixed strain rate (function) the effect of increased cross-head rate can be induced by increasing the interphase thickness (i.e. application of fiber sizing). Indeed, the experimental results reveal shorter fragment length for increased cross-head rate and/or presence of fiber sizing. Based on this premise we can write the following superposition relation for the interfacial strength:

$$\tau_c = k(a_T \, a_\delta \, v_{ref})^l \tag{10}$$

where a_t and a_δ are cross-head rate shift functions relating to environmental temperature and interphase thickness effects respectively.

II - The Effects of Cure Temperature-Time and Curing Agent Content

Experimental results [1] revealed that the increasing portion of the interfacial strength function resides between the %CA (percent curing agent content) values of 9.5% and 11.0%. As we go to higher levels of %CA above 9.0%, optimum τ_c is obtained at shorter time higher temperature levels. At and above 11.0% CA level τ_c becomes insensitive to cure temperature and time conditions due to the high level of the curing agent present. At 9.0% level, again the τ_c is considerably below the optimum value revealing the competing effects of adhesion and degradation.

These results illustrate the possibility for superposition between cure time, temperature and curing agent content on the increasing portion of an interfacial strength function. In other words, one can write:

$$\tau_c = k'(a_t \, a_{CA} \, t_{c,ref})^{l'} \tag{11}$$

where a_t and a_{CA} are cure time shift functions relating to cure temperature and percent curing agent content respectively. Also note that in the light of this consideration, equation (6) does not provide a complete description for cure effects on the adhesion strength as it lacks the effect of %CA. The authors think that equation (11) should also include interphase thickness and (ratio of) matrix and interphase material property shift functions based on the nonlinear viscoelastic work reported earlier.

III - Electric Resistive Heat Curing of the Interphase

Adhesion. Experimental results [9] revealed that application of 3.5 mA of electric heating current increases the numbers of carbon fiber fragments observed in the 50 mm gage length of specimens, in comparison to the specimens for which no electric postcure was applied. Application of higher currents does not produce additional advantage in reducing the critical fiber length, l_c and, in fact, seem to result in longer l_c values with I=6.5 mA. Application of such high amperage may cause the break number to decrease due to the forming of a highly deformable interphase. The shortest average l_c value was calculated as $l_c = (59 \pm 19) \times 10^{-2}$ mm corresponding to one hour application of 3.5 mA and the longest value was $l_c = (82 \pm 22) \times 10^{-2}$ mm corresponding to one hour application of 6.5 mA. Application of 3.5 mA or 5 mA for four hours both resulted in the same average l_c value of $(69 \pm 18) \times 10^{-2}$ mm.

Postcuring of the specimens by convective heating revealed an approximately monotonic increase in the number of fiber fragments with increasing thermal postcure temperature. This postcure procedure, of course, does not produce the severe thermal gradients produced with electric heat postcuring. It seems that the thermal postcure for 4 hours at higher temperature has some advantage in increasing adhesion over postcuring for 1 or 8 hours at the same temperature. In fact, the shortest l_c value obtained under thermal postcure conditions was $(58 \pm 20) \times 10^{-2}$ mm corresponding to four hours of thermal postcure at 74 °C. The longest l_c value was $(72 \pm 16) \times 10^{-2}$ mm corresponding to eight hours of thermal postcure at 39 °C.

The increased adhesion resulting from the applied postcure procedures is attributed to the increased "reptation" behavior of the polymer as described earlier.

The σ_m^2/E value. This value is roughly related to the G_c or the material toughness and was shown to increase with increased current heating (Fig. 4) [9]. Such higher values of σ_m^2/E mean better toughness. A similar behavior for the toughness was obtained in the thermal postcure, except 4 hours of postcure as well as 8 hours seemed to result in optimal values (i. e. they reached a maximum and then dropped off).

Cracks. Experimental results revealed that, in general, not only the number but also the size of cracks are smaller with electric postcure when compared with thermal postcure.

REFERENCES

1. A. Turgut, E. Sancaktar, The Effects of Cure Temperature-Time and Curing Agent Content on the Fiber-Matrix Interfacial Strength Clarkson Univ. Report No. MAE-219 (1990).

2. E. Sancaktar, S. C. Schenck and S. Padgilwar, Ind. and Eng. Chem. Product Res. Dev. 23, 426 (1984).

3. E. Sancaktar, P. Zhang, Failure Prevention and Reliability (ASME DE-Vol. 16, 1989), p.259.

4. E. Sancaktar, H. Jozavi and R. M. Klein, J. Adhesion 15, 241 (1983).

5. L. M. Pritykin, A. B. Zilberman, I. I. Zilberman and V. L. Vakula, private communication.

6. J. Muzzy, L. Norpoth, and B. Varughese, SAMPE Journal, 25, 23 (1989).

7. E. Sancaktar, S. K. Dembosky, J. Adhesion 19, 287 (1986).

8. L. E. Nielsen Mechanical Properties of Polymers (Reinhold Publishing Corp., New York, 1962) p. 56.

9. E. Sancaktar, W. Ma, Electric Resistive Heat Curing of the Fiber-Matrix Interphase Clarkson Univ. Report No. MIE-197 (1989).

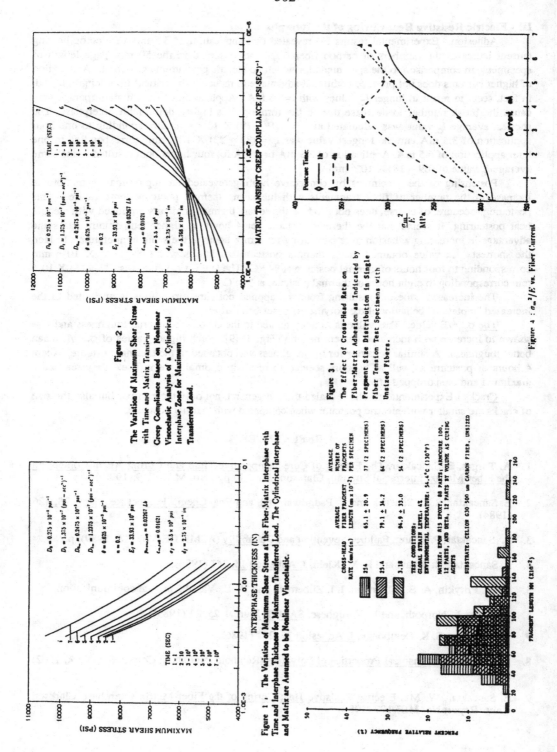

Figure 2 :

The Variation of Maximum Shear Stress with Time and Matrix Transient Creep Compliance Based on Nonlinear Viscoelastic Analysis of a Cylindrical Interphase Zone for Maximum Transferred Load.

Figure 3 :

The Effect of Cross-Head Rate on Fiber-Matrix Adhesion as Indicated by Fragment Size Distribution in Single Fiber Tension Test Specimens with Unsized Fibers.

Figure 4 : $\sigma_m{}^2/E$ vs. Fiber Current

Figure 1 : The Variation of Maximum Shear Stress at the Fiber-Matrix Interphase with Time and Interphase Thickness for Maximum Transferred Load. The Cylindrical Interphase and Matrix are Assumed to be Nonlinear Viscoelastic.

EMPIRICAL MODEL BUILDING AND SURFACE RESPONSE METHODOLOGY APPLIED TO SCLEROMETRICAL INVESTIGATION FOR REFRACTORIES CONCRETE

M.OUADOU [*], T. MATHIA [*], B. CLAVAUD [**], R. LONGERAY [***], P.LANTERI [***]

[*] Ecole Centrale de Lyon. Lab. de Technologies des Surfaces - C.N.R.S. 69131 ECULLY.

[**] Lafarge Réfractaires Monolithiques. Centre de Recherche 69804 SAINT PRIEST.

[***] Université C. Bernard Lyon I. Lab. de Synthèse Orga. Appliq. 69622 VILLEURBANNE

ABSTRACT

Recent proliferation of refractories concrete in several applications such as cementry, brick-field, petrochemistry and petroleum refining, industrial incinerator, iron and steel industry and all metals metallurgy is at least the main reason of these investigations. The motivation is to optimize mechanical features, principally to make refractories concrete more resistant to abrasive wear process.

This paper summarize a recent experimental investigation in abrasive wear phenomenon applied to refractories concretes considered as an heterogeneous composites. In spite of the industrial importance of these materials, little is known about their tribological behaviour. Consequently, for better understanding of the abrasive process concerning these materials it was necessary to develop a new experimental and analytical approach.

A a new technique based on fully computerised sclero-topometer has been adapted and applied to this type of materials. Then the application of specific strategy for experiments program and data treatments based on optimal experimental design have been undertaken for empirical model building and surface response establishment. The data issued from experimental techniques have been interpreted, explored and presented thanks to a fully computerised device. This appliance enables a standardisation of analysis techniques and logical classification of materials based on scientific knowledge in regard to abrasive processes.

The experiments were conducted on refractories concretes composed of materials as Al_2O_3 SiC, Chamotte and Bauxite.

Notation list:

A_t = Transversal section of material removed and/or displaced,

$[\text{J/m}^2]$
F_N = Normal scratch force, $[N]$
F_T = Tangential scratch force, $[N]$
E_S = Specific abrasion energy $[J/m^3]$
x_i = Input variables
y_i = Output variables

1. Experimental research methodology and response surfaces in abrasion study

The abrasion of brittle and anisotropic materials such as refractories in many industrial applications can be considered as a multi-point scratching and indenting process. In several case, the scratches and the indentations are randomly organised so it is difficult and complex to define the main features of these phenomenon.

Many works on abrasive wear consider that hardness of materials is the most important parameter determining abrasion resistance of materials. This consideration is not clearly verified for all abrasive conditions. In two or three body abrasive wear two modes can be differentiated. Ductile cutting is the first one with materials loss, and ploughing is the second with no loss of materials. The complete understanding of transition from ploughing to cutting in abrasive wear is of a great help for abrasive process optimisation.

From the scientific and technological points of view, the knowledge of the tendency for a surface either ductile or brittle to be deformed by an asperity, a protuberance or a harder particle, the forces needed and energy consumed, are very important to understand the phenomenon relative to the wear, friction and adhesion [1, 2, 8].

A common goal in may types of experimentation is to characterize the relationship between a response and a set of factors of interest to the researcher, to improve and understand the systems with which he is working. This is accomplished by constructing a model that describes the response over the applicable ranges of the factors of interest. In many applications the fitted model is referred to as a response surface because the response can be graphed as a curve in one dimension or a surface in two or more dimensions. The response surface of the process can be explored to determine important characteristics such as optimum operating conditions (factors levels that produce the maximum or minimum estimated response) or relevant tradeoffs when there are multiple response.

1.1. Experimental design strategy

An experimental design is the protocol of how to vary all the factors simultaneously, in a series of experimental runs, in order to extract the maximum amount of informations from the data in the presence of noise. In general, with experimental design, only a small numbers of experimental runs are needed.

1.2. Response surface methodology

Response surface methodology [11] comprises a group of statistical techniques for empirical model building and model exploitation. By careful experimental design and analysis of it seeks to relate a response or output variable (y) to the levels of a number of predictors or input variables (x_i) that affect it. Example for 3 variables :

$$y = b_0 + b_1x_1 + b_2x_2 + b_3x_3 + b_{12}x_1x_2 + b_{13}x_1x_3 + b_{23}x_2x_3 + b_{11}x_{12} + b_{22}x_{22} + b_{33}x_{32}$$

coefficients "b" measure the influence of the variables :
- b_ix_i is a linear term and coefficient b_i measures the main effect of variable x_i.
- $b_{ij} \cdot x_ix_j$ is a cross-product term and coefficient b_{ij} measures the interaction between the variable x_i and the variable x_j.
- $b_{ii} \cdot x_1^2$ is a quadratic term to take curvatures of the response surface into account.

1.3. Empirical model building.

Second degree polynomial model has been used to describe relationship between the studied sclerometric response (noted Y) and the experimental variables chosen (coded variables noted X_i are used) i.e. : $Y = a_0 + a_1 X_1 + a_2X_2 + a_{12} X_1X_2 + a_{11} X_1^2 + a_{22}X_2^2$. With the values of the coefficient of this model is possible to realize an optimal experimental design as proposed by BOX [11], then to draw a response surface of the phenomenon. Response surface models $Y(X_1, X_2)$ are easily established by least square fitting [12] or a polynomial of the experiment coded variables (X_i).

2. Sclerometrical approaches in material characterisations.

To simulate material removal mechanisms during abrasion in the use of scratch test, one can see many development from probably the first initiative of K. JAGGER [4] to the recent complete review by T.H. KOSEL [5] of scratch test studies performed from simplest devices in air up to sophisticated in situ in SEM.

Different systematic sclerometrical investigations on metals (T.O. MULHERN and L.E. SAMUEL [6]) than ceramics and polymers (T.G. MATHIA and B. LAMY [8]) showed the transition from ductile to brittle abrasion.

For surface coatings some specific tests to arrive at quick failure mechanism diagnosis by mean of a scratch testing device (J. VON STEBUT [7]) has been developed. Sclerometric experiments are generally ascertained, full of the informations, and time-consuming investigations.

2.1. Objectives of a fully computerised sclerometer.

The combination of a computerised 3D-topometer with a sclerometer needs an interactive software and a very performing hardware for fully automated experimentations analysis,.and show of results.

The problem concerns therefore the definition of universal techniques which should be compatible with the real needs to be taken into account for tribology, rheology and industrial purposes.

The deformation of a surface caused by abrasive particles [6] depends on several parameters such as: *the type of materials, the shape and the size of penetrators, the forces applied on the penetrators, the trajectories and speeds to be respected, the interface protuberances / surfaces, etc.*

Because of the multiplicity and the interdependence of parameters [7] which occur during test and to make the method easy of use and universal, it is essential that operator conditions should be defined so that the needs of the whole scientific and industrial communities are satisfied.

2.2. Experimental apparatus and procedure.

2.2.1. General descriptions.

The SCLERO-TOPOMETER device we developed is a fully computerised device which can be used for several kinds of mechanical surface characterisations. (Fig. 1). The device offers the possibility of three independent apparatus:
- Sclerometer for scratch test.
- Indentometer for hardness measurement.
- Topometer for topographical investigations which can be used separately for surface topography characterisation.

In sclerometrical investigations and in indentation it offers different types of tests conditions: *fixed depth scratch, fixed load scratch, variable depth scratch (with linear increasing or decreasing depth scratch and the combination of it), pendulum scratch test with fixed depth or normal load of indentation.*

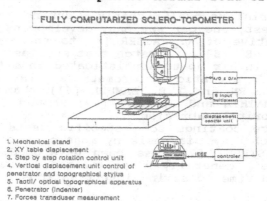

FULLY COMPUTARIZED SCLERO-TOPOMETER

1. Mechanical stand
2. XY table displacement
3. Step by step rotation control unit
4. Vertical displacement unit control of penetrator and topographical stylus
5. Tactil/ optical topographical apparatus
6. Penetrator (Indenter)
7. Forces transducer measurement

Figure 1: Composition of the SCLERO-TOPOMETER computerised unit

The main test consists of scratching the surface sample and measuring simultaneously the forces which occur during the real time scratch. 3D-topography analysis offers needed informations concerning the form and scratch dimensions with vertical sub-miconic and lateral micronic resolution. Acoustic emission signal can also be measured optionally in case of brittle materials.

The maximum length scratch is 40mm. Spatial resolution can selected from 1 to 100 ¥m. The scratch velocity can be fixed in an interval of 10¥m/s to 2 mm/s.
With reference to the above-mentioned it is obvious that when the measurements are computerised, the manipulations and the numerical treatments of the data are automatized.

2.2.2. Experimental procedure and automated treatments.

In sclerometry or indentation investigations the SCLERO-TOPOMETER offers for the operator various selection of parameters such as: *depth and speed of the indentation or scratching, controlled depth or load, acquisition step of the measured parameters, number of indentations/scratches to be performed on a same sample, choice between a single or multiple an organised or random scratches/indentation etc.*

The indentor is automatically positioned on the specimen flat surface with a precision of about 0.5 ¥m. Then in the first step of a scratch test at a specified the normal force during vertical penetration is measured and stored in a file the same way as scratch forces. The real scratch depth is measured by the tactile 3D-topometer.

Note that only the sample preparation is manual. The surface sample may be oriented for different type of test.
The main sclerometric parameters which can be accessible directly after a scratch test are: *the theoretical values of scratch normal and tangential section, dynamically hardness of a material, specific abrasion energy, three components of scratching efforts measured.*

The 3D-topographical analysis needs any manual intervention of the operator to determine the rest of parameter which can not be calculated automatically.

3. Some results of measurements and discussions.

In case of refractories concrete abrasion, it is always useful to estimate the characteristics such as: *the linear abrasion rate, the material removal rate, the volumetrically abrasion rate, the weight abrasion rate the energical abrasion rate, the abrasion coefficient and probably most important, the specific abrasion energy:* $..E_s = F_t/A_t$
This energy expressed in stress units determines the energy required to remove elementary volume of abrased material.

The penetrators used for the tests carried out were a Vickers and a Rockwell C penetrator.

However the main difficulties test are the estimation of the area of the different sections A_t and F_t which are not constant during a scratch and evaluate in different manner

depending upon the nature of material and experimental
conditions principally due to: elastic recovering of the
materials and brittle fracturation.

To calculate the specific abrasion energy Es the loss
material volume is calculated thanks to 3D-topographical
analysis of the surface after tests in relation to tangential
force recorded F_t. (Fig. 2).

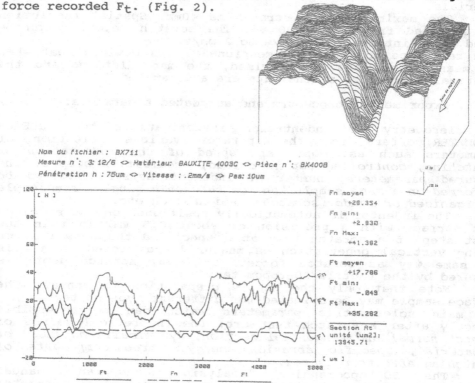

Nom du fichier : BX711
Mesure n° : 3: 12/6 <> Matériau: BAUXITE 4003C <> Pièce n° : BX400B
Pénétration h : 75um <> Vitesse : .2mm/s <> Pas: 10um

**Figure 2: Isometric diagram and scratch forces of refractory
concrete surface.**

Studying the effect of the sliding velocity and scratch depth
on the evolution of the tangential force (F_t) and the specific
abrasion energy (E_s) of some tested materials (Al_2O_3) the
existence of an optimum penetration which characterize the
maximum of resistance in abrasion can be localised. (Fig.3)

This is to confirm that not only the attack angle have a
great importance in abrasive process but in the same way the
scratch depth and sliding velocity of abrasive particle are
important.

That means, in terms of consumed energy, that for the
depth of scratches of $h_{optimum}$, abrasive resistance of material
is the best. This resistance is very sensitive to sliding
velocity too. As indicates tangential force evolution this
conclusion is not so evident.

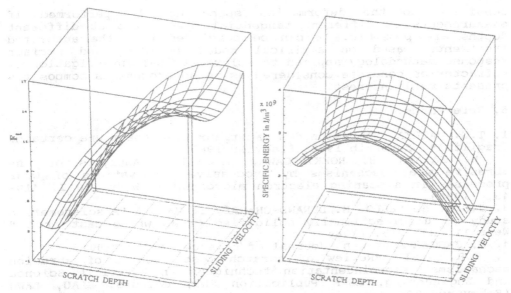

Figure 3: Response surface for the tangential force and the specific abrasion energy of (Al_2O_3).

The same observation can be made when regarding the evolution of the side flange/groove ratio or the evolution of the two flanges and groove area. In this case the transition from ploughing to cutting is clearly verified and demonstrated.

4. Conclusion.

Conceived in the basic idea to make a standardization of the scratch test it's principal feature now is that the fully computerised SCLERO-TOPOMETER is of easy use and for this reason offers large possibilities of studying materials abrasitivity. It is therefore a means of investigation which is quick, efficient and precise not only for the above-mentioned matter, but also for the rheological characterisation of composites and superficial coatings.
    Moreover, the application of cycles of successive scratches by the penetrator enables the deduction of the memory effect or the determination of the size of the area which would have been affected by the residual deformation of the nearest scratch.
    With the tests carried out on brittle areas it is possible to determine the sclerometrical or dynamically tenacity and the brittleness coefficient can therefore be defined with the ratio hardness/tenacity.
    Finally, when the geometrical parameters of the scratch caused by the indentor are known, it is possible to automatically evaluate the sclerometrical hardnesses, the specific energy of the abrasion, the different abrasitivity rates along with a dynamically approach of the abrasive wear. Characterisations of surfaces made in materials which are

sensitive to the deformation speed can be performed if measurements at different tangential speeds and at different depths are possible. It can be concluded that the automated treatments based on empirical model building and surface response methodology applied to sclerometrical investigation of refractories concrete considered as an heterogeneous composites presents real interest.

5. References.

1. T.G. MATHIA, "Abrasion des matæriaux præsentant une certaine fragilitæ", Eurotrib 1985, Elsevier 1985.
2. H. KAYABA, H. HOKKIRIGAWA, K. KATO, "Analysis of the abrasive wear mechanisms by successive observations of wear processes in a scaning electron microscope", Wear 110 I. 419-43, (1986).
3. J.D.B. de MELLO, M. DURAND-CHARRE, T. MATHIA "A sclerometric study of unidirectionally solidified Cr-Mo white cast iron " Wear 111 203-215, (1986)
4. K. JAGGER American Journal of science, 4-399,(1897)
5. T.H. KOSEL "Review of scratch test studies of abrasion mechanisms" Microindentation techniques in Materials Science and Engineering, ASTM Publication 889, editors: BLAU, LAWN (February 1986)
6. MULHEARN AND L.E. SAMUEL "The abrasion of metals: a model of the process" Wear 5 478-498, (1962).
7. VON STEBUT "Scratch testing induced surface damage of thin coatings: A study of failure mechanisms by means of appropriate surafce analytical tools" First international conference on plasma surface engineering. Sepetember 19-23, (1988).
8. T.G. MATHIA, B. LAMY "Sclerometric characterization of nearly brittle materials" Wear, 108 (1986) pp385-399
9. HOKKIRIGAWA, K. KATO, "An experimental and theoritical investigation of ploughing, cutting and wedge formation during abrasive wear", Tribology International, 1988, Vol 21, nr. 1
10. L.E. SAMUELS, "Metallographic polishing by mechanical methods", American Society for Metals, 3rd Edition, (1982)
11. P.G. FOX, I.B. FREEMAN "What does the pendulum hardness test measure" Journal of Mater. Sci. 14 (1979) pp 151-158.
12. G.E.P. BOX, N.R. DRAPER, Empirical model building and response surface, 1987, J. Wiley - New York.
13. N.R. DRAPER,G.SMITH Applied Regression Analysis, 1981, J. Wiley, New York.
14. D.H. DOEHLERT Appl. Statist. 19, 3 (1970).

DEVELOPMENT OF FATIGUE DAMAGE MECHANICS FOR APPLICATION TO THE DESIGN OF STRUCTURAL COMPOSITE COMPONENTS

P.W.R. BEAUMONT and S.M. SPEARING

Cambridge University Engineering Department,
Trumpington St, Cambridge CB2 1PZ, U.K.

ABSTRACT

The basic principle that dominates this work is the recognition that the strength of a composite changes with accumulated damage and that the composite can retain adequate strength by creating a microstructure that permits regions of sub-critical damage growth.

A structural composite laminate, carbon fibres in epoxy for example, containing a hole or notch sustains local sub-critical damage on loading: delamination and splitting of plies, fibre fracture, matrix cracking, and so forth. The precise geometry of this notch tip damage zone depends uniquely upon the level of applied stress and the laminate configuration, ply stacking sequence, fibre orientation, etc. Damage accumulation is a function of the applied stress, load cycles, temperature, and so forth, and the stress distribution in the notch tip region depends on both stress and damage state. To predict notch strength requires a knowledge of the relation between the localised critical stress distribution in the load

bearing $0°$ plies and the extent of sub-critical damage growth, split length, delamination area, for example. This involves the development of computational models that describe changes in microstructure brought about by stress and by environmental conditions which simulate those in service. Application of practical damage mechanics analysis can be made to the prediction of post-fatigue strength and stiffness, toughness, damage tolerance and fatigue-life of fibre composites. The extension of basic damage models of composite failure to design is a powerful tool in the development of design procedures for composite hardware.

OBSERVATIONS OF DAMAGING MECHANISMS

Components of Damage

An X-ray image of a typical $(90/0)_s$ notched tensile specimen loaded in tension is shown in Fig. 1 [1]. The dark images of the photograph indicate three forms of matrix crack emanating from within the notch damage zone:

1) **splitting** in the $0°$ plies;
2) **delamination** zones at 90/0 interfaces whose size is related to split length; and
3) **transverse ply cracking** in the $90°$ plies.

A concentration of fibre breaks in the $0°$ plies occurs at the intersection of a split and a transverse ply crack [2].

Damage Propagation

$(90_i/0_j)_{ns}$ Laminates

Fig. 2 shows X-ray pictures of notched $(90_2/0_2)_s$ and $(90/0)_{2s}$ laminates after 10^6 load cycles. For these cross-ply laminates, the same basic features of the damage zone are present; however, the rate of damage growth is greater for laminates having thicker plies.

In the examples above, notch tip damage has grown in a self-similar manner, with the characteristic triangular-shaped delamination zone at the $(90/0)$ interface. At the tip of the split, the delamination angle, α, is between about $4°$ and $7°$. Essentially, the damage pattern at the notch tip can be characterised by the average length, ℓ, of four splits. Fig. 3 shows a schematic view of the idealised damage pattern.

THE FATIGUE DAMAGE MODEL

Damage at notches in composite materials can be characterised by the split length ℓ, at the notch tip (Fig. 3). The starting point of the model is the fatigue-crack growth law ("Paris" law) for isotropic materials [2]:

$$\frac{da}{dN} = \lambda_1 (\Delta K)^m \qquad [1]$$

$\frac{da}{dN}$ is the crack growth rate;
ΔK is the stress intensity factor range; and
λ_1 and m are empirical constants.

This can be re-cast in terms of the split growth rate, $d\ell/dN$, of the composite and energy release-rate, G (using $K^2 = YG$, where Y is an

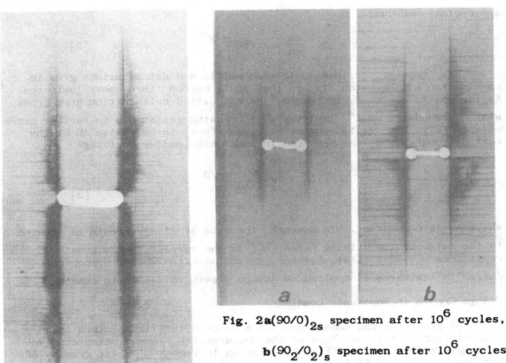

Fig. 2a $(90/0)_{2s}$ specimen after 10^6 cycles,

b $(90_2/0_2)_s$ specimen after 10^6 cycles,

$(\sigma_{max} = 208$ MPa, R = 0.1).

Fig. 1: A typical X-ray image of a $(90/0)_s$
centre-notched specimen after fatigue cycling.

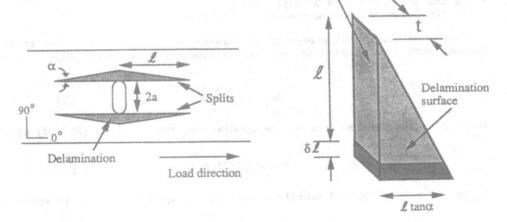

Fig. 3: Schematic of the notch-tip damage pattern in a cross-ply specimen.

appropriate modulus):

$$\frac{d\ell}{dN} = \lambda_2 (\Delta G)^{m/2} \qquad [2]$$

This, however, is inadequate when splits and delaminations grow in combination at the notch tip, as they do in carbon fibre-epoxy laminates. As the split length, ℓ, increases, the associated delamination area grows with a dependence on ℓ^2, implying an increasing resistance to further crack advance (Fig. 3). It is appropriate, therefore, to normalise ΔG by the current toughness G_c (as measured in monotonic loading), giving:

$$\frac{d\ell}{dN} = \lambda_3 \left[\frac{\Delta G}{G_c} \right]^{m/2} \qquad [3]$$

where λ_3 is a constant. In essence, the rate of damage growth is related to the ratio of the driving force, ΔG, to the current damage growth resistance G_c. But first, however, an introduction to the concept of damage in laminates as applied to damage growth in monotonic loading [1].

Damage Growth in Monotonic Loading

The growth of the damage zone in a notched tensile specimen is self-similar, i.e., the shape of the delamination front remains constant: it simply grows in scale (Fig. 1). For an increment of split growth, $\delta\ell$, the energy absorbed in forming new crack surfaces, δE_{ab}, is given by [1]:

$$\delta E_{ab} = G_s t \delta\ell + G_d(\ell \tan \alpha)\delta\ell \qquad [4]$$

G_s is the energy absorbed per unit area of split;
G_d is the energy absorbed per unit area of delamination;
t is the thickness of a 0° ply; and
α is the delamination angle at the split tip.

Some of the global energy of the system is dissipated when the split extends with a corresponding increase in specimen compliance (δC):

$$\delta E_r = \frac{1}{2} P^2 \delta C \qquad [5]$$

where P is the applied load on one quadrant of the specimen (Fig. 3) given by:

$$P = \sigma_\infty (W/2)(2t) \qquad [6]$$

σ_∞ is the remote applied tensile stress on the specimen, and W is specimen width.

For the damage zone to grow, $\delta E_r \geq \delta E_{ab}$:

$$\frac{1}{2} P^2 \delta C \geq G_s t \delta \ell + G_d \ell \delta \ell \tan \alpha \qquad [7]$$

For the limiting case $(E_r = E_{ab})$:

$$\frac{1}{2} \frac{P^2}{t} \frac{\partial C}{\partial \ell} = G_s + G_d \frac{\ell \tan \alpha}{t} \qquad [8]$$

$$(G = G_c)$$

G_c is the effective toughness of the laminate and G is the strain energy release-rate. This expression can be evaluated if $\partial C/\partial \ell$ is known. A finite element model can determine it numerically [1]. For a range of split length between $\ell/a = 0$ and $\ell/a = 6$ (2a is notch length of a centre-notched specimen), $\partial C/\partial \ell$ is constant, dependent only on the angle α and the lay-up construction [1].

Rearranging eqn. (8), the load for split initiation is given by:

$$P_i = \left[\frac{2G_s t}{\left. \frac{\partial C}{\partial \ell} \right|_{\ell=0}} \right]^{\frac{1}{2}} \qquad [9]$$

The value of G_s can be determined from a measurement of P_i.
For subsequent split growth under monotonic loading:

$$\ell = \frac{P^2 \left(\frac{\partial C}{\partial \ell} \right)}{2G_d \tan \alpha} - \frac{G_s}{G_d} \left[\frac{t}{\tan \alpha} \right] \qquad [10]$$

Dividing both sides of eqn. (10) by one-half of notch length, a, and substituting for P (in terms of applied stress σ_∞), and $\partial C/\partial \ell$, (scaled from the finite element result $\left. \frac{\partial C}{\partial \ell} \right|_{FE}$), this expression can be normalised as

follows:

$$\frac{\ell}{a} = \frac{(\sigma_\infty)^2 (t) \ (W/2)_{FE} (2t)_{FE}}{G_d \tan \alpha (2a/W)} \left. \frac{\partial C}{\partial \ell} \right|_{FE} - \frac{G_s}{G_d} \left[\frac{t}{a \tan \alpha} \right] \qquad [11]$$

In practice, $G_s < G_d$, hence the second term of eqn. (11) is small, allowing the simplification:

$$\frac{\ell}{a} = \frac{(\sigma_\infty)^2(t) \ (W/2)_{FE} (2t)_{FE}}{G_d \tan \alpha \ (2a/W)} \left.\frac{\partial C}{\partial \ell}\right|_{FE} \qquad [12]$$

Equation (12) indicates that for a given $2a/W$, the ratio ℓ/a is dependent only on applied stress, σ_∞. The normalised split length curves with applied stress curves for all $(90/0)_s$ specimens should be coincidental regardless of width. Figure 4 shows split growth data for three $(90/0)_s$ specimens $(W/2a = 3)$ under monotonically increasing load. Equation (12) produces a good fit to the data using a selected value of $G_d = 400 \ J/m^2$ and $G_s = 158 \ J/m^2$. Now we apply the model to fatigue damage.

Damage Growth in Cyclic Loading

We start with equation (3). Inserting the definition

$$\Delta G = \frac{1}{2} \frac{(\Delta P)^2}{t} \frac{\partial C}{\partial \ell} \qquad [13]$$

and the result derived from equation (8)

$$G_c = G_s + G_d \frac{\ell \tan \alpha}{t} \qquad [14]$$

gives:

$$\frac{d\ell}{dN} = \lambda_3 \left[\frac{1/2(\Delta P)^2 \ (\frac{\partial C}{\partial \ell})}{G_s t + G_d \ell \tan \alpha} \right]^{m/2} \qquad [15]$$

For an initial split length, ℓ_o, the split length after N cycles of constant load amplitude is given by the integrated form of eqn. (15):

$$\ell = \frac{1}{G_d \tan \alpha} \left[\lambda_3 (\Delta G)^{m/2} \left[\frac{m+2}{2}\right] (G_d \tan \alpha) N + (G_s t + G_d \ell_o \tan \alpha)^{(m+2)/2} \right]^{2/(m+2)}$$

$$[16]$$

where ΔG is given by equation (13). The split length in the first cycle, ℓ_o can be determined using eqn. (12). It only remains to identify values for λ_3, m, and $(\partial C/\partial \ell)$.

317

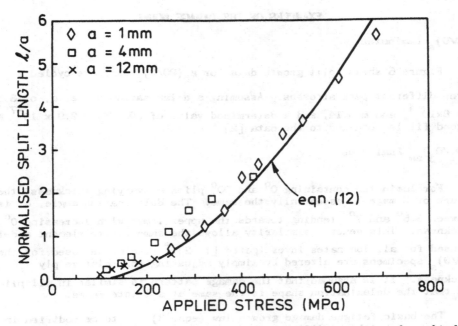

Fig. 4: Split growth for a (90/0)$_s$ laminate as a function of applied stress. The solid line represents eqn. (12).

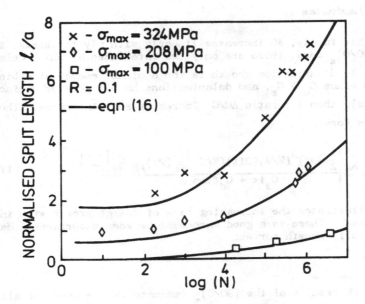

Fig. 5: Normalised split length for a (90/0)$_s$ laminate as a function of load cycles, showing the effect of varying peak stress. The solid lines represent the prediction of eqn. (16).

EXAMPLES OF THE DAMAGE MODEL

(90/0)$_s$ Laminates

Figure 5 shows split growth data for a (90/0)$_s$ laminate cycled at three different peak stresses. Assuming a delamination angle, α, of 3.5^o, $\lambda = 8 \times 10^{-4}$, and m = 14, and a determined value of $(\partial C/\partial \ell) = 2.9 \times 10^{-7}$ N^{-1}, a good fit is obtained to the data [2].

(90$_i$/0$_j$)$_{ns}$ Laminates

For laminates containing 0° and 90° plies of varying thickness, the nature of damage is essentially the same. The delamination angle, α, is between 3.5^o and 7°, tending towards the upper limit with increasing 0° ply thickness. This generic similarity allows the same finite element model to be used for all laminates investigated [1, 2].* The meshes used for the (90/0)$_s$ specimens are altered by simply adjusting for relative ply thickness. It is assumed that the damage pattern is similar in all plies and that the delamination shape is the same at all interfaces.

The basic fatigue damage growth law (eqn. 3) has to be modified in order to cope with the different conditions controlling damage growth. For convenience, the growth laws are kept in differential form but used in integrated form.

(90$_i$/0$_i$)$_s$ Laminates

For this family, ΔG increases linearly with ply thickness; i×t. As for the (90/0)$_s$ case, there are only two interfaces at which delaminations can grow. If delamination growth is the dominant energy absorbing process (and it is since $G_d > G_s$, and delaminations have a greater surface area than splits), then the ratio $\Delta G/G_c$ increases as i increases. The fatigue law has the form:

$$\frac{d\ell}{dN} = \lambda_3 \left[\frac{\frac{1}{2}(\Delta\sigma_\infty)^2 (W/2)(2it)(W/2)_{FE}(2t)_{FE} \left. \frac{\partial C}{\partial \ell} \right|_{FE}}{G_s it + G_d \ell \tan \alpha} \right]^{m/2} \qquad [17]$$

Figure 6 illustrates the increasing rate of damage growth with increasing ply thickness. There is a good agreement between experimental data and the predicted split growth curves.

---

*Thus, the FE results of the (90/0)$_s$ laminate can be used for all laminates of the (90$_i$/0$_j$)$_{ns}$ family. Values of $(\partial C/\partial \ell)$FE for specimens of various cross-ply construction are given in [1, 2].

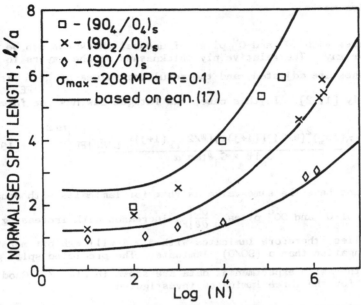

Fig. 6: Normalised split length as a function of load cycles for $(90_i/0_i)_{ns}$ laminates. The solid lines represent eqn. (17).

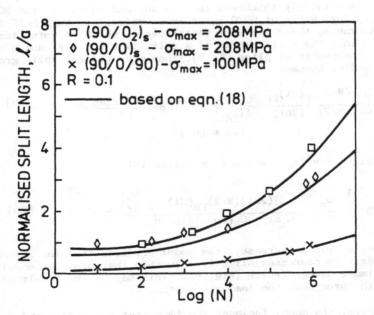

Fig. 7: Normalised split length as a function of load cycles for $(90_i/0_j)_{ns}$ laminates. The solid lines represent eqn. (18).

$(90_i/0_j)_s$ Laminates

Laminates with $90°$ and $0°$ plies of unequal thickness can be modelled in a similar way. The relative ply thicknesses of the layers in the finite element meshes are adjusted, and the resulting values of $\left.\frac{\partial C}{\partial \ell}\right|_{FE}$ are scaled appropriately [1, 2]. In this case, the fatigue law has the form:

$$\frac{d\ell}{dN} = \lambda_3 \left[\frac{\frac{1}{2}(\Delta\sigma_\infty)^2(W/2) \; ((i+j)t)(W/2)_{FE}(i+j)t_{FE} \left.\frac{\partial C}{\partial \ell}\right|_{FE}}{G_s it + G_d \ell \tan \alpha} \right]^{m/2} \qquad [18]$$

$\left.\frac{\partial C}{\partial \ell}\right|_{FE}$ does not have the same value as that for laminates with equal thicknesses of $0°$ and $90°$ plies. $\left.\frac{\partial C}{\partial \ell}\right|_{FE}$ increases with increasing thickness of the $0°$ plies; therefore laminates with $j > i$ will exhibit more rapid damage propagation than a $(90/0)_s$ laminate. The predicted split growth curves together with experimental data are shown in Fig. 7. Good agreement is achieved for the three laminates investigated.

$(90/0)_{ns}$ Laminates

In this series, ply thickness is constant and equal for the $90°$ and $0°$ laminae, only the number of 90/0 interfaces varies. Through half the laminate thickness, there are $2n-1$ interfaces at which delaminations can propagate. Since the surface area of a split is $n \times t \times \ell$, and $\partial C/\partial \ell$ scales with the number of plies [1, 2], then eqn. (8) for split growth in monotonic loading becomes:

$$\frac{1}{2} \frac{P^2}{nt} \frac{(W/2)_{FE}}{(W/2)} \frac{(2t)_{FE}}{(2nt)} \left.\frac{\partial C}{\partial \ell}\right|_{FE} = G_s + G_d \frac{(2n-1)\ell \tan \alpha}{nt} \qquad [19]$$

$$(G) = (G_c)$$

The fatigue law is, therefore, modified to:

$$\frac{d\ell}{dN} = \lambda_3 \left[\frac{\frac{1}{2}(\Delta\sigma_\infty)^2(W/2)(2nt)(W/2)_{FE}(2t) \;_{FE} \left.\frac{\partial C}{\partial \ell}\right|_{FE}}{G_s nt + G_d(2n-1)\ell \tan \alpha} \right]^{m/2} \qquad [20]$$

As the number of plies increases, the model predicts that the rate of damage propagation decreases slightly. In this instance, the model tends to underestimate damage growth (Fig. 8). This may be the result of unequal damage growth throughout the laminate thickness.

In summary, the model includes the important parameters and processes which govern fatigue damage behaviour. The ability to model fatigue damage

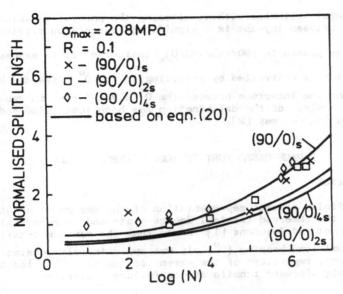

Fig. 8: Normalised split length as a function of load cycles for $(90/0)_{ns}$ laminates. The solid lines represent eqn. (20).

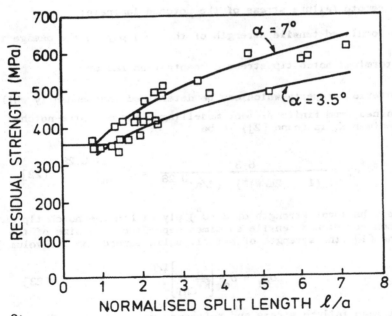

Fig. 9:

Residual strength as a function of split length for $(90/0)_s$ specimens.

growth in several laminates without altering the parameters which appear in the equations between lay-ups is a significant advance on previous methods.

For damage growth in $(90/+45/-45/0)_s$ laminates, it is assumed that damage evolution is controlled by splitting in the 0° plies and by delamination at the interface between the 0° ply and the innermost -45° plies [2]. The shape of the delamination has been identified and the model developed in a logical way [2].

DAMAGE-BASED POST-FATIGUE STRENGTH MODEL

Failure Criterion

Using a finite element representation of the damage at a notch tip, combined with the observed terminal damage-state and measured failure stress, Kortschot and Beaumont [1] showed that the maximum tensile stress in the principal load bearing (0°) ply was approximately constant at laminate failure, regardless of the extent of damage. This led them to propose a straightforward tensile stress failure criterion:

$$\sigma_{\infty f} = \frac{\sigma_{0f}}{K_t} \qquad [21]$$

$\sigma_{\infty f}$ is the remote failure stress of the notched laminate;

σ_{0f} is the localised tensile strength of the (0°) ply in the damage zone; and
K_t is the terminal notch tip stress concentration factor.

σ_{0f} is a material property which can be determined independently while K_t can be obtained from finite element modelling. For a centre-notched $(90/0)_s$ specimen K_t is found [2]) to be:

$$K_t = \frac{6.3}{(1 - (2a/W)^2) \ (\ell/a)^{0.28}} = C_1 \left[\frac{\ell}{a}\right]^{-0.28} \qquad [22]$$

Furthermore, the local strength of a (0°) ply within the notch tip damage zone of a centre-notched tensile specimen depends on the size of the damaged zone [1] (the strength of brittle solid depends on its volume):

$$\sigma_{0f} = \sigma_0 \left[\frac{V_0}{KV_{CNT}}\right]^{1/\beta} \qquad [23]$$

σ_0, V_0 is a mean failure stress and reference volume (= 1.88 GPa and 7.4

mm^3, respectively, for the 0° ply in a waisted (unnotched) $(90/0)_s$ tensile specimen [1].

KV_{CNT} is the equivalent volume of the (0°) ply in a CNT specimen; and β is the Weibull modulus (= 20) in the Weibull analysis (eqn. 23).

The stress distribution at the notch tip varies with split length, ℓ, and the delamination angle, α, at the tip of the split [1]. For $\alpha = 3.5^{\circ}$, $K = 0.0014 + 0.0025$ (ℓ/a); hence

$$\sigma_{Of} = \sigma_o \left[\frac{V_o}{(0.0014 + 0.0025 \, (\ell/a)) \, 8 \, (a^2 t)} \right]^{1/\beta} \qquad [24]$$

$$= C_2 \left[(0.0014 + 0.0025 \, (\ell/a)) \, a^2 \right]^{-1/\beta}$$

$V_{CNT} = 8 \, (a^2 \, t)$, where t is the thickness of the (0°) ply.

Strength of a Damaged $(90/0)_s$ Laminate

The notch tip stress concentration factor K_t (eqn.(22)) and the (0°) ply strength, σ_{Of}, (eqn. (24)) both depend on split length, ℓ. Combining eqns. (21)-(24), we obtain an expression for the laminate failure stress in terms of the normalised split length ℓ/a (for $\alpha = 3.5^{\circ}$):

$$\sigma_{\infty f} = \left[\frac{(1-(2a/w)^2)(\ell/a)^{0.28}}{6.3} \right] \sigma_o \left[\frac{V_o}{(0.0014 + 0.0025(\ell/a))8(a^2 t)} \right]^{1/\beta}$$

$$[25]$$

For a delamination angle of $\alpha = 7^{\circ}$, eqn. (25) becomes:

$$\sigma_{\infty f} = \left[\frac{(1-(2a/w)^2)(\ell/a)^{0.31}}{5.4} \right] \sigma_o \left[\frac{V_o}{(0.004 + 0.0065(\ell/a))8(a^2 t)} \right]^{1/\beta} \qquad [26]$$

Experimental data of residual strength of fatigue-damaged $(90/0)_s$ specimens (W = 24 mm, a = 4 mm) are plotted against the measured terminal split length (Fig. 9). These results are compared with the predicted strength (eqns. (25) and (26)). The quasi-static failure strength is about 360 MPa, which corresponds to $\ell/a = 1.3$. If higher values of ℓ/a are obtained during load cycling, higher residual strengths result. A quasi-static strength of 360 MPa represents a lower limit for the residual strength after fatigue; hence, the predicted residual strength curves are truncated by a horizontal line at $\sigma_{\infty f} = 360$ MPa, between $\ell/a = 0$ and $\ell/a = 1.3$. The strength data are bounded by eqns. (25) and (26).

Strength of Damaged $(90_2/0_2)_s$ and $(90/0)_{2s}$ Laminates

Since the shape of the notch tip damage zone is similar for all notched cross-ply carbon fibre laminates [1, 2], the model should apply successfully to other cross-ply laminates of equal thickness $(0°)$ and $(90°)$ plies. For Weibull modelling of $(0°)$ ply strength [1], the reference volume $(a^2 t)$ scales with the $(0°)$ ply thickness t. Equations (25) and (26) are, therefore, valid for all laminates of the general form $(90_i/0_i)_{ns}$. Figure 10 shows residual strength data plotted against ℓ/a for $(90/0)_{2s}$ and $(90_2/0_2)_s$ lay-ups. The $(90_2/0_2)_s$ material exhibited a delamination angle, α, of $7°$, approximately, whereas the $(90/0)_{2s}$ laminate had a narrower delamination zone. These measurements of α reflect the good agreement between data and theory. The effect of reducing the notch tip stress concentration factor is counteracted by the increased equivalent volume of material under stress. The strength model has neglected changes in material properties brought about by fatigue damage [2].

Fatigue Strength vs Cycles Diagram

In many ways, a prediction of post-fatigue notch strength is simpler than the prediction of notch strength in monotonic loading. We know that after fatigue loading there is no further damage growth during the residual strength test [2]. A successful strength prediction, therefore, depends on two distinct steps: (1) modelling damage growth; (2) equating strength to damage size at the end of a fatigue test.

The split length for a notched $(90/0)_s$ laminate after N load cycles (after eqn. (16)) is given by:

$$\frac{\ell}{a} = \frac{1}{G_d a \, \tan\alpha}\left[\lambda_3(\Delta G_t)^{m/2}\left[\frac{m+2}{2}\right](G_d \tan\alpha)N + (G_s t + G_d \ell_o \tan\alpha)^{(m+2)/2}\right]^{2/(m+2)}$$

$$= C_3\left[\Delta\sigma^m N + C_4\right]^{2/(m+2)} \qquad [27]$$

Substituting into eqn. (25) for $\alpha = 3.5°$:

$$\sigma_{\infty f} = C_5\left[1 + C_6\left(\left[\Delta\sigma^m N + C_4\right]^{2/(m+2)}\right)^{1/\beta}\frac{\left[\left[\Delta\sigma^m N + C_4\right]^{2/(m+2)}\right]^{0.28}}{C_1} \qquad [28]$$

where C_4, C_5 and C_6 are combinations of material constants which can be

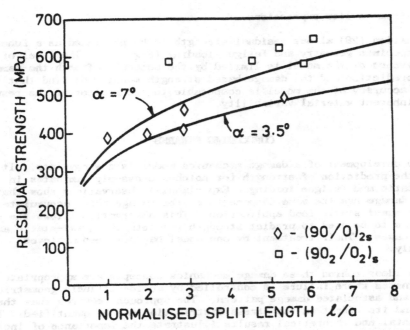

Fig. 10: **Residual strength as a function of split length in $(90_2/0_2)_s$ and $(90/0)_{2s}$ specimens.**

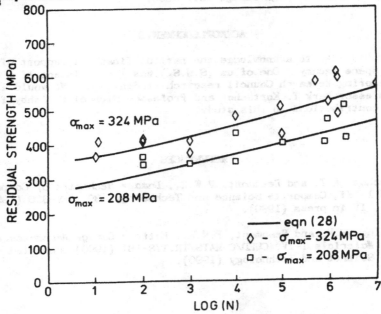

Fig. 11: **Residual strength as a function of number of load cycles for $(90/0)_s$ specimens.**

derived from earlier expressions.

Equation (28) allows residual strength to be predicted as a function of the specimen geometry and fatigue loading (Fig. 11). The level of effectiveness of the model is limited by the accuracy of both the damage growth prediction and the damage-based strength model. We find that the overall accuracy of the model is comparable to the scatter of the results due to inherent material variability.

CONCLUDING REMARKS

The development of a damage mechanics model is now complete. It allows the prediction of strength for notched cross-ply laminates in quasi-static and fatigue loading. Experimental observations show that fatigue damage has the same components as the damage which accumulates during a quasi-static load application. This observation allows the same parameters to be used to predict strength for both the quasi-static and fatigue cases. Their treatment by one model has not been achieved previously.

The ideas behind these damage mechanics concepts are appropriate for all laminates where failure is controlled by an identifiable geometrical feature and associated damage pattern. The approach requires that the damage and its effect on the stress distribution can be quantified. The experimental and theoretical results illustrate the importance of including the damage state in models for notched strength. This is called "model-informed empiricism" and is a powerful approach in the design of composite hardware of high damage-tolerance.

ACKNOWLEDGEMENTS

We would like to acknowledge the partial financial support of the European Space Agency. One of us (S.M.S.) was the recipient of a Science and Engineering Research Council research studentship. We would like to thank Professor Mark T. Kortschot and Professor Michael F. Ashby for their valuable contributions to this study.

REFERENCES

1. Kortschot, M.T. and Beaumont, P.W.R., *Damage Mechanics of Composite Materials I, II*, Composite Science and Technology, 39, 289-326 (1990). Parts III, IV in press (1990).

2. Spearing, S.M. and Beaumont, P.W.R., *Fatigue Damage Mechanics of Composite Materials I-IV*, CUED/C-MATS/TR.178-181 (1990) submitted to Composite Science and Technology (1990).

INFLUENCE OF FIBRE ORIENTATION

ON THE CRACKING AND FRACTURE ENERGY IN BRITTLE MATRIX COMPOSITES

ANDRZEJ M. BRANDT, Świętokrzyska 21, 00-049 Warsaw, Poland

ABSTRACT

The aim of the tests was to determine the relations between the fibre system orientation and fracture behaviour of brittle matrix composites reinforced with steel fibres.

An optimization problem was previously solved analytically in which the orientation angle of a parallel fibres system was determined from the condition of maximum fracture energy for a given cracking limit state. In the solution a system of approximate relations was proposed.

To verify the theoretical results a few series of tests have been executed on Portland cement concrete specimens with fibres (chopped and continuous) oriented at various angles. The results of measurements and observations confirmed theoretically determined relations between the angle of fibres orientation and the fracture energy.

The tests presented here have been executed to determine the influence of the fibre orientation on the crack propagation. Notched concrete specimens were subjected to bending and the crack propagation from the notch was closely observed and recorded. The results of the tests are presented in the form of fracture characteristics as functions of the fibre orientation and volume content.

INTRODUCTION AND REVIEW OF PRESENT KNOWLEDGE

The orientation of fibres is an important factor in the design of the composite material structure. Strong anisotropic effects may be created by fibres and there were several attempts both to determine the influence of fibre orientation on mechanical properties and to optimize it.

SANDHU [1] applied Hill's yield criterion to determine the filament orientation yielding maximum strength of transversaly isotropic material subjected to normal and transversal stresses. BRANDMEIER [2] has developed this problem for graphit/epoxy composites. COX [3] applied Michell's [4] theory to find out the disposition of fibres of least weight for a specified loading system. Several papers were devoted to optimum orientation of fibres in advanced composite materials, e.g. [5], [6], [7]. The variation of strength as a function of the fibre system orientation was studied by HAYASHI et al. [8] and KELLY and DAVIS [9]. In the last paper the variation of composite strength and also of the form of rupture with fibre orientation has been shown, which was later mentioned by many authors.

All above works concerned advanced composites with ductile matrices and the objective function for design or optimization was the composite strength. In brittle matrix composites early cracking appears because the ultimate matrix strain is much lower than that of the fibres: $\varepsilon_{mu} \ll \varepsilon_{fu}$. For technological and economical reasons the volume fraction of fibres is low: only with special technologies the fibre volume fraction may be higher than about 3% for steel fibres or 5% for other kinds of fibres.It is then obvious that the fibres and their orientation play a completely different role in cement based composites. Namely, the fibres control cracks opening and propagation and substantially improve the material toughness.The problem of the orientation of fibres in cement based matrices was presented first by MORTON [10] who has shown that the work of fracture may be considerably increased when fibres are not aligned with the direction of principal tensile strain. These results have been based on previous experimental works published by HING and GROVES [11], HARRIS [12] and MORTON and GROVES [13]. They have shown that the work of fracture calculated as the amount of work of external load absorbed by the element is the most important magnitude to be considered in the design of the brittle matrix composites. This approach was developed by BRANDT in a proposal of formulae for energy calculation, [14], and later in a simple optimization problem, [15], [16]. Analytical solution was next compared with results of the experimental studies, [17], [18].

The problem of the fibre orientation was recently examined by MASHIMA, HANNANT and KEER [19] who tested tensioned specimens reinforced with unidirectional or bidirectional systems of polypropylene fibres. Because the polypropylene fibres behave differently in the cracked matrix than the steel ones, other fracture mechanisms were considered than proposed in papers [15] and [16].

In the present paper the influence of the orientation of steel fibres on cracking characteristics is examined after experimental results.

OBJECTIVES OF REINFORCEMENT AND PRINCIPAL MECHANISMS

The fibres are introduced to brittle matrices usually neither to increase the initial cracking stress (1st crack load) nor to improve the ultimate composite strength. The first cannot be reached with low volume of fibres limited by technological reasons and high cost of fibres. Moreover, it is easy to improve the matrix by other ways, e.g. by decreasing the water/cement ratio. The increase of the ultimate composite strength is possible only by application of an efficient fibre reinforcement of adequate volume and good fibre-matrix bond. That ultimate strength is however achieved after considerable cracking, provided that cracks are sufficiently thin and correctly distributed to satisfy other requirements, e.g. impermeability and durability.

The main reason for the fibre reinforcement is usually the control of cracking process. It seems, therefore, that this reason should be clearly taken into account in the material design. For that aim in papers [15] and [16] the fracture energy has been proposed as an objective function, formulated as a sum of a few components related to two main mechanisms:
- debonding of the fibres, and
- pulling-out the fibres from the matrix subjected to cracking.

These mechanisms may be specified in different ways and consequently presented in different analytical forms. The relations proposed in [16] were

based on several simplifying assumptions, namely that:
- all fracture energy involved in the rupture of the specimen is represented by the considered mechanisms,
- the specified fracture mechanisms are not interrelated.

Among other simplifications, the multiple cracking was not taken into account. Moreover, the angle of the aligned fibre system with the direction of principal tensions was treated as the only variable.

The proposed relations may be improved by the introduction of more sophistic mechanisms, without modifying the principle of the design method, based on the conviction that the best solution corresponds to the maximum area under the load-deflection or stress-strain curve. That minimum reflects the best crack control by the fibres.

COMPARISON OF THEORETICAL AND EXPERIMENTAL RESULTS
FOR UNNOTCHED SPECIMENS

The approach developed in papers [15] and [16] was illustrated by examples of optimal solutions for an element subjected to tension and another - to bending. As an objective function the fracture energy accumulated up to a specified limit state was selected. The calculated results were obtained after derivation of proposed simplified expressions with respect to the only variable: the angle of fibre system orientation. Later, the tests of specimens with various fibre orientation were executed and tested. All details of calculation and testing may be found in papers [17] and [18] and final results are summarised in Figs. 1a) and b). From these curves certain

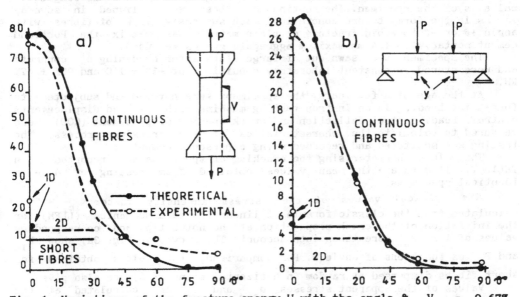

Fig. 1. Variations of the fracture energy U with the angle ϑ, V_f = 0.67%, results of calculations and tests, after [17]; a) specimens subjected to axial tension, b) specimens subjected to bending

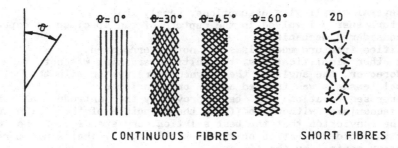

Fig 2. Fibre orientation systems in tested specimens.

confirmation of theoretical results may be concluded, at least in the general shape of the curves and characteristic numerical values. However, there were no tests of elements with small values of angle $0° < \theta < 30°$ and the existence of an extremum for $\theta \neq 0°$ was neither confirmed nor denied.

TESTS OF NOTCHED SPECIMENS

The specimens were prepared with the same mix proportions and fibre distribution as described in [17]. The reinforcement was made with mild steel round and straight wire 0.4 mm in diameter, continuous or chopped into 40 mm fibres. The results reported here concern only one volume fraction of continuous or short fibres V_f = 0.67% . The fibre orientation is shown in

Fig. 2, the angle θ being measured between aligned fibres and the longitudinal axis of the specimen. The continuous fibres were aligned in several layers in the forms before concreting, each successive layer of fibres with angle $+\theta$ or $-\theta$ to avoid fracture of plain matrix. As matrix the Portland cement mortar with 1.4 mm maximum aggregate grains was used.

The specimens were sawn out of large slabs after hardening of concrete and were stored in constant laboratory conditions of $+18° \pm 1°C$ and $90\% \pm 2\%$ RH.

At the age of a few years, the specimens were notched and subjected to four-point bending in an Instron testing mashine with the head displacements control. Load, central deflection and crack opening displacement (COD) were measured to calculate the characteristics of the cracking process. The testing was monitored and recorded using a personal computer.

The values characterising the cracking of specimens are reported in Table 1. All data are the mean values obtained from testing of 4 - 6 identical specimens.

The critical values of the stress intensity factor K_{Ic} were calculated from the classic formula of linear fracture mechanics (LEFM) for the initiation of the crack propagation at the notch tip and corresponding values of load P_{cr} were taken into account. The curves in Fig. 3a) show K_{Ic} and P_{cr} as functions of angle θ. For comparison, the values obtained for short fibres dispersed at random in horizontal planes (2D) are also given.

Values of the apparent stresses σ_{app} and σ'_{app} are calculated at the notch tip for two stages: crack opening and maximum load, respectively.

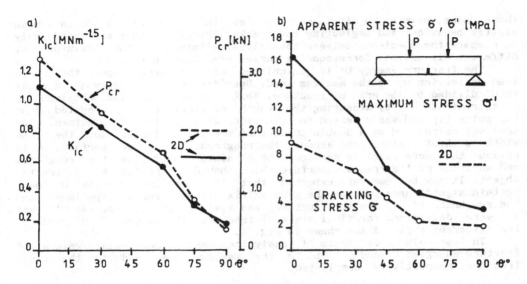

Fig. 3. Variations of characteristic values with the angle ϑ; a)variation of
P_{cr} - load corresponding to 1st crack opening and of K_{Ic};
b)variation of the apparent stresses σ_{app} and σ'_{app} corresponding
to cracking initiation and to the minimum load, respectively.

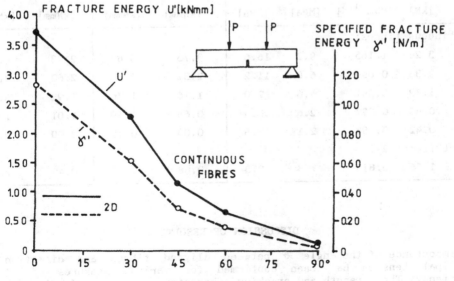

Fig. 4. Variation of fracture energy U'and specific fracture energy γ' with
the angle ϑ.

These values have no physical meaning, as they are calculated assuming elastic behaviour and neglecting the cracks. Therefore, they are used only to compare the specimens between themselves, eliminating the influence of different dimensions. Corresponding curves are shown in Fig. 3b).

The fracture energy U' is calculated as the area under the curve load–deflection up to the maximum load. Specific fracture energy γ' is equal to U' divided by the area of fracture. That last magnitude was obtained in an approximate way by measuring the length of crack which propagated from the notch tip and was observed on both sides of each tested specimen. The area was calculated as a double product of the crack length and the beam width, without taking into account the roughness of fracture surface. It is obvious that more exact values could be obtained considering the roughness and waviness of the fracture surface which should be treated as a fractal object. It may be, however, expected that the obtained results do have certain significance because the same matrix was used for all specimens and the roughness of all fracture surfaces was similar. In the same way U'' and γ'' were determined for final state of fracture. The values of U' and γ' for different angles ϑ are shown in Fig. 4.

In last columns of Table 1 analytical and experimental values of fracture energies U' and U'' up to the maximum load and to the final fracture, respectively, are listed.

TABLE 1
Calculated and experimental values of characteristic parameters

Experimental values					Experimental Values Fracture		Calculated values energy	
Angle	Load	Coeff.	Stress	Stress				
ϑ	P_{cr}	K_{Ic}	σ_{app}	σ'_{app}	U'	U''	U'	U''
[°]	[kN]	[MNm$^{-1.5}$]	[MPa]	[MPa]	[kNmm]	[kNmm]	[kNmm]	[kNmm]
continuous fibres:								
0	3.26	1.165	9.2	16.6	3.73	5.82	2.73	4.02
30	2.36	0.850	6.8	11.2	2.32	3.10	2.62	3.03
45	1.72	0.591	4.6	7.0	1.16	1.76	1.92	2.70
60	0.85	0.327	2.6	3.9	0.66	1.19	1.01	1.39
90	0.42	0.198	2.1	3.6	0.08	0.10	0.00	0.00
short fibres 2D:								
	1.66	0.815	7.3	9.3	0.88	2.82	0.49	2.69

DISCUSSION OF RESULTS

The importance of the angle ϑ between aligned fibres and direction of principal tension has been confirmed for various measures of fibre efficiency. The strength and cracking characteristics depend directly on that angle.

The shape of all obtained curves is, however, essentially different from that shown for other kinds of composites, for example by KELLY and DAVIES [9]. The difference is probably related to such mechanisms as pull-out of fibres and passing of fibres across microcracks and cracks which appear differently in ductile and brittle materials.

The shape of the curves seems to indicate that the maximum of efficiency is related to angle $\vartheta \neq 0^{\circ}$, probably around 10° - 15°. This situation was not observed because there were no specimens with such reinforcement, but it would be in agreement with theoretical considerations published in [15] and [16]. Perhaps future tests will give more information on that question.

In vue of the influence of selected angle ϑ on the efficiency of the fibre reinforcement, it seems appropriate to consider this parameter in the material design of the fibre reinforced cements and concretes. The execution of a system of short steel fibres with arbitrary angle ϑ is possible using several methods: magnetic linearisation of fibres in the fresh cement mix, casting of the mix through a kind of comb, shotcreting with appropriately designed parameters, etc.

In the case of any other kind of fibres or matrices it is also possible to determine adequate fracture mechanisms and to deduce analytical relations. Also the methods of execution should be adapted to properties of fibres and matrices.

The problems of fibre orientation examined here represent only a first step towards the optimization of the material structure of composites.

ACKNOWLEDGMENT

This research was supported by the Project CPBP 02.01, coordinated by the Institute of Fundamental Technological Research, Polish Academy of Sciences in Warsaw.

REFERENCES

1. Sandhu, R.S., Parametric study of Tsai's strength criteria for filamentary composites, AFF-TR-68-168, 1969.

2. Brandmaier, H.E., Optimum filament orientation criteria, J.of Comp. Materials, 4, July 1970, 422-425.

3. Cox, H.L., The general principles governing the stress analysis of composites. In Proc.Conf. Fibre Reinforced Materials: Design and Engineering Applications, Instn.of Civil Engineers, pap. No. 2, London, 23-24 March 1977, 9-13 and disc.

4. Michell, A.G.M., The limit of economy of material in frame structures, Phil. Magazine, 8, 47, 1904, 589-597.

5. Structural Design Guide for Advanced Composite Applications, Wright--Patterson AFB, 1968.

6. Dow, N.F., Rosen, B.W., Shu, L., Zweben, C.H., Design criteria and concepts for fibrous composite structures, N68-14176, 1967.

7. Chamis, C.C., Design of composite structural components. In Composite

Materials, Broutman, L.J., Krock, R.H., 8, Academic Press 1975.

8. Hayashi, I., Mori, K., Kaneko, S., Mogi, F., Effect of fiber orientation on mechanical properties of short fiber reinforced composite materials. In Proc. of the **1st Japan Congr. on Mat.Research**, Kyoto 1978, 287-293.

9. Kelly, A., Davies, G.J., The principles of the fibre reinforcement of metals, **Metallurgical Reviews**, 10, 37, 1965, 1-77.

10. Morton, J., The work of fracture of random fibre reinforced cement. **Materials and Structures** RILEM, 12, 71, 1979, 393-396.

11. Hing, P., Groves, G.W., The strength and fracture toughness of polycrys-talline magnesium oxide containing metallic particles and fibres. **J. of Mat. Science**, 7, 1972, 427-434.

12. Harris, B., Varlow, J., Ellis, C.D., The fracture behaviour of fibre reinforced concrete. **Cem.and Concr.Res.**, 2, 1972, 447-461.

13. Morton, J., Groves, G.W., The cracking of composites consisting of dis-continuous ductile fibres in a brittle matrix - effect of fibre orienta-tion. **J.of Mat.Science**, 9, 1974, 1436-1445.

14. Brandt, A.M., On the calculation of fracture energy in SFRC elements subjected to bending. In Proc.Conf. **"Bond in Concrete"**, P.Bartos ed., Paisley 1982. Appl. Science Publishers London 1982, 73-81.

15. Brandt, A.M., On the optimization of the fiber orientation in cement based composite materials. In Proc.Int.Symp.**"Fiber Reinforced Concrete"**, Detroit 1982, G.C. Hoff ed., American Concrete Institute 1984, 267-285.

16. Brandt, A.M., On the optimal direction of short metal fibres in brittle matrix composites. **J. of Mat.Science**, 20, 1985, 3831-3841.

17. Brandt, A.M., Influence of the fibre orientation on the energy absorp-tion at fracture of SFRC specimens. In Proc. of Int.Symp.**"Brittle Matrix Composites 1"**, Jablonna 1985, A.M.Brandt, I.H.Marshall eds, 403-420,Elsevier Applied Science Publishers 1986, 403-420.

18. Brandt, A.M., Influence of the fibre orientation on the mechanical properties of fibre reinforced cement (FRC) specimens. In Proc.**1st RILEM Congress** in Versailles, vol.2, Chapman and Hall, 1987, 651-658.

19. Mashima, M., Hannant, D.J., Keer, J.G., Tensile properties of polypropy-lene reinforced cement with different fiber orientation. Amer.Concr.Inst. **Materials Journal**, 87, 2, 1990, 172-178.

APPLICATION OF THE METHOD OF CAUSTICS
TO FIBER-REINFORCED MATERIALS

ERNST MOOR

Institute of Mechanics

Swiss Federal Institute of Technology, ETH-Zurich

CH-8092 Zurich

ABSTRACT

Cracks are studied in cross-ply carbon fiber-reinforced long, thin epoxy plates under tension while they are being loaded quasi-statically. The emphasis of this work lies on the measurement of the strain field at the crack tip. The shadow-optical technique has been applied for the first time to fiber-reinforced materials. The method is additionally used to obtain experimental results about the fracture and damage behaviour of fiber-reinforced plastics.

When the load is clearly below the fracture load, the shadow pattern shows a caustic as it is known from isotropic materials. The observed shape of the caustic is strongly dependent on the ratio of the stiffnesses in the two fiber directions. If the load is increased quasi-statically a damage zone appears at the crack tip which influences the shadow pattern. At higher loads the damage zone dominates and no caustic in the classical sense can be observed. Reducing the load one obtains a smaller but not vanishing damage zone. Once the crack starts running, disturbances due to secondary effects in the loading direction are superimposed and influence the running shadow pattern strongly.

INTRODUCTION

Pilot tests on specimens of laminates with different stacking sequences show that cross-ply carbon fiber-reinforced composites can break suddenly and catastrophically with a fast running "sharp" crack of known trajectory, if a notch is used to localize the running crack. This gives the idea to apply the shadow optical method of caustics to study fast crack propagation in fiber-reinforced plastics. The method of caustics is well established in fracture mechanics to measure the strain and stress intensity at the crack tip in the linear-elastic as in the elastic-plastic case [1], [2], [3]. Until now the technique has been applied to transparent materials as glass and some plastics and to metals. The application of the shadow optical method to fiber reinforced plastics has been presented for the first time in [4]. The present paper uses the method more extensively in conjunction with a better experimental setup and a new coating technique.

MATERIALS AND METHODS

Test Specimens

The single and double-edge-notched symmetrical test specimens usually have a width of 22 mm or 32 mm, a length between 300 and 600 mm and a thickness from 1.2 to 4 mm. All laminates have midplane symmetry. The sawed or milled slit is 4 or 5 mm deep and the groove of the notch has a radius between 0.02 and 0.25 mm.

The test specimens are sawed out of plates which are pressed using unidirectional prepreg. The prepreg consists of Torayca T300, Torayca M40 or Courtaulds Grafil XAS fibers and an epoxy matrix.

Coating technique

After the pressing the laminates are optically rough. The surface is not plane and smooth enough for the application of the shadow-optical method. Therefore the test specimens are coated with an additional film of epoxy resin of about 0.1 mm thickness. For this purpose a room-temperature curing epoxy resin is cast on the specimen and degased in a vacuum chamber. After this a smooth glass plate treated with dimethyldichlorsilane is laid on and removed after the epoxy has hardened. Furthermore a thin film of aluminium is created by applying the physical vapor deposition technique (PVD) in order to obtain a reflecting surface.

Experimental Setup

A Cranz Schardin high speed camera with 8 sparks is used. The flash duration is less than 200 ns and the sparks have a diameter of about 0.7 mm. For static measurements permanent point-like light sources with a diameter of about 0.2 mm are available. The optical arrangement is the one known for non transparent materials. The 8 sparks or point lights are imaged by a concave mirror onto 8 objectives of a camera array. The diameter of the concave mirror is 0.3 m and its focal length is 2 m. The focal length of the camera lenses is 580 mm. The distance z_0 between the reference plane and the specimen can be varied between zero and infinity. In the static case the "Kodak Technical Pan 4415" film is used and in the dynamic case the "Kodak T-Max 400 Professional".

The tensile testing machine is a very simple one. The specimen-holding jaws are free to rotate and the cross beam velocity is 0.8 mm/min.

RESULTS

Static case

Load << fracture load: When the applied load is much smaller than the fracture load, the shadow pattern shows a caustic as it is known from isotropic materials. The observed shape of the caustic is strongly dependent on the stiffness ratio E_{11}/E_{22} (Figure 1). E_{22} corresponds to the tensile stiffness in loading direction, E_{11} to the tensile stiffness in the crack plane. For a quasi isotropic laminate the shape of the caustic is nearly the same as for isotropic linear-elastic materials (Figure 1). Applying the linear-elastic fracture mechanics theory on a laminate with a sharp crack, the caustic can be calculated theoretically [5]. The experimentally observed shapes of the caustics correspond well to the theoretically calculated shapes.

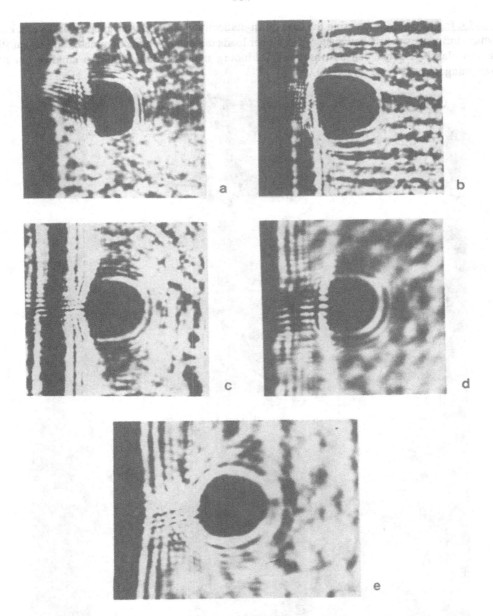

Figure 1. Typical experimental caustics of fiber-reinforced notched, long, thin plates under tension. The load is clearly below the fracture load (applied load/fracture load < 0.3).

a) cross-ply, carbon-epoxy, E_{11}/E_{22} about 0.8 (Test No. 198.1)
b) cross-ply, glass-epoxy, E_{11}/E_{22} about 1.25 (Test No. 265.3)
c) cross-ply, carbon-epoxy, E_{11}/E_{22} about 0.42 (Test No. 293.2)
d) cross-ply, carbon-epoxy, E_{11}/E_{22} about 2.0 (Test No. 211.1)
e) quasi-isotropic (Test No. 228.4)

Load < fracture load: Increasing the load quasi-statically a damage zone appears at the crack tip which influences the shadow pattern. At higher loads the damage zone dominates and no caustic in the classical sense can be observed. Reducing the load one obtains a smaller but not vanishing damage zone (Figure 2).

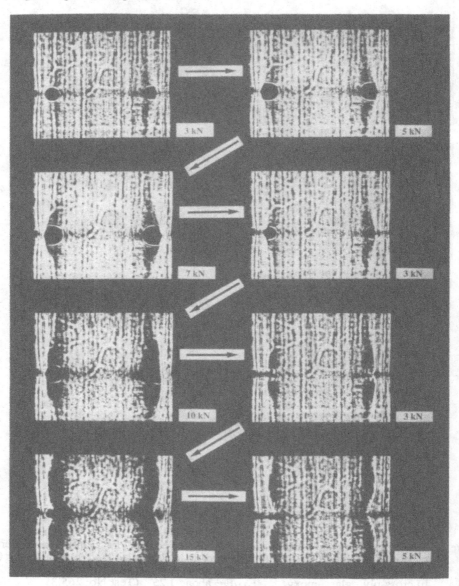

Figure 2. Typical damage process in a cross-ply carbon fiber-reinforced, long, thin plate under tension by increasing the load: E_{11}/E_{22} about 0.5; $z_0 = 0.63$ m. (Test No. 230) The specimen is double-edge-notched; width = 21.6 mm; length = 480 mm; thickness = 3.7 mm; fracture load = 50 kN.

Dynamic case

When the test specimen is loaded quasi statically up to a certain load the specimen breaks suddenly and catastrophically with a fast running "sharp" crack. In the dynamic case no caustic could be observed until now. Figure 3 shows a typically shadow pattern of a fast running crack.

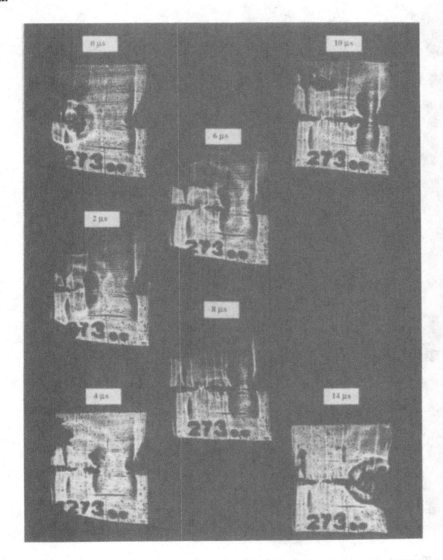

Figure 3. Typical shadow pattern of a fast running crack in a cross-ply carbon fiber reinforced, long, thin plate under tension. E_{11}/E_{22} about 2.0; $z_0 = 0.17$ m. (Test No. 273) The specimen is double-edge-notched; width = 22 mm; length = 325 mm; thickness = 1.4 mm; fracture load = 6.6 kN; crack speed about 1000 m/s.

Figure 4 shows why no caustic might be observed in the dynamic case. A damage zone around the crack tip propagates with the same velocity as the crack. This secondary effect is superimposed and influences the running shadow pattern strongly.

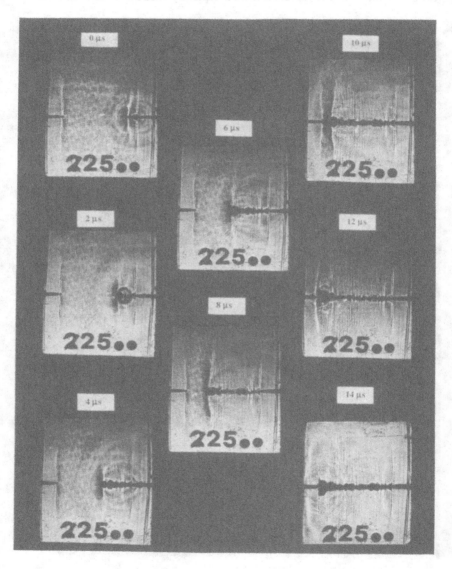

Figure 4. Typical damage zone around a fast running crack in a cross-ply carbon fiber-reinforced, long, thin plate under tension. E_{11}/E_{22} about 0.83; $z_0 = 0.07$ m. The specimen is double-edge-notched; width $= 22$ mm; length $= 440$ mm; thickness $= 1.9$ mm; fracture load $= 8.7$ kN; crack speed about 1000 m/s. (Test No. 225)

CONCLUSIONS

The presented results show clearly that the shadow optical technique is applicable to fiber-reinforced materials. Thanks to the high quality of the pictures, the technique becomes a powerful tool for the study of the fracture mechanics of laminates and the damage processes in fiber-reinforced materials.

If the applied load is much smaller then the fracture load, a caustic in the classical sense can be observed. The shape depends strongly on the stiffnesses of the anisotropic plate and is in good agreement with theoretical calculations based on the linear-elastic fracture mechanics theory. The shadow pattern technique can be used to examine the damage processes which occurs in a zone around the sharp notch if the load is increased. If a high speed camera is used the method can even be applied to study dynamic fracture behaviour.

ACKNOWLEDGEMENT

This paper was prepared in the course of research sponsored by the Swiss National Science Foundation.

The author would like to thank Prof. Dr. M. Sayir and M. Veidt for valuable discussions and comments.

REFERENCES

1. Kalthoff, J. F., Shadow Optical Method of Caustics. In Handbook of Experimental Mechanics, ed. A.S. Kobayashi, Prentice-Hall, Englewood Cliffs, 1987, pp. 430 - 500.
2. Theocaris, P. S., Elastic stress intensity factors evaluated by caustics. In Mechanics of Fracture, Vol. 7, ed. G.C. Sih, Martinus Nijhoff Publishers, The Hague, 1981, pp 189 - 252.
3. Rosakis, A. J. and Freund, L. B., Optical Measurement of the Plastic Strain Concentration at a Crack Tip in Ductile Steel Plate. J. Eng. Mat. Techn., 1982, 104, 115 - 120.
4. Moor, E., Experimental results on fast running cracks in single-edge-notched carbon fiber-reinforced plates under tension. In Euromech Colloquium no 238 on Crack Tip Modelling and Fracture Criteria, ed. F. Nilsson, Uppsala, 1988.
5. Veidt, M., Studien zum strukturdynamischen und bruchmechanischen Verhalten von Faserverbundplatten. Diss. ETH, 1991.

MODELLING OF ORTHOTROPIC LAYERED STRUCTURES CONTAINING CRACKS IN THE INTERFACES

C. PETIT
Laboratoire de Génie-Civil / E.R.I.N
I.U.T, 19300 EGLETONS
N. RECHO
E.R.I.N, Université de Haute Alsace
61 Rue A. CAMUS, 68093 MULHOUSE

ABSTRACT

To modelise the behaviour of composite layered structures containing cracks, one can distinguish different geometrical and material conditions which have an important effect on modelling. Generally, we are interested by two configurations : the crack tip is in the interface or perpendicular to the interface. Here, a cracked wood core plywood is represented by a finite element model. Adhesive interlayers behaviour are taken into account with joint finite elements. Current parameters in fracture mechanic are obtained from the Virtual Crack Extension Method (V.C.E.M) and are compared to experimental results.

INTRODUCTION

The Virtual Crack Extension Method (V.C.E.M) enables numerical users to obtain accurate results in homogeneous materials and isotropic layered structures [1-2]. In this paper, we are interested by the study of orthotropic layered materials such as wood core plywood where adhesive interlayers have a significative influence on their mechanical behaviour. As flaws exist in adhesive layers, we are studying crack growth in adhesive layers in a D.C.B specimen.

THE EXPERIMENTAL SPECIMEN

TABLE I
Different wood characteristics

	FIR	MERANTI[1]	BEECH
E_L (MPa)	12700	12700	14100
E_T (Mpa)	480	370	840
ν_{LT}	0.50	0.25	0.39
G_{LT} (MPa)	750	1100	2000
E_R (MPa)	930	1030	2100
ν_{TR}	0.35	0.33	0.31
ν_{RL}	0.03	0.02	0.05

TABLE II
Mechanical characteristics of the adhesive interlayer

	EPOXY	UREE FORMOL[1]
E (MPa)	5800	3000
ν	0.43	0.40

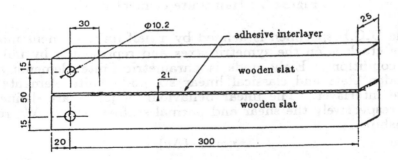

Figure 1 : Specimen D.C.B

The experimental results of energy release rates from adhesive double cantilever wood beam specimens are given by Duchanois [3]. The

---

[1] French name

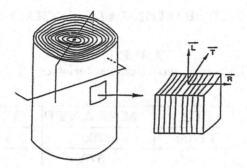

Figure 2 : Orthotropic axes

specimen size and shape is represented on figure 1. The orthotropic axes are named L, T and R according to figure 2 and the D.C.B specimen is oriented in LT plane. The elastic mechanical properties of different woods used in this study are given in table I. The mechanical properties of the adhesive interlayer is given by table II.

THE SPECIMEN MODEL

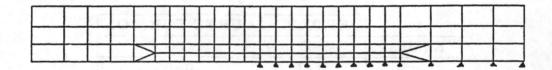

Figure 3 : Half finite element mesh

The D.C.B specimen is defined by a half finite element model (fig. 3). The crack is on the symetric axes and represented by nul vertical bound conditions. Eight node isoparametric finite elements are used for wooden slats and classical linear six node finite elements for the adhesive interlayer. The linear behaviour of joints are defined by k_s and k_n respectively the shear and normal stiffness [1][4]. We recall the relationship :

$$\tau = k_s \cdot \{\Delta u\}$$

$$\sigma = k_n \cdot \{\Delta v\}$$

$\{\Delta u\}$ is the relative displacement along the joint
$\{\Delta v\}$ is the relative displacement perpendicular to the joint.

Different values are obtained from table II according to a choice of the interlayer thickness. You notice that in the crack tip area, quarter

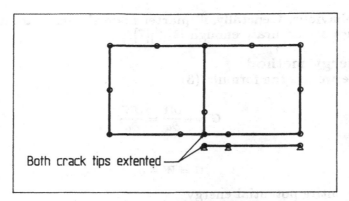

Figure 4 : The crack tip finite elements

node elements are used [5] for improved accuracy. Both finite elements represented on figure 4 are concerned.

THE DIFFERENT METHODS

Compliance method

The well known experimental method enables us to obtain the energy release rate G such as :

$$G = \frac{F^2}{2b} \cdot \frac{\partial C}{\partial a} \tag{1}$$

with $C = \frac{\lambda}{F}$ where λ is the displacement of external force F and C is called compliance of D.C.B specimen, b is the specimen width.

The formula (1) enables the experimental user to obtain G from the simple law : displacement λ against the force F easily drawn. Now, if we want to calculate G at crack tip in an other structure made of a similar material, we shall have to use numerical methods to estimate a crack parameter.

Virtual Crack Extension Method

Here, we propose the Virtual Crack Extension method (V.C.E.M) and we will compare our numerical values to experimental results. This method enables us to calculate G from formula (2) when exterior forces are constants [6].

$$G = -\frac{1}{2} \cdot \sum_{i=1}^{n} \{U_i\}^T \cdot \left[\frac{\partial K_i}{\partial a}\right] \cdot \{U_i\} \tag{2}$$

where $\{U_i\}$ and $[K_i]$ represent the displacement vector and stiffness matrix of the distorted finite element i. n is the number of distorted elements.

Several techniques enable us to have the result ; the user can choose the distorted finite element area which is used to have G value. Here, $n = 3$ according to figure 4 there are two orthotropic elements and one joint element in the crack tip area. So, in this case the result depends on local behaviour. Generally, if quarter node elements are used at crack tip, G values are accurate enough [1][2][7].

Strain energy method
In this case we use the formula (3)

$$G = -\frac{\partial \Pi}{\partial a} = \frac{\partial W}{\partial a} \qquad (3)$$

where

$$\Pi = W + \phi$$

with Π the entire potential energy
    ϕ the external force potential energy
    W the strain energy of the structure
with W defined :

$$W = \frac{1}{2} \int_{(V)} \sigma_{ij} \cdot \epsilon_{ij} \cdot dV$$

$$= \frac{1}{2} \sum_{i=1}^{N} \{U_i\}^T \cdot [K_i] \cdot \{U_i\}$$

with N the total number of finite elements
    Thus, this method takes into account nodal displacement values in all nodes of this structure. It is a global method.

The elastic bound beam method.
The D.C.B specimen can be represented by a beam such as figure 5.

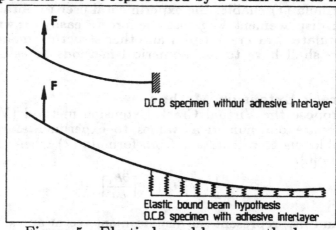

Figure 5 : Elastic bound beam method

    The analytical solution enables us to obtain the compliance C against a the crack length.

RESULTS AND COMMENTS

Global and local results

We have determined the G curves against a the crack length with two methods. Such as it has been described before, first local G values are calculated in the crack tip area with V.C.E.M which is also named finite difference method according to formula (2) (because $\frac{\partial K}{\partial a}$ is given by $\frac{\Delta K}{\Delta a}$ the finite stiffness variation).

Second, global G values are obtained from finite element calculations of the entire strain energy W (3). In fact, G is calculate from W at each step of the crack growth. The derivative values of W give us results which are quite the same as V.C.E.M. Figure 6 shows the two results of these methods on Fir wood.

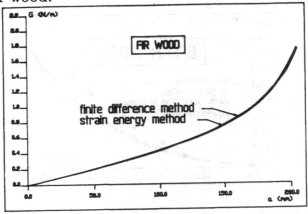

Figure 6 : G curves of Fir wood

These interesting results enable us to say that G, obtained with V.C.E.M from crack tip element informations, is accurate even if finite elements size is not very low in the near field and even if the orthotropic behaviour is considered.

Numerical and experimental results

In this part, we have compared experimental results [3] with differents ways of C compliance calculations. In figure 7 there are three curves obtained from : Compliance method (from finite element calculations), V.C.E.M and elastic bound beam method.

We can remark that Fir wood results are quite good, but Beech and Meranti [1] numerical results give lower values than experimental results. Those differences could come from some difficulties in determining orthotropic elastic parameters (table I) accurately.

---

[1] French name

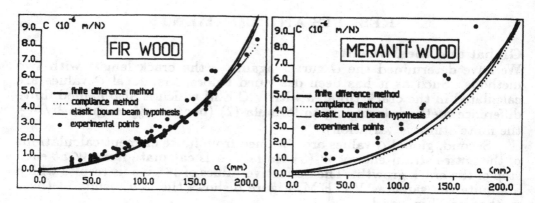

Figure 7 : Compliance values of woods without adhesive interlayer.

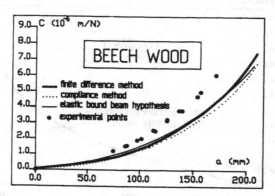

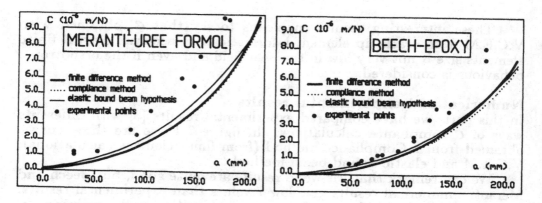

Figure 8 : Compliance values of wooden slats adhesive interlayer.

---

[1] French name

Figure 7 and 8 about adhesive wooden slats such as in figure 1 enable us to observe the same differences as before between experimental and numerical results. But, the elastic bound beam results compared with our results enable us to say that V.C.E.M associated with orthotropic finite elements and joint finite elements is quite as accurate as traditional elastic elements.

CONCLUSION

It has been proved that the Virtual Crack Extension Method is first an accurate method for homogeneous isotropic and elastic materials [6], then isotropic and elastic composite materials [2][7] and now orthotropic and elastic materials. Here, we have shown that a crack in adhesive interlayers can be modelised with finite joint elements. Now, we are interested in cracks which approach another layer separated from the cracked slat by an adhesive interlayer. A direct application of these results could be the fracture of plywood structures and more generally fractures of multilayered structures.

REFERENCES

1. C. PETIT, Modélisation de milieux composites multicouches fissurés par la mécanique de la rupture, Thesis of University of Clermont-Ferrand, Janvier 1990.
2. C. PETIT, A. VERGNE, Behaviour of Composite Materials with the Virtual Crack Extension Method., Conference proceedings of Localised Damage 90, Porsmouth (UK),26-28 June 1990.
3. G. DUCHANOIS, Mesure de la ténacité et étude du comportement mécanique des joints bois-colle, Thesis of I.N.P.L of Nancy (France), (1984).
4. A. VERGNE, Géométrie des surfaces de discontinuité et modélisation mécanique, Thesis of University of Limoges, (1984).
5. ROSHDY - S. BARSOUM, On the use of isoparametric finite elements in linear fracture mechanic, Int. J. of Num. Methods in Engineering, Vol. 10, pp 25-37, (1976).
6. H.D BUI et al., Mécanique de la rupture - Méthodes numériques pour l'ingénieur, I.P.S.I, Proceedings 18-20 Novembre 1986.
7. C. PETIT, Optimisation du calcul par éléments finis au voisinage d'une fissure, StruCoMe 1990, Conference proceedings edited by DATAID and AS & I, Novembre 1990.

A CRITERION OF MIXED MODE DELAMINATION PROPAGATION IN COMPOSITE MATERIAL

J.E. BRUNEL, D. LANG, D. TRALLERO
Louis Blériot Joint Research Center
AEROSPATIALE
BP 76, 92152 SURESNES Cedex, FRANCE

ABSTRACT

The aim of this study is to establish a propagation criterion based on Linear Elastic Fracture Mechanics for a laminated composite containing a delamination defect. This criterion is determined from two different approaches : first, calculations are achieved to perform mixed mode propagation tests : specimen geometry and load application. Second, tests are performed on T300-914 material, and Energy Release Rates are measured for different mixed mode propagation, covering pure mode I to pure mode II. Results show that, for this material and this kind of defect, a propagation criterion could be represented by the following relationship :

$$GI/GIc + GII/GIIc = 1$$

This criterion is consistent with experimental data.

INTRODUCTION

For several years, Linear Elastic Fracture Mechanics (L.E.F.M.) has been used to predict the mechanical behaviour of damaged structures in homogeneous materials. The use of this concept gives the following information :

- critical flaw length
- flaw propagation
- stability of propagation
- ...

in static loaded structures. The aim of this study is to establish a propagation criterion based on L.E.F.M. for a laminated composite containing a delamination defect. This criterion could be useful for the application of damage tolerance concept to composite structures.

STUDY METHODOLOGY

The delamination problem

The propagation of a delamination under a known load is generally complex and may be divided into a combination of "pure" modes. These basic modes are shown in figure 1 : mode 1 or opening mode, mode 2 or plane shear mode, mode 3 or tearing mode.

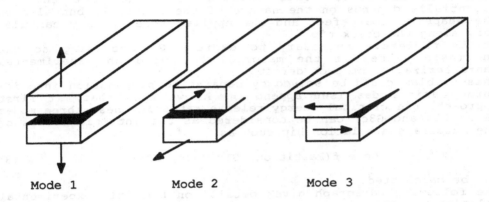

Mode 1   Mode 2   Mode 3

Figure 1. Propagation modes

Energy release rate

The propagation of a damage can be characterized by the energy release rate G, determined as follows :

$$G = \delta\Pi / \delta S \tag{1}$$

with Π : Total potential energy
     S : Area of created crack

It is worth mentioning that this figure, such as determined above, corresponds to a "virtual" crack propagation. It is obtained by the following calculation :

    Let a cracked structure be submitted to a given load. To an infinitesimal increment of the cracked area will correspond, for the same load, a variation in the potential energy of the system. The ratio of these two variations gives G.
This calculation can be made for simple cases by analytical methods. However, for more complex defects and loads, numerical methods such as the finite elements method will be used.
The critical energy release rate Gc corresponds to the energy released effectively by the structure during a defect propagation.

The comparison between G (calculated value) and Gc (experimental value) will then permit to say, on the propagation of a defect under loading :

- if G < Gc, no propagation

$$(2)$$

- if G ≥ Gc, propagation

The use of this simple concept is however difficult for laminated materials. This results from the fact that Gc essentially depends on the nature of the material, but also on the shape of the defect and the applied stress. They can also vary along the crack tip.
It is therefore necessary to express Gc according to the intrinsic values of the material to avoid an experimental characterization of each defect.
This problem could be solved by separating propagation into its three basic modes (see appendix). As we know that, at first approach, the critical energy release rates in these three modes (G1c, G2c and G3c) can be considered as the intrisic values of the materials a relationship such as

$$Gc = f(Partition, G1c, G2c, G3c) \qquad (3)$$

can be calculated.
The following paragraph gives details on how this experimental relationship can be obtained.

Experimental procedure

The scope of this study is restricted to delaminations combining modes 1 and 2. As experience and references show that for carbon/epoxy materials the value of G3c is significantly higher than the values G1c and G2c, this fracture mode will have a lower influence and its contribution is negligible. [1-3]
In order to cover the cracking modes in a field combining modes 1 and 2, it has been necessary to define several kinds of tests characterizing propagation for different partitions.
The detail of the technique used for partition calculation is mentioned in appendix.

The 4 following tests have been selected :

- DCB : Double Cantilever Beam (mode 1)
- MMF : Mixed Mode Flexure (mixed mode, see figure 2)
- CLS : Crack Lap Shear (mixed mode, see figure 2)
- ENF : End Notched Flexure (Mode 2)

According to the location of the delamination in the thickness for tests MMF and CLS, it is possible to study several mode partitions, what is shown on the following table.

TABLE 1
Field of studied partitions

Test type	G1 (%)	G2 (%)
DCB	100	0
MMF	50 < % < 60	40< % < 50
CLS	4 < % < 50	50 < % < 96
ENF	0	100

For each test, the critical energy release rate Gc is determined. The values of Gc according to the mode partition are plotted on a graph. A propagation criterion is then adjusted on these points, using G1c and G2c for these different partitions.

Material, test pieces

Tests have been performed on a unidirectional 32 ply-tape laying of T300-914.

RESULTS

Numerical results

Mode partition

The selected tests in mixed modes are shown in figure 2.

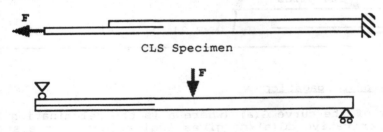

CLS Specimen

MMF Specimen

Figure 2. CLS and MMF tests, loading conditions

In the following table are quoted the 3 CLS and the 3 MMF configurations selected, as well as the combined partition for a crack length of 50mms.

TABLE 2
Partitions calculated for the different selected configurations

Test specimen		Mode 1 %	Mode 2 %
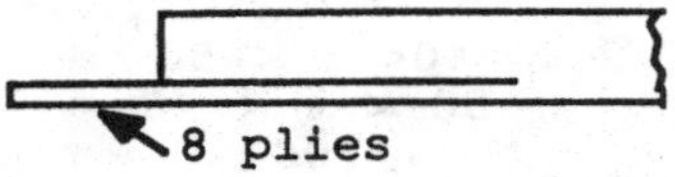 8 plies	CLS 8	4	95
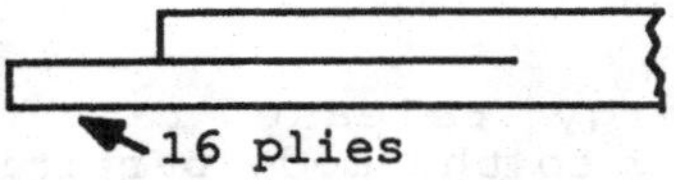 16 plies	CLS 16	21	78
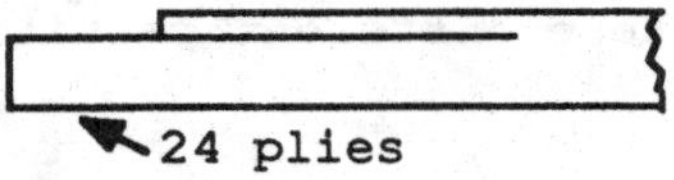 24 plies	CLS 24	42	58
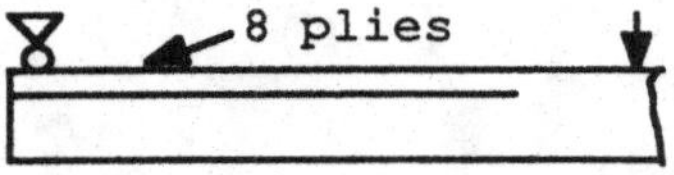 8 plies	MMF 8	51.5	48.5
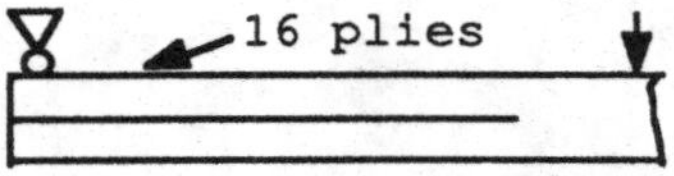 16 plies	MMF 16	58	42
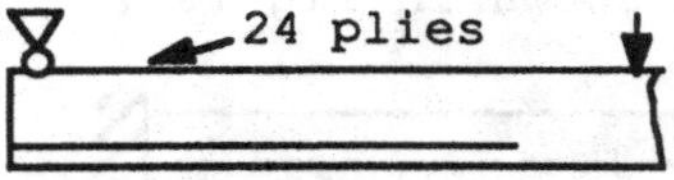 24 plies	MMF 24	61	39

Stability of propagation

The study of the curve G(a) (where a is the delamination length) and the derivative dG(a)/da gives indications on the stability or the unstability of the delamination propagation, if Gc is considered as independant from a. The following reasoning is therefore applied :

Let a cracked structure be submitted to a loading such as constant crack opening. For this structure, the energy release rate G is calculated at different crack lengths (a), and can be plotted on the following curve.

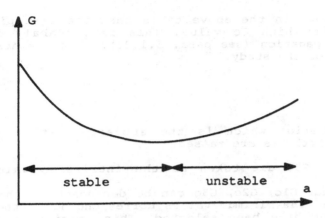

Figure 3. Curve G(a), fields of stable and unstable propagation

For a constant critical energy release rate Gc the analysis of the slope leads to the following conclusions .

- if dG/da ≤0 ; the crack releases less energy than necessary for propagation. Therefore, this energy must be supplied to the system to continue propagation (by increasing loading). This propagation is stable.
- if dG/da>0 ; in such case, during propagation, energy, higher than that required for propagation is released. Cracking is then not controllable, and therefore unstable.

This has been applied to the selected MMF and CLS configurations. The results of this study and the collation with experimental observations show an excellent consistency between numerical predictions and behaviour during testing.

Experimental results

All experimental results are summarized in the following table.

TABLE 3
Gc for the different partitions - material T300/914

Test specimen	Mean Gc (J/m2)	Standard Deviation	Mode 1 amount (%)
ENF	440	36	0
CLS8	475	10	4
CLS16	305	51	21
CLS24	289	27	42
MMF8	244	39	52
MMF16	237	37	58
MMF24	315	47	61
DCB	180	14,5	100

It can be seen in the above table that the test piece MMF 24 shows a quite high Gc value. This is probabaly due to the unstable propagation (see para. 3.1.2.), to the contrary of the other tests of the study.

Criterion

<u>Hypotheses</u>

In the criterion which is the subject of this study, the following hypotheses are raised.

- G1c, G2C, G3c are taken as the intrinsinc values of the material
- Knowing that G1c, G2c, G3c can be dependant on the length of the damage, the minimum values corresponding to the beginning of cracking have been selected. This permits to get rid of fibre bridging or structure effects, which can affect measurements and ensures that the proposed criterion can be maintained.
- In the same way, with a conservation goal, those values are obtained on unidirectional laminates test pieces. Other tests show that propagation values measured for delaminations located between 2 plies in the same direction are lower than those obtained for delaminations located between 2 differently orientated plies.

<u>Adjustment of a criterion</u>

A first idea to set up a propagation criterion is to draw up the ratio between the contribution of each mode and the critical propagation value, which leads to plot the curve G1/G1c + G2/G2c, according to the percentage of mode 1 (figure 4).

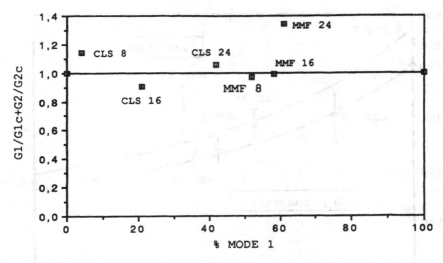

Figure 4. Representation of G1/G1c +G2/G2c = f (% of mode 1)
(G1c = 180 J/m² : G2c = 440 J/m²)

A first approximation show that propagation is governed by a criterion such as :

$$G1/G1c + G2/G2c = 1 \qquad\qquad (4)$$

This criterion may be differently expressed, according to the total energy release rate and the percentage of mode 1.

Knowing that :

$$G = G1 + G2 \qquad\qquad (5)$$

and stating that :

$$tg\alpha = G1/G2 \qquad\qquad (6)$$

The criterion (4) may be written as follows :

$$Gc = (1 + tg\alpha)*G1c*G2c/(G1c + tg\alpha*G2c) \qquad\qquad (4bis)$$

This formula evidences that knowing G1c and G2c is sufficient to deduce the delamination propagation value Gc for which a mode partition has been determined.

On figure 5 are represented the curves obtained by the previous formula (4bis) by using respectively the minimum and maximum values of G1c and G2c. On the same graph are plotted the experimental values with their standard deviation.

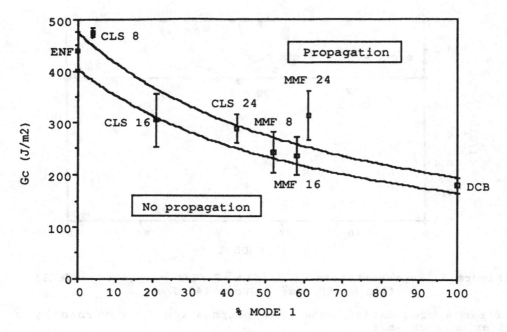

Figure 5. Curve Gc (J/m²) according to the percentage of mode 1.
Experimental values and surrounding curves.

It is worth noting that the value corresponding to the test MMF 24 is out of the criterion for the above mentioned reasons. Although the experimental dispersion is high, the selected criterion seems satisfactory.

CONCLUSIONS

The study presented focused on the delamination propagation in cross ply laminate composite materials.
Tests have been developed and performed in order to determine the critical energy release rate Gc combined with propagation of delamination in mode 1, mode 2 and mixed mode (1+2).
Tests performed on unidirectional material T300/914 permitted to adjust a delamination propagation criterion, expressed as follows :

$$G1/G1c + G2/G2c = 1$$

The results obtained all over the study work show the validity of the adopted approach, of partition calculations and of calculations of energy release rates carried out by various methods, among which the finite elements methods.
The analysis of the evolution of the energy release rate as a function of the delamination length leads to determine the stability or unstability of the propagation fields.

An excellent consistency between the numerical predictions and the experimental behaviour is observed.

REFERENCES

[1] D. BERTON
"Mise au point d'un essai de Mode III sur stratifié carbone-époxyde"
AEROSPATIALE, Rapport de stage, 08/88

[2] J.L. DONALDSON
"Interlaminar fracture due to tearing (mode III)". Materials laboratory, Air Force Wright Aeronautical Laboratories, Dayton, Ohio 4J433 -USA.

[3] G. BECHT, JW. GILLEPSIE
"Design and Analysis of Crack Rail Specimen for Mode II Interlaminar fracture" Center for Composite Materials, Department of Mechanical Engineering, University of Delaware, Newark, DE 19 716, USA, 1987.

APPENDIX

EVALUATION OF MODE PARTITIONS BY THE FINITE ELEMENTS METHOD.

When a crack grows, from a small length da, the released energy is equal to the work necessary to close the crack back to its initial length.

Let be :

$$G(a) = \lim_{\Delta a \to 0} \frac{1}{2\Delta a} \int_0^{\Delta a} \vec{\tau} . \Delta \vec{u} \, da$$

$\vec{\tau}$: Stress vector

$\Delta \vec{u}$: Vector of necessary diplacements to close the crack.

Formulated in finite elements, the expression becomes :

$$G(a) = \frac{1}{2b\Delta a} F_1 (U_2 - U_1) + \frac{1}{2b\Delta a} F_2 (V_2 + V_1) + \frac{1}{2b\Delta a} F_3 (W_2 - W_1)$$

with $\begin{cases} F_1; F_2; F_3 = \text{Node forces at crack tip} \\ U_i; V_i; W_i = \text{Displacement of nodes of crack tip} \end{cases}$

METHOD FOR DETERMINING NODE FORCES AND MOVEMENTS (three-dimensional modellisation)

First step :

—Modelling the test piece with a delamination length a+da (BACON module). See figure A hereunder.

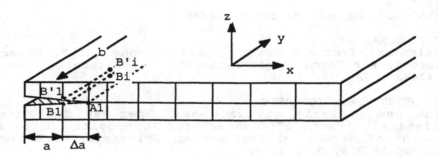

Figure A.

- Linear calculation with the multilayer three-dimensional element T11 (ASEF module)
- Movements released
- U_i, V_i, W_i of nodes B_i and B_i', in structural directions (x, y, z).

Second step

- Modellisation of the same test piece, but the crack is closed on a length a by connecting together the freedom degrees of nodes B_i and B_i' (Figure B)

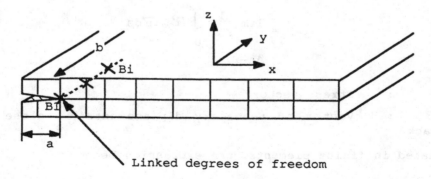

Linked degrees of freedom

Figure B.

- Linear calculation (ASEF module) with the multilayer three-dimensional element T11.

- Released reactions F1i, F2i and F3i of crack tip nodes (B_i, B_i')

Third Step

- Formatting and creating files containing the movements and forces at nodes Bi and Bi'.

Fourth step

- Calculation of the energy release rate and separation of the three crack propagation modes (Partition programme).

We then have :

$$G(a) = \frac{1}{2b\Delta a} \sum_{i=1}^{j} F_{1i} (U_i - U_i') + \frac{1}{2b\Delta a} \sum_{i=1}^{j} F_{2i} (V_i - V_i') + \frac{1}{2b\Delta a} \sum_{i=1}^{j} F_{3i} (W_i - W_i')$$

The three modes can be easily differentiated :

$$\text{Mode I} \quad \rightarrow \quad G_I(a) \quad = \frac{1}{2b\Delta a} \sum_{i=1}^{j} F_{3i} (W_i - W_i')$$

$$\text{Mode II} \quad \rightarrow \quad G_{II}(a) \quad = \frac{1}{2b\Delta a} \sum_{i=1}^{j} F_{1i} (U_i - U_i')$$

$$\text{Mode III} \quad \rightarrow \quad G_{III}(a) \quad = \frac{1}{2b\Delta a} \sum_{i=1}^{j} F_{2i} (V_i - V_i')$$

VALIDITY OF THE METHOD

At stages 1 and 2 above described, the linear static analysis modules (ASEF) provides for the potential energy P of the total structure, it is therefore easy to calculate $\Delta P = P$ (propagated) $- P'$ (not propagated), as we know the propagated surface dS we have directly :

$G(a) = \Delta P / \Delta s$
compared to G(a) calculated with the above formulas, we have :

$G(a) = G_I(a) + G_{II}(a) + G_{III}(a)$

This confirms the validity of the partition mode calculation.

FRACTURE AND FATIGUE DAMAGE IN A CSM COMPOSITE

J. J. L. MORAIS
Departamento de Física
Univ. de Trás-os-Montes e Alto Douro
Quinta dos Prados, 5000 VILA REAL, PORTUGAL

A. T. MARQUES, P. M. S. T. DE CASTRO
Faculdade de Engenharia da Universidade do Porto
Rua dos Bragas, 4099 PORTO, PORTUGAL

ABSTRACT

This paper presents a study on the mechanical properties, mode I fracture toughness and on the kinetics of fatigue crack growth in a CSM (chopped strand mat) E glass reinforced polyester resin composite. The environmental influence was studied through the immersion of the composite in an aquous solution of merthiolate tincture. The absorption kinetics of the composite when fully immersed in the aquous solution was also examined.

INTRODUCTION

The CSM (chopped strand mat) glass reinforced polyester resin composites have a wide range of engineering applications. The components and structures in which they are used, are, in general, subjected to variable service loads and are in contact with agressive environments. Moreover, they contain flaws, introduced during the fabrication process or in service. The characterization of the fracture and fatigue behaviour, and the evaluation of the environmental effects is obviously necessary, if reliable and safe structures are to be made.

This paper summarises the results of a study on (1) the absorption kinetics of a CSM E glass fibre reinforced polyester resin composite when fully immersed in an aquous solution of merthiolate tincture, (2) the effect of the fluid content on the mechanical porperties, (3) the effects of fluid absorption on the notched strength in mode I, (4) applicability of several predictive models for the notched strength in mode I and (5) the kinetics of mode I fatigue crack growth, including the effect of the fluid absorption.

MATERIALS AND METHODS

The material tested is a CSM E glass fibre reinforced CRYSTIC 272 resin composite. The resin CRYSTIC is a polyester isophtalic resin. The cure system employed was a MEKP catalyst (Butanox M50) and a cobalt accelerator (TP 395 VZ), in a weight fraction of 2% and 0.1%, respectively. Plates with four layers and a nominal weight fraction in fibre content of 40%, were made by hand-lay up process. A post cure of 3 hours at 80±2°C was used. All the specimens tested were cut from these plates in a milling machine.

The specimens used for the absorption tests were plates with $60 \times 60 \times 3.5 mm^3$ nominal dimensions. The specimens were dried in an oven at 60±2°C during 15 days, before placing them into an aquous solution of merthiolate tincture, at 25±1°C. They were periodically removed from the bath and weighted.

The tensile tests were carried out using straight sided specimens with $290 \times 25 \times 3.5 mm^3$ nominal dimensions [1]. The fracture and fatigue tests were conducted on DEN (double edge notched) specimens with $290 \times 50 \times 3.5 mm^3$ nominal dimensions, for which the finite-width isotropic correction factor is [2]:

$$Y = (1-2a/W)^{-1/2} \left[1.12-0.56(a/W) - 0.015(a/W)^2 + 0.091(a/W)^3 \right]. \qquad (1)$$

The notches were cut with a jeweler's saw of 0.3mm thickness. For the fracture specimens the a/W ratio was in the range 0.05 to 0.35; for the fatigue specimens a nominal a/W ratio equal to 0.1 was used.

Before testing some tensile specimens were immersed in the aquous solution of merthiolate tincture for a variable period up to 383 days, and some DEN specimens were immersed for 5 months. The bath temperature was maintained at 21±3°C.

All tests were conducted at 21±2°C in an INSTRON 1125 machine. The tensile and fracture specimens were tested at a crosshead displacement rate of 2mm/min. An instrumented 100 kN load cell and a displacement transducer with a 50mm gage length were used in order to assess the load versus displacement curve. The tension to tension fatigue tests were carried out under constant load amplitude and the loads were applied using a triangular wave forme with a frequency range of 0.5Hz to 1.0Hz. At specified moments the fatigue tests were stopped in order to obtain a load versus displacement curve up to the maximum load of the fatigue test and at a crosshead displacement rate of 2mm/min. Some fatigue specimens were tested in the presence of the liquid environment in the region of the crack extension.

RESULTS AND DISCUSSION

Absorption Kinetics Properties

The experimental results of weight change are presented in figure 1, as a function of square root of time. The weight change is defined as:

$$M = \frac{\text{Weight of moist material} - \text{Weight of dry material}}{\text{Weight of dry material}} \times 100. \qquad (2)$$

In the figure 1 the experimental results are also compared with the approximate Fick's law solution for a homogeneous plate of thickness h, small when compared with the other dimensions (a and b), and initially dry [3]:

$$M/M_m = (4/h)(Dt/\pi)^{1/2} \quad , \quad Dt^2/h^2 < 0.07 \tag{3}$$

$$M/M_m = 1-(8/\pi^2)\exp(-D\pi^2 t/h), \quad Dt^2/h^2 \geq 0.07 \tag{4}$$

In the equations (3) and (4), M_m is the maximum liquid content and D is related with the material diffusivity, D_0, by [3]:

$$D = D_0(1+h/a+h/b)^2. \tag{5}$$

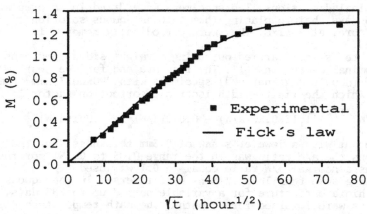

FIGURE 1. Weight change versus square root of time.

The maximum liquid content, taken equal to the maximum experimental value, and the calculated material diffusivity are:

$$M_m = 1.20\% \quad \text{and} \quad D_0 = 26.1 \times 10^{-8} \text{mm}^2/\text{s}. \tag{6}$$

Tensile Properties

The evaluation of the 0.5% secant modulus ($E_{0.5}$) and of the tensile strength (σ_0) with the liquid content (M) are presented in figure 2. The data obtained suggest that the examined tensile properties are not

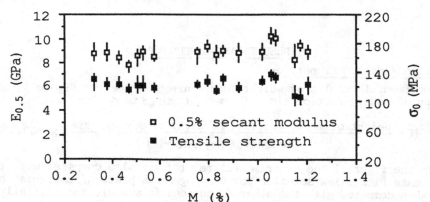

FIGURE 2. Tensile properties as a function of liquid content.

influenced by the immersion of the material in the aquous solution of merthiolate tincture at the room temperature, at least for immersion periods of up to 1 year.

Fracture Analysis

The experimental values of notched strength (σ_N), normalized by the unnotched strength (σ_0), are shown in figure 3 as a function of crack half-length (a). The notched strength is defined as the quotient between the maximum load and the unnotched transversal section area of the specimen. The notched strength of the satured material is lower than the notched strength of the dry material, and both fall below the notch insensitive line (figure 3).

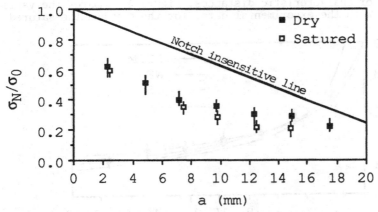

FIGURE 3. Notched strength of the dry and the satured material.

Four predictive models for the notched strength have been considered and compared with experimental data. LEFM is based on a single parameter, the stress intensity factor:

$$\sigma_N = K_{IQ}/[Y(\pi a)^{1/2}]. \qquad (7)$$

Waddoups, Eisenmann and Kaminski [4] introduced a two parametric model which predictive equation was:

$$\sigma_N = (\sigma_0/Y)\left[a_0/(a_0+a)\right]^{1/2} \qquad (8)$$

where a_0 is a characteristic distance assumed to be a material property. Nuismer and Whitney [5] proposed two models based also on the unnotched strength and on characteristic distances, the Point Stress (PS) model and the Average Stress (AS) model, for which predictive equations were respectively:

$$\sigma_N = (\sigma_0/Y)\left[1-a^2/(a+b_0)^2\right]^{1/2} \quad ; \quad \sigma_N = (\sigma_0/Y)\left[c_0/(2a+c_0)\right]^{1/2}. \qquad (9,10)$$

For the material tested, in the dry or satured condition, it is not possible to obtain a strictly valid critical value of the stress intensity factor, since the load versus displacement curves are not linear up to

rupture. Even so, the maximum load stress intensity factor was calculated
and the averaged value is presented in table 1, conjointly with the

TABLE 1
Characteristic parameters of the strength models.

Material	a_o (mm)	b_o (mm)	c_o (mm)	K_{IQ} (MPa m$^{1/2}$)
Dry	2.059	0.962	4.123	9.767
Satured	1.388	0.655	2.792	7.946

averaged values of characteristic distances. Figure 4 compares the various
strength models with the experimental data, for the dry and the satured
material.

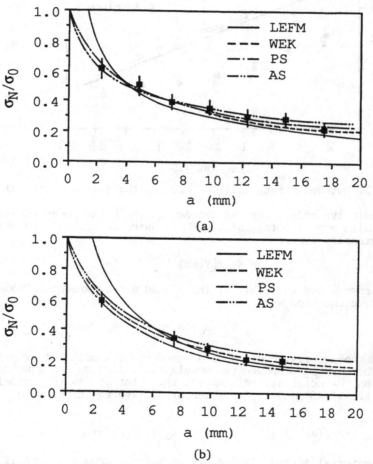

FIGURE 4. Comparision between strength models and experimental data for the
(a) dry and (b) satured material.

Fatigue Crack Growth

The variables examined and their effect on the fatigue life (number of cycles up to rupture, N_r) are summarised in table 2. It was observed that an increase in R (minimum/maximum load ratio), or in the maximum applied stress (σ_m), as well as the immersion in the aquous solution of merthiolate tincture up to saturation, imply a reduction of fatigue life.

TABLE 2
Fatigue variables and fatigue life

Material	Fatigue Cycle	Specimen	Fatigue Life
Dry	$R = 0$ $\sigma_m = 0.80\sigma_N$	1-A 2-A 3-A	2425 1407 6500
	$R = 0$ $\sigma_m = 0.65\sigma_N$	1-B 2-B	61500 21200
	$R = 0.5$ $\sigma_m = 0.80\sigma_N$	1-C 2-C	150 40
Satured	$R = 0$ $\sigma_m = 0.80\sigma_N$	1-D 2-D 3-D	512 751 477

The extension of the initial crack was measured using the compliance method [6]. The compliance calibration curves are shown in figure 5, for the satured and dry material. During the fatigue tests an equivalent crack

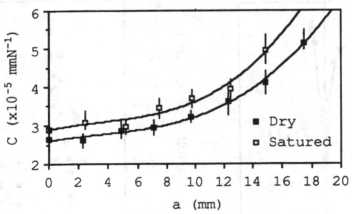

FIGURE 5. Compliance calibration curves.

length (a_e) was calculated from the compliance calibration curves and plotted against the cyclic ratio (actual number of cycles/fatigue life

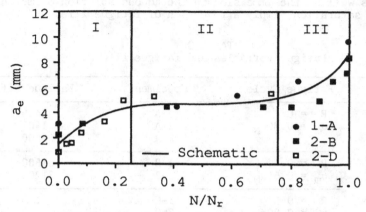

FIGURE 6. Fatigue equivalent crack growth.

ratio, N/N_r), as illustrated in figure 6. The shape of these propagation curves is not influenced by the variables examined and suggests the existence of three propagation phases (figure 6).

The dependency of the equivalent crack propagation rate (da_e/dN) with the range of the stress intensity factor (ΔK) is illustrated in figure 7(a). The Paris'law is applicable to the two latter propagation phases (figure 7(b)) which correspond to an effective growth of initial crack.

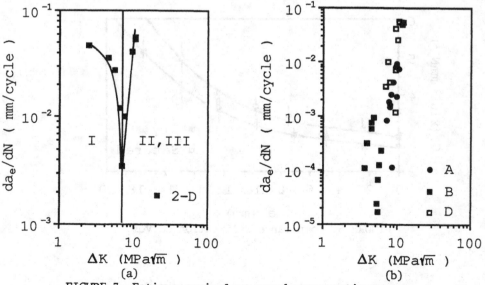

FIGURE 7. Fatigue equivalent crack propagation rate.

CONCLUSIONS

From this study, the following conclusions can be drawn:
- The Fick's law is adequate to describe the weight change of specimens fully immersed at 25°C.
- The 0.5% secant modulus and the tensile strength are not influenced by the liquid content, at least for immersion periods of up to 1 year.
- The mode I notched strength of the satured material is lower than the notched strength of the dry material. The models of Waddoups et al. and Nuismer et al. are adequate to describe the dependency of the notched strength with the crack length.
- An increase in R or in the maximum applied load, as well as the immersion up to saturation, imply a reduction of fatigue life. The equivalent crack propagation curves displayed three distinct regions, the Paris'law being applicable to the two latter phases.

REFERENCES

1. Lima, A. V., Simões, J. A., Marques, A. T. and de Castro, P. M. S. T., The influence of specimen geometry and test conditions on the tensile and fracture mechanics properties of GRP. In Developments in the Science of Composite Materials, ed. A. R. Bunsell, P. Lamicq, A. Massiah, Elsevier Applied Science, 1989, pp. 705-711.

2. Stress Intensity Factors Handbook, ed. Y. Murakamy, Pergamon Press, Oxford, 1987, (repr. 1988).

3. Whitney, J. M. Moisture diffusion in fibre reinforced composites, ICCM 2, Elsevier Applied Science, 1978, pp. 1584-1603.

4. Waddoups, M. E., Eisenmann, J. R. and Kaminski, B. E., Macroscopic fracture mechanics of advanced composite materials. J. Comp. Mat., Vol. 5, 1971, pp. 446-454.

5. Whitney, J. M. and Nuismer, R. J., Stress fracture criteria for laminated composites containing stress concentrations. J. Comp. Mat., Vol. 8, 1974, pp. 253-265.

6. Redjel, B. and De Charentenay, F. X., Résistance à la fissuration des matériaux composistes SMC. Part 2: Influence de la fraction volumique de fibres de verre, du temps de post-cuisson et de la temperature. Composites, Vol. 6, 1987, pp. 12-19.

NON LINEAR BEHAVIOUR OF CERAMIC MATRIX COMPOSITES

R. EL GUERJOUMA, B. AUDOIN, S. BASTE, A. GERARD

Université de Bordeaux I, Laboratoire de Mécanique Physique, URA CNRS 867
351 Cours de la Libération, 33405 TALENCE Cedex (France)

ABSTRACT

In the general framework of thermodynamics irreversible processes, the damage phenomenon is described by some internal variables taking into account differents non-linearities (micro-cracks propagation, closing-opening threshold) during load-unload cycles of ceramic-ceramic composites. An ultrasonic method provides the needed stiffness variations, inaccessible by classical strain measurements, required to identify the anisotropic damage. When it is possible, the confrontation of damage evaluation by these two metrologies shows a good agreement.

1.INTRODUCTION

Continuum damage mechanics has been an active subject of research in the last ten years [1,2]. It constitutes a phenomenological approach to the progressive material strength deterioration which induces rupture of volume representative element in the common sense of continuum mechanics or, more precisely, crack initiation [3]. The damage parameter takes into account, on a macroscopic scale, the microscopic deterioration (voids, microcraks, decohesions) of the material, by using the general framework of the thermodynamics of irreversible processes (TIP) with internal variables [4].

For composite materials, this decrease of strength phase, which precedes the macroscopic cracks propagation, shows itself by different manner, according to the relative stiffness between fiber and matrix [5]. Thus, the ceramic-ceramic composites (brittle-brittle materials) present a strongly non-linear behaviour [6]. Micrograph observations carried out at various stages of stress-strain curves allow to detect a severe multiple micro-cracking of the matrix [7].

Different macroscopic measures of damage are then suitable like measures of the stress-strain behaviour. These are the most appropriate from the mechanical point of view [8]. They can be obtained in terms of residual strain and absorbed energy [7] or elastic constants (Young's modulus and poisson's ratios) [9]. These quantities are measurable by conventional mechanics metrology. We emphasize that these metrologies authorize assessment of phenomenon only in the axis of load and do not allow a direct measure of the Young's modulus perpendicular to the loading direction and of the shear modulus requisites for identification of an anisotropic damage [10]. Also, another method must be used to characterize ceramic-ceramic composite mechanical behaviour and to measure the damage during a tensile test [11].

The objectives of this paper are multiple. In the framework of TIP (section 2) the damage phenomenon is described by some internal variables and the anisotropic elastic coupled with anisotropic damage behaviour law is given. We propose also an unidimensional behaviour law of ceramic-ceramic during load-unload cycles. The spreading of micro-cracks, is described as previously [9], by a damage internal variable and we introduce a new internal variable in the purpose to describe the threshold of closing and opening micro-cracks which was experimentally observed on strain hardening curves. This model is validated (section 3) using an experimental apparatus consisting of an ultrasonic immersion test device coupled with a loading machine [12]. The ultrasonic characterization of anisotropic solid is classically based on the measurement of the phase velocities of longitudinal and shear waves crossing the sample for various propagation directions [13]. The velocities decrease continuously with the applied stress. These variations (section 4) of the identified stiffness components are those required to assess the damage of the investigated plane. The global damage parameters are increasing differently with stress and are affected by the state of possible pre-existing microcracks. The unidimensional type of behaviour of these materials during load-unload cycles is validated also by ultrasonic method.

2. DAMAGE MODELS

The method developped here proceeds of a phenomenological point of view. This consists in the estimation of the evolution of microstructure by its result on the stress-strain behaviour of material. In the framework of TIP we postulate an energy equivalence between virgin and damaged material which automatically impose the symmetry of elastic tensor [14]. The technique based on the free energy (Helmholtz energy) gives an expression of the behaviour law which is not well suited to ultrasonic methods [15]. Also, it is necessary to search a dual formulation. We need to specify the damage tensorial nature of D and the transformation law $M(D)$ of the strain elastic tensor. An extension of the preceding step [9] is made with the free enthalpy. Introducing the effective elastic strain tensor ε^e_{eff}, connected with the strain tensor ε^e by:

$$\varepsilon^e_{eff} = (M^t(D))^{-1}[\varepsilon^e] \tag{1}$$

the energy equivalence [16] gives the relation between the stiffness tensor C (resp. C_o) of the damaged (resp.virgin material) :

$$C(D) = (M(D))^{-1} : C_o : (M^t(D))^{-1} . \tag{2}$$

Then,we obtain the constitutive law for the damaged material :

$$\sigma = C(D)[\varepsilon^e] = (M(D))^{-1} : C_o : (M^t(D))^{-1}[\varepsilon^e] . \tag{3}$$

For D, we choose a fourth rank diagonal tensor which privileges the composite symmetry axes. The non zero C_{ij} terms are :

$$C_{11} = C^o_{11}(1-d_1)^2 \; ; \; C_{12} = C_{21} = C^o_{12}(1-d_1)(1-d_2) ;$$

$$C_{22} = C^o_{22}(1-d_2)^2 \; ; \; C_{13} = C_{31} = C^o_{13}(1-d_1)(1-d_3) ;$$

$$C_{33} = C^o_{33}(1-d_3)^2 \; ; \; C_{23} = C_{32} = C^o_{23}(1-d_2)(1-d_3) ;$$

$$C_{44} = C^o_{44}(1-d_4)^2 \; ; \; C_{55} = C^o_{55}(1-d_5)^2 ; \; C_{66} = C^o_{66}(1-d_6)^2. \tag{4}$$

For any loading, these relations allow the whole determination of damage tensor components d_i by :

$$d_1 = 1 - \left(\frac{C_{11}}{C_{11}^0}\right)^{1/2} , \qquad d_4 = 1 - \left(\frac{C_{44}}{C_{44}^0}\right)^{1/2} ,$$

$$d_2 = 1 - \left(\frac{C_{22}}{C_{22}^0}\right)^{1/2} , \qquad d_5 = 1 - \left(\frac{C_{55}}{C_{55}^0}\right)^{1/2} , \qquad (5)$$

$$d_3 = 1 - \left(\frac{C_{33}}{C_{33}^0}\right)^{1/2} , \qquad d_6 = 1 - \left(\frac{C_{66}}{C_{66}^0}\right)^{1/2} .$$

We note that the off-diagonal components variations of the stiffness tensor are not essential for the evolution of damage but they should be verified to consolidate the model validity.

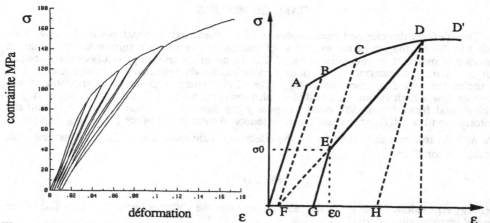

Figure. 1 Experimental stress-strain curve

Figure.2 Schematic stress-strain curves and notation during load-unload process

The constitutive law (3) and (4) is in good agreement with the global behaviour of ceramic-ceramic composites in the case of monotonous loading, but this equation is not verified during load-unload processes. During the loading processes a hardening of the material corresponding to a closing of the micro-cracks is observed (fig.1). These micro-cracks tend to reopen during the loading processes of subsequent cycles. The spreading of the micro-cracks resumes when the stress level exceeds the one of the previous cycle (fig. 1). Three different parts are distinguishable on the schematic curves (fig.2) of stress-strain evolution. In the first part of the curves (GE fig.2), the slope is parallel to the one of the initial loading; the damage is equal to zero while the stress is lower than the critical load of closing microcracking (E fig.2). In the second part, we have another slope (ED fig.2) corresponding to constant damage and reopening of the micro-cracks created during the previous cycle. In the third part the damage increases when the maximal stress of the previous cycles is exceeded (DD' fig.2). In an unidimensional case, according to the framework of TIP [4], we introduce a scalar internal variable D corresponding to damage and the proportion of opened micro-cracks β, which is a new state variable. Evidently this quantity should satisfy the relation :

$$0 \le \beta \le 1. \tag{6}$$

Moreover, we choose as state variable the strain ε and preserve the assumption of its partition into an elastic ε_e and anelastic ε_{an} strain ($\varepsilon = \varepsilon_e + \varepsilon_{an}$). We note ε_o, σ_o the strain and stress on the threshold (E fig.2) and we choose as free energy $\rho\psi$ given by :

$$\rho\psi = \frac{1}{2} \beta \, E \, \varepsilon_e^2 + \frac{1}{2}(1 - \beta) \, E(1 - D) \, \varepsilon_e^2 + (\varepsilon_e - \varepsilon_o) \, I(\beta) . \tag{7}$$

Noting E the Young's modulus of virgin material, the first and second terms are the volumetric free energies of opened and closed micro-cracks and the third term is the free energy which takes into account the possible interactions between opened and closed micro-cracks. This last term depends on the strain difference relative to the strain threshold and on β. This quantity β must verify relation (6) and consequently justifies our choice where I is the indicator function of the interval [0,1] (convex funcfion : $I(x) = 0$ if $x \in [0,1]$, $I(x) = +\infty$ if $x \notin [0,1]$).

We assume that the behaviour is not dissipative, or elastic, with respect to the proportion β. It results :

$$\frac{\partial \rho\psi}{\partial \beta} = 0 \ \Rightarrow \ ED\varepsilon_e^2 = (\varepsilon_e - \varepsilon_o)\frac{\partial I(\beta)}{\partial \beta} . \tag{8}$$

The first term of this equality is positive, also we deduce that : if $\beta = 0$, $\varepsilon_e - \varepsilon_o < 0$ (because $\frac{\partial I(\beta)}{\partial \beta} < 0$), if $\beta = 1$, $\varepsilon_e - \varepsilon_o > 0$ (because $\frac{\partial I(\beta)}{\partial \beta} > 0$) and $0 < \beta < 1$, $\varepsilon_e - \varepsilon_o = 0$.

Thus, on the path GE (fig.2) all micro-craks are closed ($\beta = 0$), on ED all micro-cracks are opened ($\beta = 1$) and on the threshold (E fig.2) we have a β proportion opened micro-cracks ($0 < \beta < 1$). By dealing with the intrinsic dissipation only, the Clausius - Duhem inequality derived from the second principle of thermodynamics must be satisfied by any velocities $\dot{\varepsilon}$, $\dot{D}$ with :

$$\rho\psi(\varepsilon,D,\beta) = \frac{1}{2}\beta \, E(1 - D) \, (\varepsilon - \varepsilon_{an}^o - \varepsilon_{an}^c)^2 + \frac{1}{2}(1 - \beta) \, E(\varepsilon - \varepsilon_{an}^o)^2 + (\varepsilon_e - \varepsilon_o) \, I(\beta) \tag{9}$$

and we deduce classically :

$$\sigma = \beta \, E(1 - D) \, (\varepsilon - \varepsilon_{an}^o - \varepsilon_{an}^c) + (1 - \beta) \, E(\varepsilon - \varepsilon_{an}^o) \Leftrightarrow \sigma = E(1 - \beta D) \, (\varepsilon - \varepsilon_{an}^o - \beta\varepsilon_{an}^c) , \tag{10}$$

$$Y = -\frac{\partial \psi}{\partial D} = \frac{1}{2}\beta \, E(\varepsilon - \varepsilon_{an}^o - \varepsilon_{an}^c)^2 + \sigma\frac{\partial \varepsilon_{an}^o}{\partial D} + \beta \, E(1 - D)(\varepsilon - \varepsilon_{an}^o - \varepsilon_{an}^c)\frac{\partial \varepsilon_{an}^c}{\partial D} , \tag{11}$$

where σ is the actual stress, Y the dual variable associated with D, ε_{an}^o anelastic strain on threshold (OG fig.2) and ε_{an}^c the complementary strain after this (GH fig.2). The intrinsic dissipation is only :

$$YD = \sigma\dot{\varepsilon}^o_{an} + \frac{1}{2}\beta E(\varepsilon - \varepsilon^o_{an} - \varepsilon^c_{an})^2\dot{D} + \beta E(1 - D)(\varepsilon - \varepsilon^o_{an} - \varepsilon^c_{an})\dot{\varepsilon}^c_{an} , \qquad (12)$$

where the dot denotes the derivative with respect to time. The evolution of damage appears only if σ is maximum and the constitutive law allows to write :

$$YD = \sigma\dot{\varepsilon}^o_{an} + \frac{1}{2}\sigma\dot{\varepsilon}^c_{an} - \frac{1}{2}(1 - \beta) E(\varepsilon - \varepsilon^o_{an})\dot{\varepsilon}^c_{an} . \qquad (13)$$

The last term depends on $(1 - \beta)$ i.e. the proportion of closed micro-cracks which characterizes the threshold opening micro-cracks. By using the constitutive equation in this case it results :

$$YD = \sigma\dot{\varepsilon}^o_{an} + \frac{1}{2}(\sigma - \sigma_o)\dot{\varepsilon}^c_{an} . \qquad (14)$$

On this dissipative form we recognize with the integral of the first term the area OACGO (fig.2) corresponding to the anelastic strain increase coupled with threshold and the second one the area CED (fig.2) corresponding to the damage increase. This interpretation of each terms shows that the Clausius-Duhem inequality is indeed satisfied $(YD \geq 0 \ \forall \ \dot{D})$.

3. ULTRASONIC UNDER LOAD CHARACTERIZATION DEVICE

The ultrasonic characterization of anisotropic solids is based on the measurement of the phase velocity of longitudinal and shear waves transiting in the sample for varying directions [17]. A complex device was elaborated to have access under load to the velocities of propagation as function of transmission angle and stress. This device consists of an immersion ultrasonic bench and his data collection system associated to a small volume machine test [12]. Velocities are obtained from the measurement of the transit-time of the wave by filtering correlation techniques and normal modes tracking. Data handling of pulsed waves, angular scanning and transit time measurements are fully computer assisted. Ultrasonic measurements are made in pulsed through transmission, using two transducers. The coupling medium is water thermoregulated at 25°C with high stability.

For not yet known materials, the main problem is to choose experimental conditions which simulate in the best way the hypothesis of finite homogeneous medium. In fact, each material is homogeneous at a given scale depending on the size of a characteristic material dimension. The study of transfer function of considered material (longitudinal wave at normal incidence) shows a cut of frequency around 2.5 MHz. This justifies the use of 2.25 MHz central frequency transducers.

The exploitation of experimental data necessites the knowledge of the relations between velocities and stiffnesses [17]. These relations are obtained when we introduce the strain field associated to a monochromatic plane wave in the equation of motion. The composite is homogeneous and continuous at the frequency used. The expression of the velocities of the two propagated modes are obtained by resolving the equation of propagation. The materials is assumed orthotropic (9 independent stiffnesses) and is supposed to stay orthotropic during loading (3), (4). In the considered principal plane (1,3), the relations between velocities, stiffnesses and direction of propagation are [12]:

$$2\rho V^2_{L,S} = A \pm (A^2 - 4B)^{1/2} ,$$

with :

$$A = C_{33}\cos^2 \theta + C_{11}\sin^2 \theta + C_{55}$$
$$B = (C_{33}\cos^2 \theta + C_{55}\sin^2 \theta)(C_{55}\cos^2 \theta + C_{11}\sin^2 \theta) - (C_{13} + C_{55})^2\cos^2 \theta \sin^2 \theta ,$$

θ is the transmission angle, L (resp. S) designate the longitudinal (resp. shear) mode. An identification process using the Newton method allows to calculate the whole stiffnesses at every level of stress from the set of experimental velocities data. Evolution of identified stiffnesses as function of applied stress is typical of an anisotropic damage.

An important loss of stiffness is noted along the tensile axis 1 and a more weak variation for other stiffness tensor components. In this paper investigation in (1,3) plane only is related.

4. DAMAGE EVOLUTION, ULTRASONIC AND STRAIN METROLOGIES CONFRONTATION.

Four components C_{11}, C_{33}, C_{55} and C_{13} can in a simple manner quantitatively characterize damage evolution in the (1,3) investigated principal plane. By this ultrasonic evaluation and using relations (5) we obtain three components D_1, D_3 and D_5 of the fourth rank damage tensor. The evolution of these damage parameters with stress is shown in figure 3.

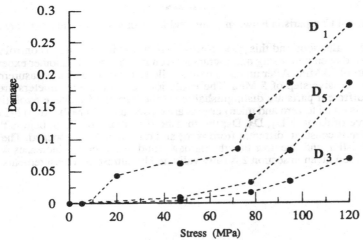

Figure 3. Damage tensor components evolution with stress.

These three internal variables are increasing with stress and three different evolutions are revealing an anisotropic damage. We note that D_1, damage in the loading direction 1, is the largest while D_3, the damage in the direction 3 is the smallest. Moreover, the component D_5 has a faster increase. This let us predict a more ruining damage in the (1,3) interlaminar shear plane.

In paper [9] we have defined the Young's modulus and damage by :

$$E_1 = E_1^o(1 - D_1)^2, \ \nu_{12} = \nu_{12}^o(1 - D_1)(1 - D_2)^{-1} , \ G_{12} = G_{12}^o(1 - D_6)^2 ,$$
$$E_2 = E_2^o(1 - D_2)^2, \ \nu_{13} = \nu_{13}^o(1 - D_1)(1 - D_3)^{-1} , \ G_{13} = G_{13}^o(1 - D_5)^2 ,$$
$$E_3 = E_3^o(1 - D_3)^2, \ \nu_{23} = \nu_{23}^o(1 - D_2)(1 - D_3)^{-1} , \ G_{23} = G_{23}^o(1 - D_4)^2 .$$

Therefore, it is interesting to compare ultrasonic measurements with those obtained by strain measurements. But we have observed in [10] that only are reliable the damage evaluation by classical mechanical mesurements in the load axis (here D_1). These measurements compared with ultrasonic ones are shown in figure 4. We constate a good agreement between these two types of metrologies.

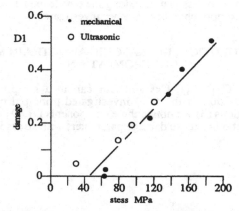

Figure 4 Comparison between strain and ultrasonic damage measurements.

After these evaluations and this good confrontation it is interesting to control the model validity of threshold opening-closing micro-cracks given in section 2. In another experiment we load the sample until 75 MPa. After unload a new tensile test and ultrasonic measurements have been made by successive step of 5 MPa. The evolutions of damage parameters are given in figure 5. Three different parts are distinguishable on the curves of damage evolution. In the first, the damage is equal to zero and micro-cracks are closed until 20 MPa. At 20 MPa we have a gap on the curve of damage D_1, D_3, D_5; the threshold is attained. Then, begins the second part where damage is constant (different from zero) and microcracks are opened. The third part appears about 75 MPa and we note that the damage (and micro-craks) increases with stress. Also the modelisation given in section 2 is fully validated by ultrasonic measurements.

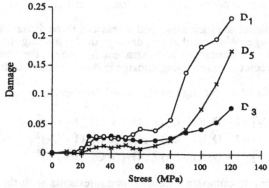

Figure 5. Damage evolution and opening-closing micro-cracks threshold.

5 CONCLUSION

The continuum damage mechanics find a priviliged application field in ceramic-ceramic composites. In the framework of the thermodynamics of irreversible processes with internal variables we have given the behaviour modelisation of these brittle-brittle composites. The constitutive equation coupling anisotropic damage with anisotropic elasticity takes into account in a simple manner the threshold of opening-closing micro-cracks observed in load-unload cycles.

This assumption currently used, for example see [18,19], and coming from physical considerations is, here, a direct consequence of general framework of thermodynamics irreversible processes applied to the materials with inner connection.

This cycle executed during the tensile test allow measurement of the elastic properties variations. The confrontation of damage measurements with classical mechanical metrology and ultrasonic one are in good agreement. The threshold of opening-closing micro-cracks, the constitutive law coupling anisotropic damage - anisotropic elasticity are fully verified and validated by the present ultrasonic metrology.

References.

[1] CHABOCHE J.L., "Le concept de contrainte effective appliqué à l'élasticité et à la viscoplasticité en présence d'un endommagement anisotrope." Comportement mecanique des solides anisotropes, Ed. J. P. Boehler, EuroMECH, Villard de lens. 1979.

[2] KRAJCINOVIC D., "Continuum Damage Mechanics". Applied Mechanics Update.ASME, New York, pp.397-406,1986.

[3] CHABOCHE J.L., "Phenomelogical Aspects of Continuum Damage Mechanics". Nuclear Engng. and Design, 79, pp. 309-319, 1984.

[4] GERMAIN P., "Cours de Mécanique des milieux continus". Tome I, Ed. Masson, Paris.1973.

[5] PERES P., ANQUEZ L., JAMET F., "Analyse micromécanique de la rupture des composites céramiques." Rev. Phys. Appl.,23,pp. 213-228, 1988

[6] AVESTON J., COOPER G.A., KELLY A., "Single and multiple fracture". Properties of fiber composites , IPC science and technology press, Guildford, Surrey, pp. 15-24. 1971.

[7] INGHELS E., LAMON J., "Caractérisation de l'endommagement de composites tissés à matrice SiC". Revue Phys. Appl. , 23, pp. 193-200, 1988.

[8] CHABOCHE J.L., "Anisotropic Creep Damage in the Framework of Continuum Mechanics". XVIIth Inter. Congress of Theoretical and Applied Mechanics, Grenoble, August 21-27.1988.

[9] BASTE S., GERARD A., "Endommagement anisotrope des composites à matrice céramique". C. R. Acad. Sci. Paris, t. 305, Serie II, pp. 1511-1516.1987.

[10] BASTE S., GERARD A., "Off-diagonal elastic constants and anisotropic damage" 13th conf. Energy Sources Technology; New-Orleans 14-18 Jan. 1990, Proc. ASME Composite Material Technology Ed. D. HUI pp.71-76 1989.

[11] EL GUERJOUMA R., BASTE S., "Evaluation of anisotropic damage in ceramic-ceramic (SIC-SIC) composite by ultrasonic method". Ultrasonics International, Madrid Jul. 1989 pp.895-900.

[12] BASTE S., EL GUERJOUMA R., GERARD A., "Mesure de l'endommagement anisotrope d'un composite céramique-céramique par une méthode ultrasonore". Rev. Phys. Appl., 24,pp721-731.1989.

[13] ROUX J., HOSTEN B., CASTAGNEDE B., DESCHAMPS M., "Caractérisation mécanique des solides par spectro-interférométrie ultrasonore". Revue Phys. Appl. , 20, pp. 351-358, 1985.

[14] CORDEBOIS J.P., SIDOROFF F., "Endommagement anisotrope en élasticité et plasticité". J. Méca. Théor. Appl. , numéro spécial, pp.45-60, 1982.

[15] GERARD A., "Mesure de l'endommagement par méthodes ultrasonores". Revue Franç. Méca. ,82, pp. 49-56,1982.

[16] SIDOROFF F., "Damage mechanics and its application to composite materials". Mechanical characterization of load bearing fiber composites, Ed. A. H. Cardan, G. Verchery, pp. 21-35,1985.

[17] HOSTEN B., CASTAGNEDE B., "Optimisation des constantes élastiques à partir des mesures de vitesses d'une onde ultrasonore". C. R. Acad. Sci. Paris, t. 296, Série II, pp. 297-300,1983.

[18] LADEVEZE P., LEMAITRE J., "Damage effective stress in quasi-unilateral conditions." Proc. of IUTAM Congress, Lyngby (Denmark) 1984.

[19] LA BORDERIE C., BERTHAUD Y., PIJAUDIER-CABOT G., "Crack closure effects in continuum damage mechanics-numerical implementation." Proc. SCIC-C 1990 pp. 975-987.2th Inter. Conf. Austria Ed. N. BICANIC, H.MANG

Notched Strength and Damage Mechanism of Laminated Composites with Circular Holes

Jiayu XIAO, Dawei LAI, Claude BATHIAS

Conservatoire National des Arts et Métiers
Département des Matériaux Industriels
292, Rue Saint Martin, Paris 75141
FRANCE

ABSTRACT

Two fracture criteria proposed by Nuismer and Whitney and two models proposed by S. C. Tan are used in this paper for the strength prediction of quasi-isotropic composite laminates containing circular holes. The comparisons show that the minimum strength model can provide a best agreement with experimental data. An improved method that can be used for studying the notched composite laminate damage mechanism is proposed. Some relevant results calculated are verified with uniaxial traction, x-ray photography, and acoustic emission.

I-INTRODUCTION

Since the point stress criterion and the average stress criterion were established by Whitney and Nuismer (1) (2), a number of articles have been published (3)-(6) for the strength prediction of notched composite laminates under uniaxial load. All the criteria associate with a characteristic dimension that expresses the characteristic damage zone size (7). The characteristic dimension was considered as a universal constant (1); then it was treated as a function of the hole diameters (5). Although these criteria containing the characteristic dimension can provide a good agreement with experimental data for the strength prediction, they leave much to be done for the explication and study of the damage mechanism in notched laminates.

In the present study, comparisons of strength prediction for notched laminates are made among the two criteria (1) (2) and the two models (6). In particular, an improved method that can be used for studying the notched laminate damage mechanism is proposed. This method is a modification for Tan's two models in combining with the classical method of lamination. Some relevant results obtained theoretically are respectively compared with those from uniaxial traction, x-ray photography, and acoustic emission.

II-THEORETICAL CONSIDERATION

2-1 Point stress criterion (WN-A) and average stress criterion (WN-B) from
Whitney and Nuismer

The two criteria are no more than based on the stress distribution along the ligament of notched laminates. These two criteria can be expressed respectively by equations [1] and [2]:

$$\frac{\sigma_N}{\sigma_0} = \frac{2}{2+\xi_1^2+3\xi_1^4-(K_T-3)(5\xi_1^6-7\xi_1^8)}, \qquad \xi_1 = \frac{R}{R+d_0} \tag{1}$$

$$\frac{\sigma_N}{\sigma_0} = \frac{2(1-\xi_2)}{2-\xi_2^2-\xi_2^4+(K_T-3)(\xi_2^6-\xi_2^8)}, \qquad \xi_2 = \frac{R}{R+a_0} \tag{2}$$

where, d_0 and a_0 are the characteristic dimensions. The other symbols are explained in (6) - (8).

2-2 Point strength model (Tan-A) and minimun strength model (Tan-B) from S. C. Tan

The two models are based on the stress distribution around the hole (see fig.1). In the first model, the ratio of two first ply failure (FPF) strengths is proposed to be the same as the ultimate strength reduction factor (SRF) (b_1 and b_0 are the characteristic dimensions), i.e.

$$SRF = \frac{\sigma_N}{\sigma_0} = \frac{\text{FPF notched strength at } (b_1,0)}{\text{FPF unnotched strength}} \qquad [3]$$

In the second model, the strength distribution along a curve which is parallel to the hole contour with a distance b_0 apart is considered. The curve is called as the characteristic curve. Along it, the FPF strength is calculated point by point, the ratio of the minimum FPF notched strength to the unnotched strength is proposed to be the same as the strength reduction factor, i.e.

$$SRF = \frac{\sigma_N}{\sigma_0} = \frac{\text{Minimum FPF notched strength at } c}{\text{FPF unnotched strength}} \qquad [4]$$

where, σ_N and σ_0 are the notched ultimate strength and the unnotched ultimate strength respectively; c is the characteristic curve of which the equation is as follows:

$$\frac{x^2}{(a+b_0)^2} + \frac{y^2}{(a+b_0)^2} = 1 \qquad [4.a]$$

FPF strength can be determined with the Tsai-Wu quadratic failure criterion (9) containing the stress interaction term $F_{12}^* = -0.5$. The expressions of the stress distribution can be found in (6).

2-3 Modification for Tan's models

The characteristic dimensions in Tan's models are determined only in laminate. But, since a first ply failure strength is considered, it is more reasonable to calculate them in this ply. After the first ply failure, one can recalculate the laminate rigidity and determine a new characteristic dimension in the second failure ply. With this modification for Tan's models one can study the notched laminate damage mechanism, including the failure sequences, the failure locations, and the failure moments.The calculation related is carried out with the VAX.

III-EXPERIMENTS

3-1 Materials and specimens

T300/Epoxy is used.The stacking sequences are as follows, X: $(0/0/45/45/-45/-45/90/90)_s$ and J: $(0/90/45/-45/-45/45/90/0)_s$. All specimens are cut from a 300x600 plate. They have a fixed ratio of width to diameter, equal to 5. They are respectively 15, 25, 40, 50 mm wide by 250 mm long including 75 mm of each end tap. Each ply has the same thickness of 0.125 mm. The fabrication of specimens is carried out with diamond tools. The elastic properties of each ply in the notched laminates used in the present study are shown in table 1.

3-2 Uniaxial Traction and damage detection

All specimens are tested at a relatively low cross-head speed of 0.4 mm/mn with MAYES (25 tons) and at room temperature. The charge-deformation curves are recorded with an acquisition controlled by the Apple-II and the acoustic emission count rate is recorded. At the charge ratio desired, the specimens are decharged and impregnated into an opaque liquide to x-ray, Iodure Zinc, for 45 minutes. The liquide can filter into cracks and delaminations. Using x-ray photography, one can see the appearances and evolutions of the damages in these specimens.

IV-RESULTS AND DISCUSSIONS

4-1 Ultimate strength

The comparisons between the ultimate strength predictions and the experimental data are illustrated in Figs.2 and 3. All the predictions agree reasonably well with the experimental data. In particular, the prediction of the minimum strength model is more accurate than that of the others. It can be seen from these figures that the characteristic dimension is a function of stacking saquences.

After the first failure ply (90° ply) fails, the following +45° and -45° plies can be retreated as the so-called first failure plies, and a new SRF-Radius relation that can give a good prediction result can be calculated. Based on the same treatment for 0° ply, an other new SRF-Radius relation can also give a good prediction after +45° and -45° plies fail, as shown in Fig.4. From the figure it can be seen that the characteristic dimensions of each ply in a laminate are not identical. The results given in (8) show that the characteristic dimensions have a less influence on the minimum strength model than on the others.

4-2 Damage mechanisms

1) Failure sequences

C. Campello et al have studied the experimental failure process of a notched laminate whose configurations are the same as those in the present laminate (10). Their experimental results show that the failure sequences are as follows: first, the 90° ply fails; next, +45° and -45° plies fail; finally, failure occurs across the 0° ply and the final fracture occurs. Using the method given in this paper, one can calculate the security coefficients of each ply at every damage stage and give prediction results (8). The results calculated can explicate theoreticaly these experimental phenomena obtained by them.

In fact, the real damage process contains: first, cracks appear in the pre-damaged regions; secondly, they appear in 90° ply and at the same time the cracks in the pre-damaged regions evolue; next, there appear cracks in +45° and -45° plies at the almost same time across 0° ply; delaminations around the hole and near the border appear; finally, the final fracture occurs. The process can be illustrated by a series of acoustic emission charts and X-ray photos (8). Photo A extracted from (8) is presented in this paper. Although it is much more complex than that of prediction, the latter can give a good result for several main damage stages.

2) Failure locations

The failure initiation locations, θ, the angles between the 1-axis and the normal vector of the tangent at the hole contour (see Fig.1) are calculated and shown in Fig.5.

From Fig.5, It can be seen that the predicted final failure locations in the 0° ply (Fra.) are almost identical with those in the 90° ply. This result is similar to that given by Tan (6). From photo A, it can be seen that the path followed by final fracture in 0° plies is influenced both by cracks in 45° plies and by those in 90° plies.

In 90° ply, the results measured experimentaly for specimen-X agree well with those calculated at larger radius (R=4 mm, θ is about 10°; R=5mm, θ about 0°) but are not very good at smaller radius because of very deep pre-damage around the hole. The agreements are better in +45° and -45° plies than in 90° ply (R=1.5 mm, θ=20-25°; R=2.5 mm, θ about 20°; R=4.0 mm, θ=10-20°; R=5.0 mm, θ=0-10°). It is difficult to measure precisely these angles. The failure locations are significantly influenced by the pre-damage around the hole, especially in 90° ply. A detail discussion has been given in (8).

3) Failure moments

The stresses at the first damage P_1 (cracks in 90° plies) and at the final fracture P_u for stacking sequence x are calculated and shown in Fig.6. It can be seen from the figure that the two curves calculated are almost "parallel". In fact, the ratio calculated of P_1 to P_u can be treated as a material constant. As shown in fig.7, the ratio is independent of the hole radius and of the models used. It can be also seen that all experimental points are smaller than those from calculations because of the pre-damage (see fig.7). The results related for the other stacking sequence can be found in (8).

V-CONCLUSIONS

1) These criteria and models used in the present study can give good results for the strength prediction of notched laminates, but the minimum strength model is the most accurate.

2) The characteristic dimensions are not only a function of the stacking sequences but are different in different plies of a laminate.

3) The so-called combining method proposed in the present study can make Tan's models useful for a laminate of which the rigidity has been progressively degraded and can be used for studying the notched laminate damage mechanism, including the failure sequences, the failure locations, and the failure moments.

4) The failure sequences of notched and unnotched laminates for the same configurations are identical, and several main damage stages predicted agree well with the experimental results; The failure angles around the hole in each ply change with the hole size and their predictions can give a good result for a specimen fabricated well; The ratio calculated of the stress at the first damage P_1 to that at the final fracture P_u is independent of the hole radius and of the models used. It can be treated as a material constant.

Table 1 Elastic properties of each ply in the notched laminates

Properties	Data
E_1	140 000 MPa
E_2	10 000 MPa
G_{12}	5 700 MPa
X+	1 500 MPa
X-	1 500 MPa
Y+	60 MPa
Y-	220 MPa
S	110 MPa
v_{12}	0,30

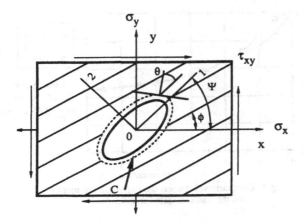

Fig.1 Coordinates of an infinite anisotropic laminate with an elliptical notch

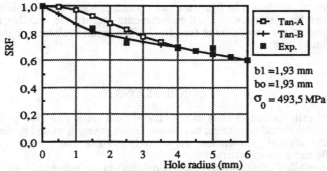

Fig.2a Comparaison between the experimental results
and the Tan's models for specimen-X

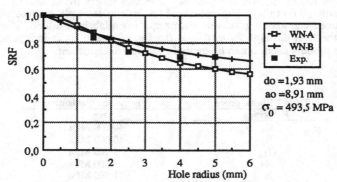

Fig.2b Comparison between the experimental results
and the criteria WN for specimen-X

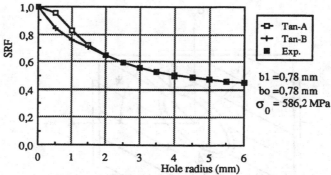

Fig.3a Comparison between the experimental results
and the Tan's models for specimen-J

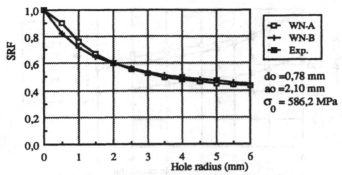

Fig.3b Comparison between the experimental results
and the criteria WN for specimen-J

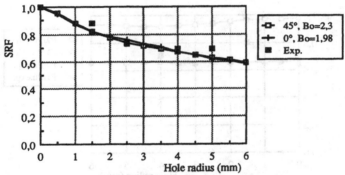

Fig.4 Comparison between the prediction according to the characteristic
dimensions in 45° and 0° plies (Tan-B) and the experimental results.

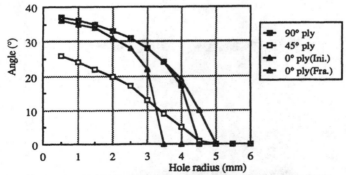

Fig.5 Failure initiation locations around the hole in 90°, 45°, and 0°
Plies and final fracture location in 0° ply for specimen-X.

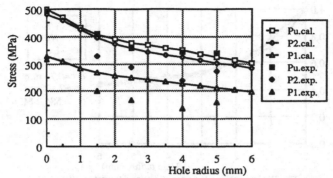

Fig.6 Comparisons of P1, P2 and Pu between the results from experiments and from calculations with the model Tan-B for specimen-X.

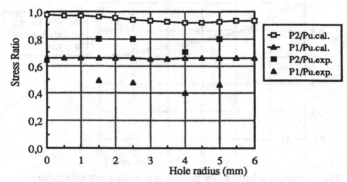

Fig.7 Values from experiments and from calculations for ratio of P1 to Pu, and of P2 to Pu for specimen-X.

A1 (90% of Pu)

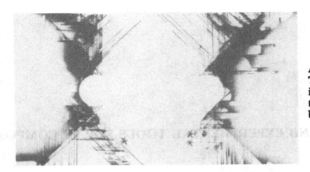

A2 (100% of Pu)
The final fracture path in 0° ply
is influenced both by that in
the +45° and -45° plies (*) and
by that in the 90° ply (**).

Photo A. Specimen-X10 (ϕ10)

REFERENCES

(1)   J. M. Whitney and R. J. Nuismer, Stress fracture criteria for laminated composite containing stress concentration, J. Composite Materials,Vol.8 (July 1974) pp253-265.
(2)   R. J. Nuismer and J. M. Whitney, Uniaxial fracture of composite laminates containing stress concentrations, Fracture Mechanics of Composites, ASTM STP 593, pp117-142 (1975).
(3)   H. J. Konish and J. M. Whitney, Approximate stress in an orthotropic plate containing a circular hole, J. Composite Materials, Vol.9 (April 1975) pp157-166.
(4)   R. Byron Pipes et al, Notched strength of composite materials, J. Composite Materials, Vol.13, (1979), pp148-160.
(5)   Karlak, R. F., Hole effects in a related series of symmetrical laminates, in proceedings of failure modes in Composites, IV. The Metallurgical Society of AIME, Chicago, pp105-117 (1977).
(6)   Seng Chuan Tan, Fracture strength of composite laminates with an elliptical opening, Composite Science and Technology, 29 (1987), 133-152.
(7)   S. C. Tan, Laminated composites containing an elliptical opening II: Experment and model modification, J. Composite Materials, Vol.21 (Oct. 1987), pp949-968.
(8)   Jiayu Xiao, Rapport de DEA, Université de Technologie de Compiègne, Sept. 1989.
(9)   S. W. Tsai and H. T. Hahn, Introduction to composite materials, Technomic (1980).
(10)   C. Campello, C. Bathias et al, Etude de l'endommagement de plaque stratifiée contenant un trou circulaire sous chargement quasi-statique, JNC5, Paris, 9-11 Septembre 1986, pp149-161, Pluralis, 1986.

DAMAGE MODELING AND EXPERIMENTAL TOOLS FOR 3D COMPOSITES

C. CLUZEL   E. VITTECOQ
Laboratoire de Mécanique et Technologie
(ENS Cachan/CNRS/Université Paris VI)
61 Av du Président Wilson 94235 CACHAN CEDEX
FRANCE

ABSTRACT

In this paper we define a mechanical behaviour modeling and experimental tools for the characterization of 3D composites and especially for their deterioration. In the framework of damage mechanics a simple model of the behaviour of the material has been define at a Mesoscopic scale that is the scale of fibre-yarns, matrix blocks and interfaces. This allows the prediction of the behaviour of different geometry knowing the behaviour of the elementary constituents. This approach was carried out on AEROLOR, a 3D massive composite from AEROSPTIALE. The application on an evolutive 3D composite is at present in progress. The method and the experimental precautions are explained below with the first results obtained on the latter material.

INTRODUCTION

This text deals with an approach of 3D composites. Such materials have fibres in the three space directions at least. So we introduce the meso level, at this scale the different elementary components are fibre-yarn (composed of thousands of fibres), matrix-blocks (homogeneous material which fills the necessary space without any fibre) and their respective interfaces. From the macroscopic point of view, 3D composite materials have a complex non-linear damageable behaviour. The mechanisms which lead to such a behaviour are manifold. We can have fibre rupture ; cracks in the matrix in a fibre-yarn ; micro-cracks in a matrix-block ; micro-cracks in matrix-blocks/fibre-yarn interfaces . . . The aim of our approach consists in working at the scale where those phenomena are easy to understand and to model. This approach, which was used for the 3D composite AEROLOR from AEROSPATIALE [1],[2],[3] , is hereafter described on a new 3D material. The structure of this new material is evolutive. It means that the directions of fibre reinforcement may change according to the local strength requirement in the whole structure. The aspect is laminate with normal reinforcements. The main interest of these normal reinforcement is to avoid but it is important to know their effect on the plane behaviour. The multi-scale approach is being defined to understand and to model the damage behaviour of such a material. The first results of the plane modeling are described below.

MULTI-SCALE PRESENTATION OF THE MATERIALS

The two investigated materials are 3D composites.
The first 3D composite the L.M.T. worked on was the 3D massive composite from Aerospatiale called AEROLOR. Its structure consists in fibre-yarns arranged in the three space directions (fig 1) with the empty volumes being filled with Matrix. The material studied at present (fig 2) is a plate composite with normal reinforcements. It is an evolutive structure. The approach to such material is carried out at different scales. At the macro-scale the composite is seen as a homogeneous material. At the micro-scale, the characteristic size is the fibre diameter. The "meso" scale is used to describe the structural effect given by such a material. At the meso scale, the constituents are : fibre-yarn, matrix blocks and their interfaces.

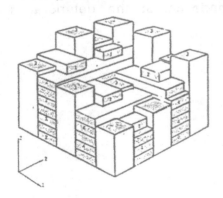

fig 1. 3D massive

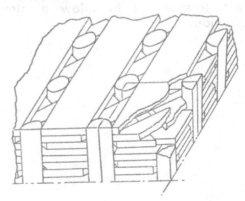

fig 2. 3D evolutive

Description of each different scale

Macro-scale

It is the scale of the structure. But if we just approach the material at this scale, the behaviour is complex to understand and thus to model. The second point is that there is no understanding of the structural local effect of the different constituents. Damage mechanisms are not really taken into account but just their global effect.

Micro-scale :

It is a natural scale to approach the composites with, because at this scale, the elementary constituents appear i.e fibre and matrix. It is interesting to study this scale because the damages of the matrix and the fibre/matrix interface are visible and consist in cracks and micro-cracks. When it is possible to study these phenomena, this scale is well adapt to understand them. The experimental values are not really representative of the local behaviour in the structure because the building method may change the properties and it is actually not possible to take each mechanism into account in such a large homogenization tool.

Meso-scale :

It is an intermediate scale where the structure of the material is visible. For the 3D massive, the meso constituents are fibre yarns, matrix blocks and their interfaces. For the 3D evolutive, two kinds of meso elementary cells are chosen according to the needs (fig 3). Its main interest is to allow a simple modeling of the deterioration behaviour.

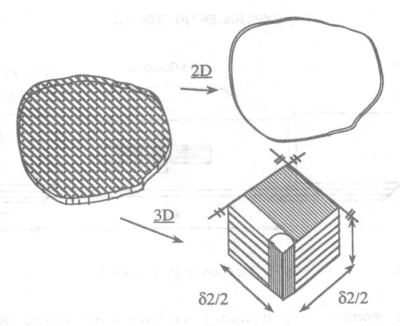

FIG 3. Choice of Elementary Cell

A first 2D elementary cell is defined to model the behaviour in the middle of the plate as simply as possible. Considering that the normal reinforcements represent a low proportion of the total fibre reinforcements, one can neglect effects on the plane direction. The meso constituent is then a single layer and homogenization consists in using the plate theory. This hypothesis is only acceptable if we have a global plane stress state. So it is no longer true near the edge of the structure. To validate the first approach and to understand the normal reinforcement effects, a second meso approach is defined. Considering the periodicity of the normal reinforcements, a 3D elementary cell is defined. The link between macro scale and meso scale is obtained by a homogenization tool using asymptotic development [4],[5].

The multi scale approach tries to use the data collected at each scale analysis. That is why the test will take place at macro level as long as it is possible to get all the information on the meso constituents. The model is simple at the meso level, and we use the micro level to understand damage mechanisms. This approach is particularly interesting for a material whose the structure is evolutive (many different geometries using the same meso constituents). The first 2D approach of the 3D evolutive is presented with experimental precautions.

EXPERIMENTAL TOOLS

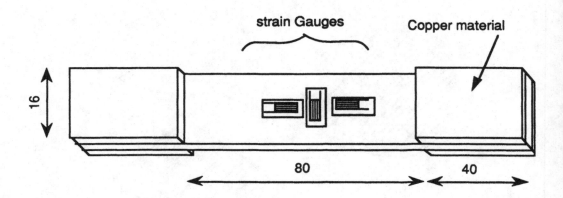

fig 4. Traction test specimen

For the moment, only traction tests have been carried out. The specimens described (fig4) are used with several geometries and tested on a M.T.S. tension machine. Piloting and data-acquisition are carried out by an Apollo computer. A cell measures the axial loading and gauges are used to measure the strain. The difficulty to test such a heterogeneous material is that a little variation of some parameters may modify the whole result. The philosophy of our test consists in using the simplest one and in making enough measurements to be able to post-analyse the true loading and test conditions [6][7]. Then we can use criterion to define the bending by comparison with strain measurements on each side of the composite. Shear stress may be appreciated on a $[0;90]_{4s}$ specimen by comparing a strain measurement in the 45° direction and the half sum of longitudinal and transverse strains (fig 5). The error on the measurement due to an error of gauge position is estimated by calculation. Knowing the homogeneous anisotropic behaviour , it is possible to calculate [8] the theoretical strain in the different directions. We then define, (fig 6) for each measurement in the direction "φ", an error "η" associated with an error on gauge position "θ" such as:

$$\eta(\varphi) = |\{ \varepsilon(\varphi+\theta) - \varepsilon(\varphi)\} / \{\varepsilon(\varphi) \}|$$

For a $[0;90]_{4s}$, we obtain the following curve (fig 6) where we can notice a sure value in the longitudinal direction and an only acceptable value, with a careful gauge positioning, for the 45° and 90° directions.

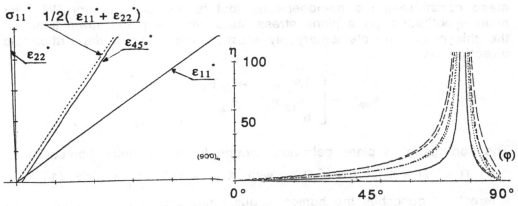

fig 5. Test with shear control   fig6: Error as a function of the gauge position

Another important checking procedure consists in testing the homogeneity of the macro strain by comparing the strain on different points of the specimen. If the response is the same all along the specimen, we can conclude that the gauge size is appropriate to the material and that the damage is diffuse throughout the specimen and not confined as for local macro-cracking or delamination. An example of a local phenomenon is given (fig 7) comparing total displacement and gauge response.

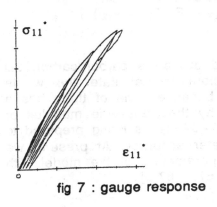

fig 7 : gauge response

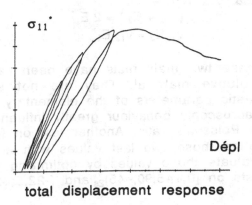

total displacement response

IDENTIFICATION OF ELASTIC PARAMETERS AND DAMAGE FUNCTIONS OF AN ELEMENTARY PLY (2D MESO CONSTITUENT)

The identification of the elementary ply will use tests on specific composites and homogeneous tools. It is to be notice that it does not exist unidirectional material with normal reinforcements. For the 2D meso constituent, the homogeneous tool is the plate theory [9]. The main hypotheses are a plane stress state and the plane strain linear in the thickness. The elementary ply elastic compliance in the orthotropic direction is :

$$
K_{cp}^{-1} = \begin{bmatrix} 1/E_1 & -v_{12}/E_1 & 0 \\ -v_{12}/E_1 & 1/E_2 & 0 \\ 0 & 0 & 1/\ 2G_{12} \end{bmatrix}
$$

The homogeneous plane behaviour written for a symmetric composite is:

$$ \Pi\ \varepsilon^*\ \Pi = K_{cp}^{-1*}\ (\Pi\ \sigma^*\ \Pi) \quad \text{with}\ K_{cp}^{-1*} = 1/2h \int_{-h}^{h} K_{cp}\ dz \quad (1) $$

where " * " describes the homogeneous characteristics.
On a given test, we measure : σ_{11}^* ; ε_{11}^* ; ε_{22}^*
The macroscopic characteristics are E_{11}^* and v_{12}^*

$$ \text{with}\ E_1^* = \sigma_{11}^* / \varepsilon_{11}^{*e}\ \text{and}\ v_{12}^* = -\ \varepsilon_{11}^{*e} / \varepsilon_{22}^{*e} $$

To get information on the elementary ply, we need different tests which allow us to identify some elastic parameters.
Using the expression (1) and the first experimental result which is $E_1 >> E_2$ and $E_1 >> G_{12}$, we have a relation between test results and elementary ply characteristics for each composite :

Test on $[0;90]_{4s}$

$$ (E_1 + E_2) = 2\ E_1^* $$
$$ v_{12}E_2 = v_{12}^* E_1^* $$

Test on $[-45;+45]_{4s}$

$$ G_{12} = E_1^*/2(1 + v_{12}^*) $$

These two main tests have been carried out on a carbon/carbon 3D evolutive material. They are not sufficient to calculate the whole elastic parameters of the elementary ply because none of them has a macroscopic behaviour greatly influenced by the transverse modulus or by Poisson's ratio. Another test on $[-67.5;+67.5]_{4s}$ is being prepared to define those two last values with a greater accuracy. At present, we evaluate those values by optimizing the parameters of the model with tests on $[0;+45;90;-45]_{2s}$ and $[-22.5;+22.5;67.5;-67.5]_{2s}$.

DAMAGEABLE BEHAVIOUR

As for elastic coefficients, the main tests to understand damage mechanisms are $[0;90]_{4s}$; $[45;+45]_{4s}$ and $[-67.5;+67.5]_{4s}$.

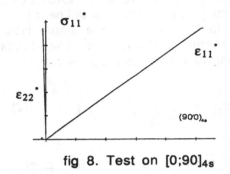

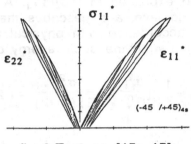

fig 8. Test on $[0;90]_{4s}$ fig 9.Test on $[45;+45]_{4s}$

On the (fig 8), we can notice that with traction tests in the fibre direction, there is a brittle elastic macroscopic behaviour whereas, with a test outside the fibre direction (fig 9), the behaviour is damageable with residual strain. To write a first damage kinematics, an optical observation of the structure (fig 10) leads to notice that the porosity of material is mainly due to the matrix blocks. As for 3D massive, the non-linear behaviour must be due to the matrix blocks and matrix blocks/fibre-yarn interfaces.

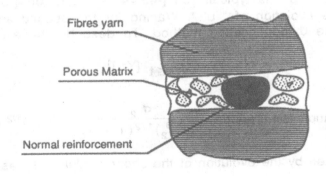

fig 10. Optical observation of the matrix blocks

In accordance with the first experimental tests, the main elementary ply characteristics which can be affected by such a mechanism are shear modulus and transverse modulus. To model damage behaviour, we use the evolution of the elastic moduli as damage indicators. The kinematics used to describe anisotropic damage is due to P. LADEVEZE and explained in [10][11]. A simple damage kinematics is define at the meso level and describes the damageable behaviour of the single layer in accordance with physical and experimental observations. That leads us to write the strain energy of the damaged elementary ply as :

$$E_D = \frac{1}{2} \left[\frac{\sigma_{11}^2}{E_1^0} + \frac{\sigma_{22}^2}{E_2^0(1-d')} - \left(2 \frac{\nu_{12}^0}{E_1^0} \right)\sigma_{11}\sigma_{22} + \left(\frac{\sigma_{12}^2}{G_{12}^0(1-d)} \right) \right] \qquad (2)$$

The conjugate quantities associated with d and d' for positive dissipation are :

$$Y_d = \frac{\delta E_D}{\delta d} \Big|_{\sigma \,:\, cst} = 1/2 \left(\frac{\sigma_{12}^2}{G_{12}^0(1-d)^2} \right) \qquad (3)$$

$$Y_{d'} = \frac{\delta E_D}{\delta d'} \Big|_{\sigma \,:\, cst} = 1/2 \left(\frac{\sigma_{22}^2}{E_2^0(1-d')^2} \right)$$

The identification of the two functions $d(Y_d)$ and $d'(Y_{d'})$ needs appropriate tests. As for the elastic parameter, a good test to describe the evolution of " d " is typically on $[-45;+45]_{4s}$, and on $[-67.5;+67.5]_{4s}$ to describe the evolution of " d' ". Waiting for the second specimen, we just present the d identification method. To describe the evolution of d, we introduce :

$$Y = Y_t = \sup_{\tau \,\le\, t} [Y\tau]^{1/2}$$

Using the equation (3): $[Y_{d\tau}]^{1/2} = \dfrac{\sigma_{12}}{(2G_{12}^0)^{1/2}(1-d)} = (2G_{12}^0)^{1/2} \, \varepsilon_{12}^e$

" d " is measured by the evolution of the shear modulus G_{12} as :

$$d = 1 - G_{12}/G_{12}^0$$

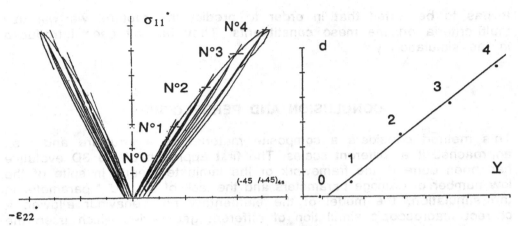

fig 11.Traction test loading-unloading   fig 12. Damage evolution

The informations of ε_{12}^e ; G_{12} are obtained with each unloading during the test (fig 11). The first point N° 0 corresponds to the initial value of the elastic parameters, and for the point N° 1 to N° 4, we get the evolution of "d" (fig 12). An approximation of the evolution of "d" under a linear form allowed a simple description of damage :

$$d = < \underline{Y} - \underline{Y}^0 > / \underline{Y}_c$$

A simulation with the damage evolution of each ply according to the stress it supports taken into account, allowed a description of the global evolution of different composites. A comparison between simulation and elastic part of the test is presented (fig 13) and (fig 14). The simulation, of the evolution of damage in a 45° oriented ply on $[90;-45;0;+45]_{2s}$ shows a value of 0.24 just before the fracture although the global curve is almost linear.

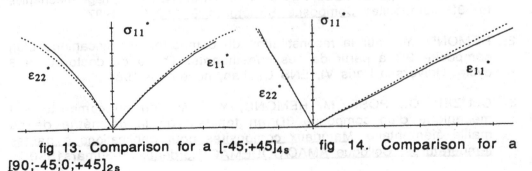

fig 13. Comparison for a $[-45;+45]_{4s}$ fig 14. Comparison for a
$[90;-45;0;+45]_{2s}$

------- Simulation ; ——— Experimental

It has to be noted that in order to predict the rupture, we will use multi-criteria on the meso constituents. They haven't been introduced in the simulation yet.

CONCLUSION AND PERSPECTIVES

This method considers a composite material as a structure and then approaches it at different scales. The first approach of the 3D evolutive has been done in the framework of the laminate theory. In spite of the low number of damage parameters and the lack of the " d' " parameter in the simulation, the model of the elementary ply behaviour allowed a correct macroscopic simulation of different geometries which used the same meso constituent. The identification of linear and non linear parameters is carried out on specimens which involve lots of precautions in order to have good loading conditions and uniform damage. Test results are coherent through the different geometries for the elastic and the non linear parts, except for the rupture because it is rather an edge effect which imposes the maximum loading instead of the expected plane behaviour. New experimental precautions will have to be used for the next tests. Those results show the limit of the model which is no longer true on the edge of the structure. The next step for the study of such a material is the use of the 3D meso-model. It will improve our knowledge of the normal reinforcement effect and will allow the study of delamination.

REFERENCES

1 . DUMONT, J.P.; LADEVEZE, P.; POSS, M.; REMOND, Y. Damage mechanics for 3D composites <u>Composite Structures 8 119-141</u> 1987

2 . REMOND, Y. Sur la reconstitution du comportement mécanique d'un composite 3D à partir de ses constituants. Thèse de doctorat de 3 cycle, Université Paris VI, ENS Cachan, novembre 1984

3 . CLUZEL, C., POSS, M, REMOND, Y. Prévision du comportement mécanique d'un composite 3D en fonction de la géométrie de sa maille élémentaire. <u>Matériaux composites pour applications à hautes températures. Colloque AMAC/CODEMAC Bordeaux.</u>29-30mars1990

4. SANCHEZ-PALENCIA, E. Comportement local et macroscopique d'un type de milieu physique hétérogène. Int. Engineering Science 12 331, 351 1974

5. SALHA, B. Elements pour une théorie des plaques hétérogènes. Rapport EDF 1988

6 . LE DANTEC, E. Modélisation des composites stratifiés. Thèse de doctorat de 3 cycle, Université Paris VI, ENS Cachan, novembre 1989

7. LADEVEZE, P. ,REMOND, Y. ,VITTECOQ, E. Essais mécaniques sur composites à hautes performances : difficultés et critères de validité. Revue française de Mécanique. 2 1989

8. LEDANTEC, E. , REMOND, Y. Etude et qualification d'un essai de compression sur un composite stratifié. Rapport interne LMT/ENS CACHAN oct 1986

9 .LADEVEZE, P. Homogénéisation et matériaux hétérogènes. Cours de DEA PARIS VI

10 .LADEVEZE, P. Sur la théorie de l'endommagement anisotrope. Rapport interne n° 34, Laboratoire de Mécanique et Technologie, Cachan 1983

11.LADEVEZE, P. About a Damage Mechanics Approach. Mechanics and Mechanisms of damage in Composite and Multimaterial: International Seminar of MECAMAT St ETIENNE nov 1989.

12.PERES, P. Caractérisation en cisaillement interlaminaire de composites thermostructuraux. Matériaux composites pour applications à hautes températures. Colloque AMAC/CODEMAC Bordeaux 29-30mars 1990

13.LEQUERTIER, Les composites thermostructuraux carbone/carbone inox, Comportement mécanique et tectonique de la protection anti-oxydation. AMAC -JNC7 LYON 6-8nov 1990

INCREASING ENERGY ABSORPTION BY "ROPE EFFECT" IN BRITTLE MATRIX COMPOSITES

STEPHANE FAIGNET, FRANK THEYS and JAN WASTIELS

COMPOSITE SYSTEMS AND ADHESION RESEARCH GROUP OF THE FREE UNIVERSITY OF BRUSSELS (COSARGUB).
VUB - TW - Kb - Pleinlaan 2 - B1050 Brussels - Belgium

ABSTRACT

Mineral polymers (MiP) have, like all ceramic type matrices one major disadvantage, i.e. a very low failure strain in tension. Toughness can be increased by addition of fibres. Thanks to the low hardening temperature of the matrix, organic material like bundles of jute fibres can be added to MiP.

Due to the special, rope like behaviour of the used bundles of fibres, the energy absorption capacity of the obtained composite is much higher than predicted when considering only the fibre characteristics.

MINERAL POLYMERS (MiP)

At the Department of Constructions of the Brussels University (V.U.B.) new ceramic materials, so called "Mineral polymers (MiP)", are being developed [1].

Although their properties can be compared in general to porcelain, ceramics, pottery or glasses, they are not manufactured in the same way : MiP are inorganic materials that are obtained by chemical polymerisation or polycondensation of different oxides from H, Li, Na, K, Mg, Ca, Al, Si, P, ... at atmospheric pressure and low temperature (between room temperature and 100°C). MiP can be prepared from powders, grains, pastes or liquids, and thus can be manufactured using processes as pressing, extrusion, casting and most resin composites processing techniques.

Due to their chemical nature, MiP present some attractive properties : chemical inertia, high temperature resistance, low energy input and, thanks to the low process temperature, organic materials like natural fibres or insulation materials can be added to MiP.

Mineral polymers have, like all ceramic type materials, one major disadvantage, i.e. a very low failure strain in tension which means not so much that the material is weak in tension but that it is unreliable. A small over-strain, for example resulting from impact or restrained shrinkage, results in catastrophic failure. Toughness can be increased by addition of fibres, but this does not change the failure strain of the matrix in most cases.

Most applications and further developments of MiP are focused on their high temperature resistance (up to 1100 °C), the possibility of net shape technology at low process temperatures, and the increase of the energy absorption capacity through the use of fibres, whiskers and other filler materials.

JUTE REINFORCED MINERAL POLYMER

To assess the possibilities of reinforcing MiP with long fibres, a series of jute reinforced laminates was fabricated. The used jute consisted of an orthogonal woven fabric with a surface density of approximately 210 g/m^2. To make unidirectional reinforced specimens the transverse fibre bundles were pulled out before laminating.

The laminates were made with the hand lay-up technique, similar to a resin matrix composite. Care was taken to avoid misalignment of the fibre bundles during the lamination. Curing at a temperature of 60°C during 20 hours was performed.

THE NPL THEORY

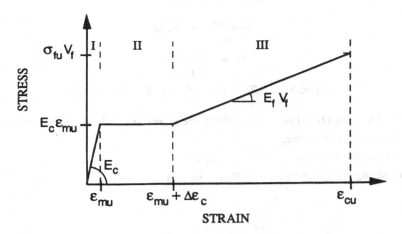

Figure 1 . Idealized tensile stress-strain diagram for brittle matrix composites.

During a tensile loading, three stages of mechanical behaviour can be distinguished, corresponding to the NPL theory [2] (also called the Aveston, Cooper, Kelly (ACK) theory [3], [4]) for brittle matrix composites with continuous aligned fibres with frictional bond between fibres and matrix :

I   Initial Stage before Matrix Cracking.
Having a perfect bond between fibres and matrix, the law of mixtures can be applied with sufficient accuracy to determine the stiffness of the composite :

$$E_c = E_m V_m + E_f V_f \qquad (1)$$

II   Stage of Crack Formation (multiple cracking) during which load is transferred from the matrix to the fibres in the vicinity of the cracks. This is occurring at a constant composite stress, which is equal to the first crack stress :

$$\sigma_c = E_c \, \varepsilon_{mu} \qquad (2)$$

where ε_{mu} is the cracking strain of the matrix, supposed to be also the cracking strain of the composite.

However this stage can only take place if the strength of the fibres bridging the cracks is sufficient to withstand the total stress on the composite expressed in equation (2), i.e. when :

$$V_f > \frac{E_c \, \varepsilon_{mu}}{\sigma_{fu}} = V_f \text{ (crit)} \qquad (3)$$

where σ_{fu} is the fibre strength, and V_f (crit) is called the critical fibre content
Out of the strain distribution after cracking we can calculate the average additional strain "$\Delta \varepsilon_c$" in the fibres, due to the formation of cracks :

$$\Delta \varepsilon_c = k \, \alpha \varepsilon_{mu} \qquad (4)$$

where :

$$\alpha = \frac{E_m \, V_m}{E_f \, V_f} \qquad (5)$$

and : $0.5 < k < 0.75$, (statistically $k = 0.659$ [3])

III   Stage of Crack Opening during which the fibres are sliding relative to the blocks of matrix and stressed up to failure.
The tangent modulus becomes : $E_f V_f$ (6)
The ultimate strength is given by : $\sigma_{cu} = \sigma_{fu} V_f$ (7)

EXPERIMENTAL RESULTS

The specimens, 25 mm wide and approximately 200 mm long, were cut out of the laminate parallel to the fibre bundles with a diamond saw. The tensile tests were performed on an INSTRON 1195 testing machine at a displacement rate of 0.2 mm/min.

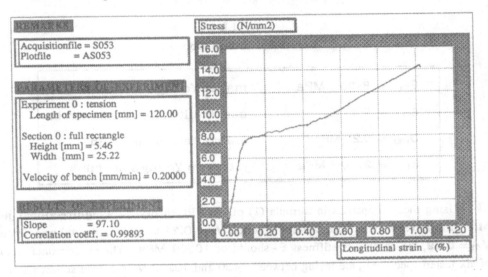

Figure 2 . Tensile test on a Jute reinforced MiP laminate ($V_f = 0.11$, $V_m = 0.89$)

The fibre volume fraction "V_f" of the laminate was determined, knowing the average weight and unit weight of the fibre bundles, by measuring the thickness of the laminate and the number of fibre bundles used in the laminate ($V_f = 0.11$, $V_m = 0.89$).

At first sight the NPL-theory seems to be valid for the Jute reinforced MiP laminate.
Let's take a closer look :

I   Initial Stage before Matrix Cracking.

The mechanical properties of the constituent materials, needed to calculate the composite characteristics, were determined in the following way :

Matrix : bulk Mip matrix specimens were cast and cured in the same way as the laminates. Young's modulus E_m, tensile strength σ_{mu}, and tensile strain ε_{mu} at failure (m = "matrix",u = "ultimate") were determined during a strain-gauge instrumented, displacement controlled four point bending test on an INSTRON 1195 testing machine.

Fibres : before matrix cracking in the composite, a perfect bond is supposed between fibres and matrix, so to determine the appropriate Young's modulus of the jute fibre a polyester-jute laminate was fabricated. Young's modulus of this laminate was measured during a strain-gauge instrumented tensile test on the INSTRON machine and application of the law of mixtures, knowing Young's modulus of the polyester and the fibre volume fraction, yielded the jute fibre modulus.

The results are given in table 1 :

TABLE 1

Constituent material properties

E_m = 8400 MPa		matrix Young's modulus
ε_{mu} = 0.08 %		matrix ultimate tensile strain
σ_{mu} = 6.2 MPa		matrix tensile strength
E_f = 27400 MPa		jute fibre Young's modulus

Using these properties in formula (1) of the NPL-theory for a **unidirectional jute-MiP laminate** with a fibre volume fraction $V_f = 0.11$ and a matrix volume fraction $V_m = 0.89$, the composite stiffness E_c should be 10509 MPa. The experimental values, of six tested specimens, are laying between 9400 and 10200 MPa : those somewhat lower values are probably due to insufficiently impregnated fibre bundles, and thus the existence of air voids.

II   Stage of Crack Formation (multiple cracking)

The condition that the fibre volume fraction V_f, here 0.11, is greater than the critical fibre content V_f (crit) = 0.56 (formula 3) is fulfilled.

Formula (2) of the NPL-theory predicts that the multiple cracking is occurring at a constant composite stress of $\sigma_c = 8.4$ MPa. Experiments showed as usually a slightly increasing stress, with a mean value between 7.5 and 9.4 MPa, and a mean crack distance of about 2 mm.

As the number of cracks is increasing and bonding between fibres and matrix is deteriorating, the fibres are not working in the same way any more; they have now to be looked at as fibre bundles : the composite seems more like a number of blocks connected to each other by fibre bundles.

The rope effect : the modulus of a rope or a twisted bundle of fibres in a tensile test is not a constant, but is function of the strain. In the beginning of the test the individual fibres constituting the rope are orientating towards the stress direction. Depending on the construction of the rope, this reorientation can continue to relatively high stress levels.

The result of this all is a rather low "bundle" modulus at low stress levels, which is increasing with strain up to a certain strain level, were it becomes a constant : "E_{bf}" (called Young's modulus of the bundles of fibres).

TABLE 2

Bundle of jute fibres ("bf"):

characteristics obtained by tensile tests on individual bundles

E_{bf} = 9160 MPa	bundle modulus for $\varepsilon_{bf} > 3\%$
ε_{bfu} = 3.6 %	ultimate bundle strain
σ_{bfu} = 148 MPa	bundle strength

In the case of a composite the adherence between fibres and matrix in the initial stage leads to the use of fibre characteristics, but when the matrix is cracking the effective modulus of the fibres is dropping towards the "bundle" modulus. The increase with strain of the "bundle" modulus combined with the dropping effect results in an effective modulus of the fibres close to E_{bf}.

The measured additional strain $\Delta \varepsilon_c$ ranges between 0.28 and 0.40%, predicted (4) between 0.29 and 0.44 % (statistically $\Delta \varepsilon_c$ = 0.38 % [3]).

When there would have been no "rope effect", i.e. when the effective fibre modulus is constant and equal to the fibre Young's modulus E_f, the maximum additional strain would have been only 0.15 % (statistically $\Delta \varepsilon_c$ = 0.13 % [3]).

This "rope effect" leads to a higher straining in the crack formation stage, and so also in a increase of the energy absorption of the composite.

III   Stage of Crack Opening

The effective fibre modulus to be considered in formula (6) is equal to the fibre bundle modulus of table 2, so the tangent composite modulus becomes 1017 MPa. The experimental values are laying between 840 and 1220 MPa.

Again, thanks to the "rope effect", there is an increase of the energy absorption : calculated with the fibre modulus E_f, the tangent modulus would be equal to 3041 MPa and, the composite strength being the same, the corresponding ultimate strain and so also the energy absorption would have been smaller.

The composite strength ranges from 14.4 to 18.0 MPa for the specimens and, according to formula (7), is equal to 16.4 MPa.

In figure 3, the calculated stress-strain curves (statistical value for $\Delta\varepsilon_c$), with and without "rope effect", are plotted over the experimental curve of figure 2.

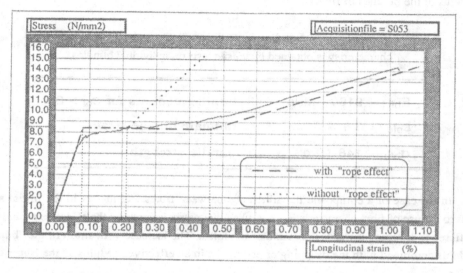

Figure 3 . Predicted stress strain diagrams plotted over the experimental curve of figure 2.

CONCLUSIONS

Energy absorption can be further increased by using bundle reinforcement instead of fibres. What does a bundle or rope reinforcement change in the tensile behaviour of a brittle matrix composite, compared to a fibre reinforcement :

- In the initial stage before matrix cracking the modulus to be considered is that one of the fibres; using a high volume fraction of stiff fibres this will lead to a significantly increased stress level of matrix cracking.
- During the multiple cracking stage the effective modulus of the rope is dropping towards the "bundle" modulus, as bonding between fibres and matrix is deteriorating. This leads to a higher straining in the crack formation stage, and so also in an increase of the energy absorption.
- In the final stage, during which the fibres are stressed up to failure, again due to the fact that the fibres are working at the lower "bundle" modulus.

This "rope effect" combines the beneficial effects of stiff fibres (before matrix cracking) and flexible fibres (during and after multiple cracking), and may be advantageously used in the design of specific composites.

ACKNOWLEDGMENTS

The authors gratefully acknowledge the support of the Research Council (OZR) of the Vrije Universiteit Brussel, and of the Belgian National Fund for Scientific Research, through contract FKFO 2.9007.88.

REFERENCES

1. Patfoort, G., Wastiels, J. , Bruggeman, P. and Stuyck, L., Mineral Polymer Matrix Composites. In Brittle Matrix Composites 2, Elsevier Applied Science, London, 1989, pp. 587 -92

2. Aveston, J., Mercer, R. A. and Sillwood, J. M., The Mechanism of Fibre Reinforcement of Cement and Concrete (Part I), National Physical Laboratory, publ. n°SI 90/11/98, January 1975

3. Aveston, J., Mercer, R. A. and Sillwood, J. M., Fibre reinforced cements - scientific foundations for specifications. In Composites - Standards Testing and Design, Proc. National Physical Laboratory Conference, IPC Sci. and Technology Press , UK, 1974, pp.93-103

4. Bentur, A. and Mindess, S. , Fibre Reinforced Cementious Composites , Elsevier Applied Science, London, 1990, pp. 89 - 135

SPECIFIC DAMPING OF A CARBON/EPOXY LAMINATE UNDER CYCLIC LOADING

M. NOUILLANT, M. CHAPUIS
Laboratoire de Génie Mécanique
Université de Bordeaux I
IUT A - 33405 Talence Cedex

ABSTRACT

This work summarizes several experimental results of tests performed in alternate tension/compression and alternate torsion on symmetric filament wound tubes. Depending on stress level, three domains of behavior are considered : linear viscoelastic, nonlinear viscoelastic, and damage domains. Most measurements show that damping increases with stress and temperature. Damaged specimens exhibit an irreversible increase of damping, which mainly depends on the maximum stress level experienced rather than on the number of load cycles. Early occurence of damage during torsion test has not yet been elucidated. Using Adams and Bacon's theory, damping may be split into three terms. In the linear viscoelastic domain, measurements lead to $\Psi_L = 1.4\%$ (longitudinal), $\Psi_T = 5.6\%$ (transverse) and $\Psi_{LT} = 8.5\%$ (shear damping). In the non linear viscoelastic domain, Ψ_L and Ψ_T are almost stress independant, while the $\Psi_{LT}-\sigma_{LT}$ relationship is strongly non linear.

INTRODUCTION

This work is intended to summarize the previous experimental results (1), (2) and (3) obtained with tubular carbon/epoxy laminate specimens under various sinusoidal loadings. The specimens were made of an epoxide matrix Ciba-Geigy LH/Y 5052 and carbon fibers Hercules IM6 G12 K symmetrically wound around a mandrel with orientations $(\pm 15°)_s, (\pm 30°)_s, (\pm 45°)_s, (\pm 60°)_s, (\pm 75°)_s$ to cylinder axis. Fiber volume fraction was about 60 % according to fiber orientation. The main parameters taken into account are fiber orientation, temperature and stress level.

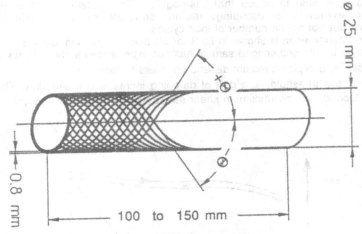

Figure 1 . specimen geometry

DOMAINS OF BEHAVIOR

According to various stress levels, one can consider three domains of behavior :

1 - Linear viscoelastic domain : at 25°C, the laminate behaves as approximately linear viscoelastic for shear stresses σ_{LT} lower than 10 MPa. (see fig2) However, this linear viscoelastic domain reduces considerably with increasing temperature.

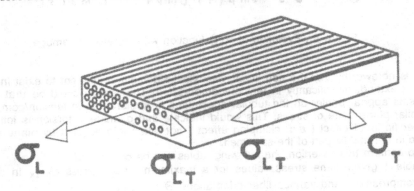

Figure 2 . Plane state of stress in a lamina

2 - Non linear viscoelastic domain : within this domain, damping is a function of stress level, but it recovers its initial value when stress decreases.

3 - Damage domain : this domain corresponds to an irreversible evolution of the Specific Damping Capacity (SDC $\Psi = \Delta w / w$) .

It is advisable to notice that "damage", which is taken here in the sense of an irreversible evolution of damping, mainly depends on the maximum stress level experienced, but not on the number of load cycles.

This phenomenon is shown in fig 3 for an alternate torsion loading :

- as far as the undamaged sample has not experienced a shear stress level higher than the threshold τ_s, damping does not depend on stress history.

- over the threshold τ_s, a part of damping increase is irreversible. The new function $\Psi(\tau)$ only depends on the maximum shear stress experienced previously, viz.τ_{mp}

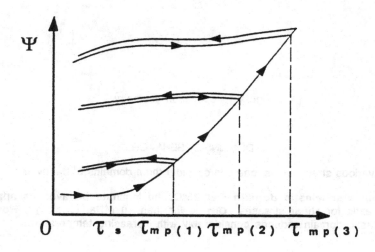

Figure 3 . Evolution of $\Psi(\tau)$ function with increasing damage

It is noteworthy that the non linear viscoelastic domain seems not to exist in torsion, whereas it extends significantly in tension/compression. The reason could be that damage mechanisms appear earlier on the tubular specimens in torsion than in tension/compression for a similar plane state of stress. This would indicate that damage in torsion is initiated by some specific local effect (e.g. clamping effect) not taken into account by plane stresses computed in the middle part of the specimen.

To confirm this assertion, the following tables can be considered:

- Table 1 gives plane stress values for a maximum tensile stress σ_{max} in alternate tension/compression, and various fiber orientations Θ .

At this level σ_{max}, tension/compression specimens were undamaged.

- Table 2 gives the same values σ_L, σ_T and σ_{LT} for the threshlold τ_s defined above.

For instance it arises from table 1 that the $[\pm15°]_s$ specimens should be significantly damaged, because stresses σ_L, σ_T and σ_{LT} have the same sign and are much higher than stresses related to τ_s shown in table 2 for the same fiber orientation.

Indeed, these samples were undamaged, which shows that the plane stress values in the middle part of the specimen cannot account for damage initiation in torsion.

TABLE 1
Plane state of stress in tension/compression for various fiber orientations (MPa)

θ	σ_{max} (MPa)	σ_L (MPa)	σ_T (MPa)	σ_{LT} (MPa)
15°	110	115	-5.5	5.5
45°	35	33.5	1.4	17.5
75°	14.5	2	12	5.7

TABLE 2
Plane state of stress in torsion for damage threshold shear stress τ_s (MPa)

θ	τ_s (MPa)	σ_L (MPa)	σ_T (MPa)	σ_{LT} (MPa)
15°	10*	115	-5.5	5.5
45°	15	31	-1.2	0
75°	17*	48	-1.35	5.4

* These values should be equal if we consider that damage is initiated only by plane stresses, which are the same for (±15°)$_s$ and (±75°)$_s$ specimens at a given τ_s.

INFLUENCE OF STRESS, TEMPERATURE AND OTHER PARAMETERS

Temperature and stress
Tension/compression tests have been performed in the viscoelastic domain, either linear or non linear, while most of the torsion tests have been performed in the damage domain.
In the non-linear viscoelastic domain as in the damage domain, damping increases with stress level σ and with temperature T (e.g. see fig 4 and 5 for (±45°)$_s$ specimens in tension/compression).

Generally, the effects of σ and T are such that $\dfrac{\partial^2 \Psi}{\partial\sigma \cdot \partial T}$ be positive.

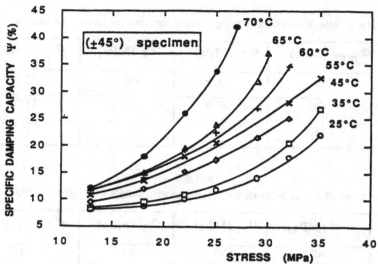

Figure 4 . Damping capacity versus stress for various temperature levels
(tension/compression)

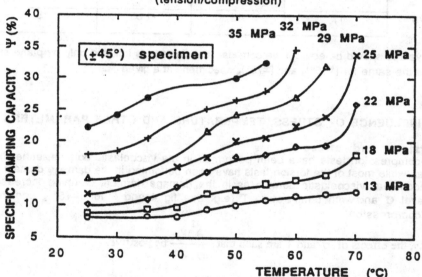

Figure 5 . Damping capacity versus temperature for various stress levels
(tension/compression)

Other parameters

At this day, the influence of frequency seems to be small within a range of 10 Hz to 1000 Hz.
The influence of manufacturing parameters such as following could be more significant :
- the proportioning and state of conservation of the epoxy resin and the curing agent
- the moisture content
- the curing cycle, which governs the glass transition temperature of the matrix and the
residual stress level
- the statistical distribution of fibers inside the matrix
- the fiber treatement

INFLUENCE OF FIBER ORIENTATION

For practical reasons, the influence of fiber orientation on damping has been studied only for a few angles of winding (±15°, ±45° and ±75°, plus ±30° and ±60° for some tests).

In the linear viscoelastic domain, we can write from Adams and Bacon's work (5):

$$\Psi = \Psi_L \cdot \frac{W_L}{W} + \Psi_T \frac{W_T}{W} + \Psi_{LT} \frac{W_{LT}}{W}$$

With $W = W_L + W_T + W_{LT}$

Where W_L, W_T and W_{LT} are the strain energies related to axial stress, transverse stress and pure shear respectively.

In tension/compression, the Ψ_L, Ψ_T and Ψ_{LT} contributions to global strain energy Ψ computed from laminate theory are shown on fig 6 (6).

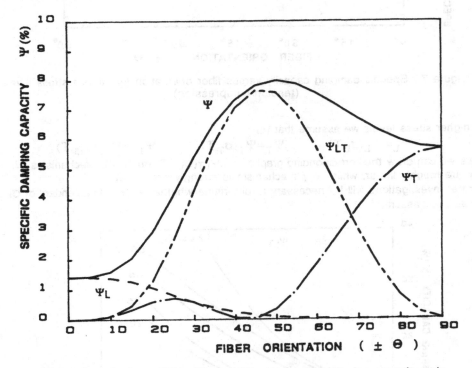

Figure 6 . Contributions of W_L, W_T and W_{LT} to specific damping capacity Ψ in tension/compression.

From the global damping capacity (curve obtained at low stress level in fig 7) we derive :

$$\Psi_L = 1.4\% \qquad \Psi_T = 5.6\% \qquad \Psi_{LT} = 8.5\%$$

These values denote a preponderant contribution of shear to global damping capacity.

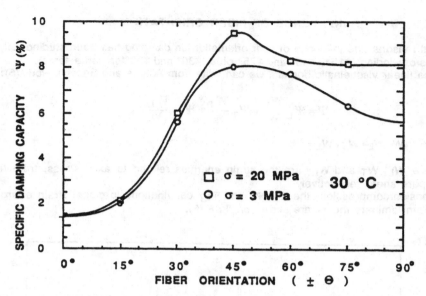

Figure 7 . Specific damping capacity versus fiber orientation for various stress levels (tension/compression).

At higher stress levels we assume that (5) :

$$\Psi_L=\Psi_L(\sigma_L,T) \qquad \Psi_T=\Psi_T(\sigma_T,T) \qquad \Psi_{LT}=\Psi_{LT}(\sigma_{LT},T)$$

Thus we can draw the corresponding graphs (e.g. $\Psi_{LT}(\sigma_{LT},T)$ on fig 8). Mechanisms W_L and W_T are roughly linear, while W_{LT} mechanism is strongly non linear.
Further investigations will be necessary to determine whether W_{LT} is independant of σ_L and σ_T as it was assumed.

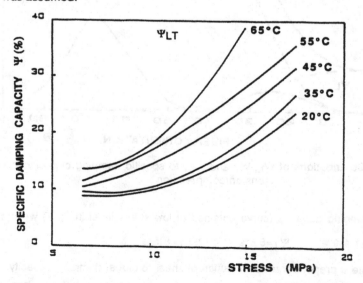

Figure 8 . Influence of stress level on damping mechanism Ψ_{LT} for various temperatures.

In alternate torsion, damage leads to damping capacities very different from the values predicted by Adams and Bacon's theory (fig. 9) when the mean shear stress exceeds 8 MPa (see fig.10)

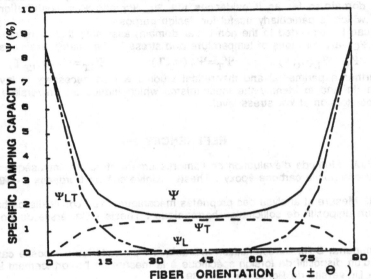

Figure 9 . Contribution of Ψ_L, Ψ_T and Ψ_{LT} to specific damping capacity in torsion.

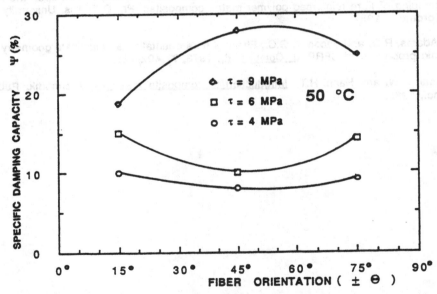

Figure 10 . Specific damping capacity versus fiber orientation for various stress levels (torsion)

It is likely that the convex pattern of the upper curve in fig 10 is due to local cracks within the material.

CONCLUSION

Adams and Bacon's theory is an interesting approach of damping prediction in the linear viscoelastic domain,so far as it evidences the Ψ_L, Ψ_T and Ψ_{LT} contributions to global damping Ψ ,which is particularly useful for design purpose.

This theory can be extended to the non linear domain, assuming that damping mechanisms Ψ_L, Ψ_T and Ψ_{LT} are functions of temperature and stress in the related directions only ,v.i.z :

$$\Psi_L = \Psi_L(\sigma_L, T) \qquad \Psi_T = \Psi_T(\sigma_T, T) \qquad \Psi_{LT} = \Psi_{LT}(\sigma_{LT}, T)$$

However, further experimental and theoretical studies will be necessary to investigate the damage domain and to identify the mechanisms which induce an irreversible increase in damping under torsion at low stress level.

REFERENCES

1   Delas, J.M., Méthode d'évaluation de l'amortissement structural intrinsèque d'un matériau composite carbone-époxy . Thesis - Université de Bordeaux I - 1989

2   Amlil, M., Mesure et analyse des propriétés mécaniques d'un composite carbone-époxy à partir d'un dispositif de sollicitation harmonique . Thesis - Université de Bordeaux I - 1990

3   Rakotomanana, C.R., Caractérisation mécanique d'un matériau composite carbone-époxy à l'aide d'un dispositif de torsion spécifique à la mesure de l'amortissement interne. Thesis - Université de Bordeaux I - 1989.

4   Morisson, W.D., The effects of moisture loss and elevated temperature upon the material damping of fibre reinforced polymer matrix composites. Ph. D Thesis. University of Toronto - 1987

5   Adams, R.D. and Bacon, D.G.C., Effect of fibre orientation and laminate geometry on the dynamic properties of CFRP, J. Comp. Mat., 1973, 7, 402-428.

6   Tsai, S.W. and Hahn, H.T., Introduction to composite materials, Technomic Publishing Co. Inc., 1980.

THE APPLICATION OF AUTOMATED ULTRASONIC INSPECTION FOR DAMAGE DETECTION AND EVALUATION IN COMPOSITE MATERIALS

F. DE ROEY, D. VAN HEMELRIJCK, L. SCHILLEMANS,
I. DAERDEN, F. BOULPAEP & A. H. CARDON
Composite Systems and Adhesion Research Group of the Free University of Brussels
VUB, Faculty of Applied Sciences, Building Kb, Pleinlaan 2, B-1050 Brussels, Belgium

ABSTRACT

Among the many methods available for non-destructive testing of composite materials, ultrasonic inspection is certainly one of the most widely used. Many of the inspection problems posed by composites can be dealt with effectively using ultrasonics. Signal processing techniques have furthermore advanced the state-of-the-art of non-destructive evaluation and testing [1]. In this paper we will describe the system built at the V.U.B. [2]. It will be shown that the use of digital signal processing and modern spectral analysis can solve some problems concerning the determination of flaws in a composite laminate.

INTRODUCTION

Various kind of flaws can be introduced into a fibre reinforced composite laminate during its fabrication and processing. Some of these are contaminations, voids, delaminations, resin-rich and resin-starved areas, ply gaps and surface scratches.

Delaminations near edges may also develop due to drilling and cutting after laminate cure. Void content is important because of its influence on the interlaminar shear strength (ILLS) and can ultrasonically be estimated by the measurement of the attenuation and reduction in the wave amplitude on passing through a specimen.

The major in service defect to be found is the presence of delamination. Small cracks are generally not to be found because they lead to delamination before a critical stage is reached.

Delaminations are normally ideally suited for the detection by ultrasonics. Because delaminations are good reflectors, they can be detected by measuring the amount of energy reflected or the corresponding loss of transmitted energy.

Foreign bodies are likely to be found by looking for a reflected signal or by measuring the attenuation. The sound velocity can also be correlated with incorrect cure, incorrect fibre volume fraction and voids.

Other defects, like fibre and ply misalignment are only readily detected using special techniques (polar scan) not normally applied and these will not be mentioned here.

In this article we will concentrate mainly on the use of signal processing techniques to classify the different flaws in a composite laminate.

MATERIALS AND METHODS

Ultrasonic scanning equipment

Figure 1 shows the ultrasonic scanning system constructed for non-destructive detection of flaws and flaw growth in composite laminate specimens.

The immersion scanning was selected because it is more suitable to be automated and of its better reproducibility. Pulse-echo was selected as the ultrasonic scanning mode.

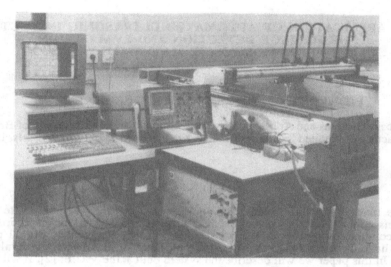

Figure 1. Ultrasonic scanning system.

The system consists of an immersion tank 110cm long, 90cm wide and 36cm high, resting on a supporting table. An aluminium frame surrounds the tank and it supports the mechanical system for scanning the specimen. This frame supports the mechanical system for scanning the specimen. It consists of two stepmotors with a resolution of 0.0125mm/step for the X-axis and 0.014mm/step for the Y-axis. The stepmotor controller has a GPIB interface.

The ultrasonic signals are generated and received by the USIP 12 manufactured by Krautkrämer [3]. This pulser-receiver is a device which can readily operate transducers in broad-band in a range of 1 to 20MHz and in narrow-band from 0.5 to 15MHz. It has an overall gain of 120dB. It provides a time domain output for reviewing the radio frequency signal on an oscilloscope and has two gated peak analog outputs from 0 to 10Volt.

To be able to measure the transit-time between different pulses, the USIP 12 has been extended with a DTM 12 module. It has a resolution of 1nsec or 0.01mm.

The data storage and processing is based on a HP vectra 80386 computer. The microcomputer is equipped with a digital to analog convertor card with a resolution of 12bits and a sampling frequency of 27.5kHz. An IEEE-488 interface card serves for the connection with the stepmotor controller.

A waveform digitizer card from Sonotek was installed to digitise the whole time signal at a frequency of 100MHz and a resolution of 8bit.

The Software package

The software written in TURBO C is completely menu-driven and very user-friendly [4].

In the input-menu the user has to specify the geometry to scan, the starting point, the measuring resolution and the choice between a continuous or a discontinuous scanning sequence. One also has to specify what will be collected as raw data; there are three basic parameters, namely the echo in gate 1, the echo in gate 2 and the transit time. A typical input-menu is shown in Figure 2.

C-SCAN : CARBON 15.G1R

<u>General parameters of scan</u>

Name of file: carbon15
Date of scan: 26-09-1990
Time of scan: 12: 26: 38
Type of scan: Discontinue
Startpoint: 1 = Lower left corner
X-length (mm): 400.
Y-length (mm): 400.
X-step (mm):   1.
Y-step (mm):   2.
Measure gate 1: Yes
Measure gate 2: Yes
Measure DTM12: No
Real time on screen: Gate 1
Autosave: Yes
Shift: 1
Comment: Carbon failure testplate met 15MHz taster van Panametrics

<u>DTM-12 parameters</u>

Velocity in water (km / s): 1.480
Velocity in specimen (km / s): 2.000
Focus length (mm): 50.8
Thickness (mm): 2.25
Scan depth (mm): 1.12

<u>USIP-12 settings</u>

Probe 1: V313
Probe 2: none
Gain (dB): 48
Pulser voltage: Hi
Frequency selection: 15 MHz BB
Echo representation: HWP
Backwall attenuation: 0
Filter: 0
Damping: 0
Echo reject: 0
Repetition rate: AUTO
Repetition rate range: * 1.0
Gate triggering by: I.F. echo
Echo in gate 1: Glasswall echo
Echo in gate 2: Defect echo

Figure 2. Input-menu.

The respons of the inspected structure is shown in real time on a VGA display. An amplitude distribution curve shows the percentage of occurrence of an energy portion. A smoothing algorithm is used to filter the raw data from accidental errors. The user also has the possibility to divide the range logarithmically and can even choose for a black and white plot. For visualisation an iso-contour plot can be generated, as shown in Figure 3.

Figure 4 shows a 3-D view of the defects in a plexiglas specimen. The depth is assessed by the DTM 12 module. The software enables the user to select the viewing angle as well as an option for magnification.

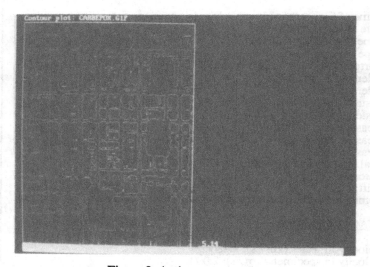

Figure 3. An iso-contour plot.

Figure 4. 3-D view of defects.

Application: a composite test plate

As an example a testplate with a thickness of 2.3mm in carbon fibre reinforced epoxy (T300, CIBA GEIGY 914) was prepared [5]. The lay-up was (0-90)9 and the different defects embedded are given in Figure 5. A focused shock-wave probe of 10MHz and a near field length of 15mm was used. The results for the backwall attenuation (gate 1) are given in Figure 6.The results for the defect echo are shown in Figure 7.

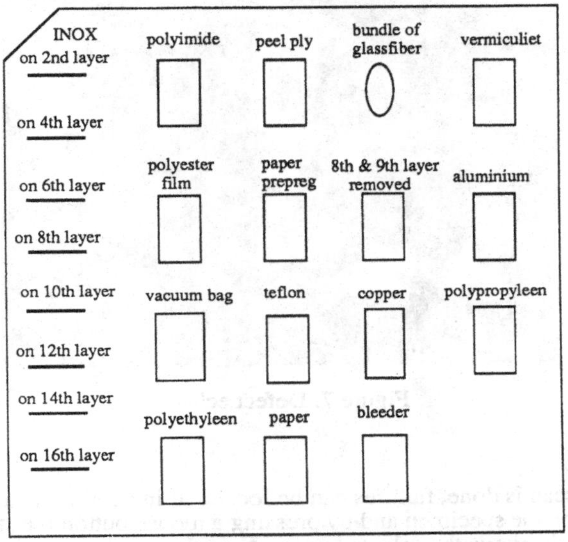

INOX on 2nd layer	polyimide	peel ply	bundle of glassfiber	vermiculiet
on 4th layer				
on 6th layer	polyester film	paper prepreg	8th & 9th layer removed	aluminium
on 8th layer				
on 10th layer	vacuum bag	teflon	copper	polypropyleen
on 12th layer				
on 14th layer	polyethyleen	paper	bleeder	
on 16th layer				

Figure 5. Localisation of defects.

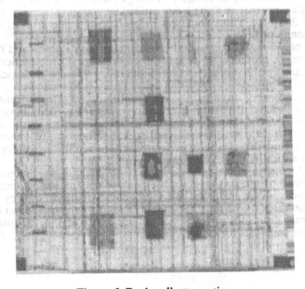

Figure 6. Backwall attenuation.

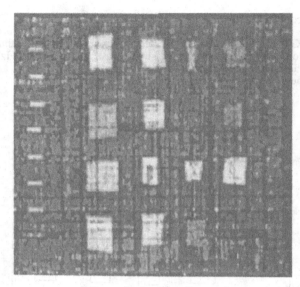

Figure 7. Defect echo.

Signal processing

When a complete c-scan is done, failures can be looked at in detail. By moving the mouse we can select any point on the specimen and by pressing a mouse button the stepmotors will place the transducer precisely above the selected point. Now the whole time signal will be digitized by the waveform digitizer card at a sampling frequency of 100MHz. The time domain already offers us a lot of information. For instance, the knowledge of the amplitude of the signal enables us to gauge the intensity or severity of the physical process or event that produced it. The activity level of a process may also be determined by counting the number of times a signal exceeds a particular threshold. This critical level may correspond to a key change in the state of the system being monitored [6].

The knowledge of the moment at which a particular peak occurs and of the time period between different peaks provides an awareness of the signal propagation time and delays within a given medium. Furthermore, the pulse width of a particular peak indicates the duration of a given incident.

In most signal processing applications, an important waveform transformation called Fourier Transform is employed. This transformation translates a time domain signal into what is known as a power domain signal. It can also be refered to as a frequency domain signal because it actually shows the power of the signal at each frequency; its effect is a plot of power versus frequency. Using this domain improves classification considerably since many phenomena occur at a specific range of frequencies.

In a manner analogous to that employed in the time domain, the intensity of a particular mechanism is indicated by the amplitude of the peak of the frequency domain signal. In this case, however, the peak will indicate the intensity at a particular frequency rather than at one instant in time. In the cases where the phenomena of interest are spread out over a large range of frequencies, we may divide the power spectrum into zones and use the percentage of power in each of these zones as a basis for classification [7].

RESULTS AND DISCUSSION

Four time signals were taken on different places of each embedded failure. Figure 8 shows the four signals taken from the paper prepreg.

Afterwards the FFT was calculated for the signals between the two vertical lines. Figure 9 gives the results for the paper prepreg. We can see that for the first three peaks the amplitude and the location is the same for the four signals.

At last we compared the frequency spectrum of three different failures like paper, polyethyleen and teflon. In Figure 10 we see that the signals differ a lot for each spectrum.

This proves that the ultrasonic signals contain a lot of information to identify a certain defect.

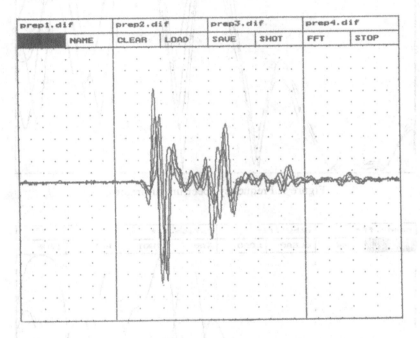

Figure 8. Time signal of paper prepreg.

422

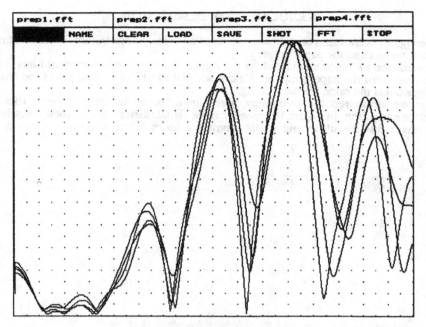

Figure 9. Frequency spectrum of prepreg.

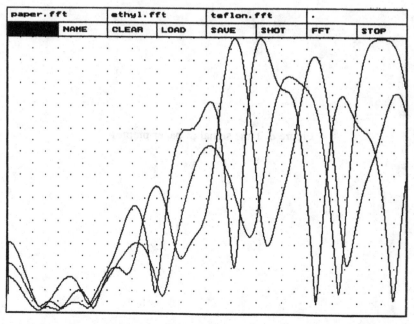

Figure 10. Frequency spectrum of paper, polyethyleen and teflon.

CONCLUSIONS

An approach to improve the testing capabilities of ultrasonic inspection is to extract more information from the ultrasonic signal in order to increase our knowledge of the hidden failures.

The results show that by using FFT algorithm, we can easily distinguish between the different failures.

The next step is to extend this method to an automated classifier with a large quantity of signals. The ultrasonic signals will be coupled with a pattern recognition system to identify the unknown signals.

ACKNOWLEDGEMENTS

The authors wish to thank Miss M. Verheyden for the poster arrangement and the typing and Mr. A. Vrijdag and Mr. F. Theys for the beautiful pictures.

REFERENCES

[1]   Chen C. H., "High resolution spectral analysis NDE techniques for flaw characterization, prediction and discrimination", 1988.

[2]   Van Hemelrijck D. , "Automation of ultrasonic test equipment", Thesis 1987.

[3]   Krautkrämer,"Operating manual USIP 12 and DTM 12".

[4]   De Roey F., "Software optimization for ultrasonics", Thesis 1989.

[5]   Van Hemelrijck D., Sol H., De Wilde W. P., Cardon A. H., De Roey F., Schillemans L., Boulpaep F., "Automatization of C-scan data acquisition". Cadcomp '90.

[6]   Hay R., Chan R., "Non-destructive testing handbook", Speedtronics LTD, 1990.

[7]   Roy C., Maslouhi A., Gaucher D., Piasta Z., "Classification of acoustic emission sources in CFRP assisted by pattern recognition analysis", Canadian aeronautics and space journal, 1988.

THE USE OF VIBROTHERMOGRAPHY FOR MONITORING DAMAGE EVOLUTION IN COMPOSITE MATERIALS

L. Schillemans, D. Van Hemelrijck, F. De Roey, I. Daerden, F. Boulpaep, A. Cardon
Composite Systems and Adhesion Research Group of the Free University of Brussels
COSARGUB - VUB - TW - Kb - Pleinlaan 2 - 1050 Brussels - Belgium

ABSTRACT

For the non-destructive evaluation of materials, several methods are well known, for instance the ultrasonic, the holographic, the radiographic method and others. For the visualization of stress concentrations on the other hand, Moire interferometry and photoelasticity are common practice. In the past few years however, a new method, called Stress Pattern Analysis by Thermal Emission (SPATE), is more and more known and used. A number of papers about the research on the value and interpretation of this method are already available but they are all especially concentrating on isotropic homogeneous materials. The basic thermoelastic theory for isotropic materials will be explained. However, because of the fact that the use of SPATE on composite materials has not been examined very well, the research being done at the Free University of Brussels is focusing on this domain. In this case, the deduction of the thermal emission - stress law must be reviewed and will be presented. A difficulty arises when at certain load levels, a phase shift is encountered between the upper and lower part of the hole. A basic theoretical explanation of this phenomenon will be put forward. Notification will be made that this is a general effect, valid for both isotropic and anisotropic materials. However, the magnitude of this effect is in general not noticeable on isotropic materials. On the other hand, this effect is in laminate materials enhanced and even of the same magnitude as the expected SPATE signal itself. As a result, SPATE scans of stressed laminated materials are much more difficult to interpret and evaluate correctly.

ISOTROPIC MATERIALS

Theory

Based on the thermomechanics for isotropic materials [7], a relation between the temperature change according to a stress state can be written as

$$T - T_0 = - \frac{\alpha_T T_0}{\rho\, c_p}\, \sigma_{kk}$$

(1)

where T stands for temperature, T_0 for the ambient temperature, α_T for the thermal expansion coefficient, ρ for the density, c_p for the heat capacity and σ_{kk} for the sum of the principal stresses. So, by measuring minute temperature changes, an evaluation of the sum of the principal stresses can be made.

Experimental results

Rectangular specimens of aluminium (Al.Mg.Si.0,5(F.22)) alloy, steel and PVC were cyclically loaded in a 10 kN capacity MTS servo-hydraulic testing machine. The Spate 9000 scan unit was positioned about 0.5 meters from the specimen surface. The surface temperature along a selected line across the width of the specimen was examined. An average of these results was calculated and plotted in function of dynamic stress amplitude. Figure 1 gives the results of the measurements on AL and PVC for a loading frequency of 5 Hz, 10 Hz and 15 Hz. Linear fits to these data are very good and we see no frequency dependence.

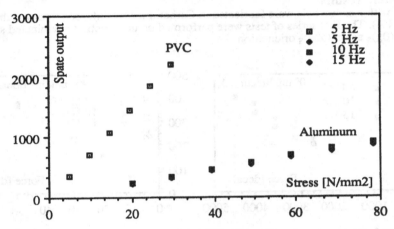

figure 1. Spate output as a function of the applied stress for PVC and aluminium

A rectangular steel specimen with a hole drilled in the middle was loaded uniaxially. The presence of this hole produces a complex non-uniform stress field in the surrounding material. Since the solution of the stress field in an uniaxially loaded infinite plate is analytically known, and since we know that the Spate output is proportional to the sum of the principal stresses, we can compare the analytical results with the Spate output. A computer program was written to compute this complex stress field [2]. The results are given in the form of a colour map and it has been noticed that the agreement is remarkably good [13].

ANISOTROPIC MATERIALS

Theory

The basic law governing the thermoelastic effect must now be expanded to anisotropic materials. The basic difference is due to the directionality of the thermal expansion coefficient.

$$\Delta T = - \frac{T}{\rho\, c_p} \left(\alpha_j \Delta \sigma_j + \alpha_k \Delta \sigma_k \right)$$

(2)

This formula is derived and valid in principal axes. But since the principal axis in most cases do not coincide with the material axis and since we do not, a priori, know the thermal expansion coefficient in principal axes, the formula is transformed to material axes, cfr. Potter [14].

$$\Delta T = - \frac{T}{\rho \, c_p} \left(\alpha_1 \Delta \sigma_1 + \alpha_2 \Delta \sigma_2 \right)$$

(3)

Again, there are only two terms, but here for a different reason. In principal axis we do have a thermal expansion coefficient coupling the shear stress with the variation in temperature, but by definition, there is no shear stress in principal axis. On the other hand, in material axis there does exist a shear stress but in this case the respective thermal expansion coefficient is zero. Notice however that in another axis system, the formula will have three terms and thus there will be a thermoelastic effect associated with shear stress.

Experimental results

Specimens with a centre hole were fabricated in Carbon Fibre Reinforced Epoxy (Fibredux 914 C TS 6K 34%). The first series of tests were performed on unidirectional laminated specimens, as well on $(0_8)_S$ as on $(90_8)_S$ orientation.

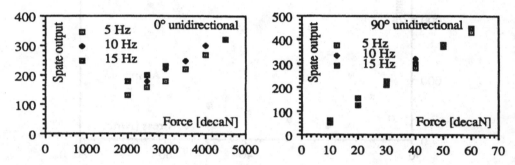

Figure 2. Spate output for 0° and for 90° unidirectional laminate (dim: 2 x 40 mm)

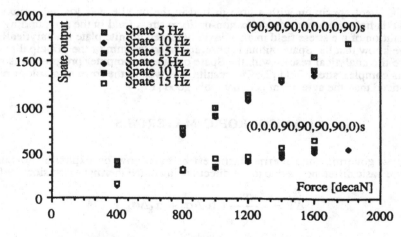

Figure 3. Spate output for type 1 and type 2

Unlike the results on isotropic materials, we noticed, across the specimen, a substantial standard deviation of the stress of about 20% to the averaged value. But, on both orientations linear fits are still very good. However, there is a frequency dependence for the 0° specimen, seen on figure 2. A frequency dependence can be very well explained when we consider the effect of the underlying layers when adiabatic conditions are not fulfilled. There will be a frequency dependent conduction of each layer leading to a different contribution to the resulting surface temperature. With respect to the applied stress, the 90° specimen establishes a much greater Spate output then the 0° specimen. This is explained by the anisotropic character of the thermoelastic expansion coefficient, which is much greater perpendicular to the fibre direction as in the fibre direction itself. Linearity tests were also done on a more complex laminate, namely $(90°_3,0°_4,90°)_s$ and $(0°_3,90°_4,0°)_s$. This lay-up was especially chosen because the macroscopical stress-strain relationship is exactly the same in both cases according to the composite laminate theory. Looking at figure 3, one does notice a different level of Spate output. This shows that Spate does not give an averaged signal, through the thickness, but that it is much more sensitive to the stress field on the surface layer then to the stresses in the underlying layers. Future tests will investigate the contribution of each layer dependant on their depth. Next, investigation was done on the same kind of specimens but now with a hole. Based on the laminate theory, it is possible to calculate, analytically [1], the stresses on the edge of a hole in a laminated plate, of an infinite size. The input consists of the layer properties, the number of layers, the stacking order and the loadcase. Based on the expanded anisotropic formula (formula 3), the theoretical Spate output can be calculated. Results for the quasi-isotropic case, for the 0° unidirectional case and for the 90° unidirectional case conform very well with practical Spate scans [13]. Work is now being done to predict the stresses, in each layer, and Spate output on the edge of a hole for more complex lay-ups, based on a non-adiabatic thermoelastic theory presented by Wong [5].

MONITORING DAMAGE EVOLUTION

Figure 4 shows a series of spate scans of a CFRP specimen $(90°_3,0°_4,90)_s$, with a hole drilled in the middle to initiate the damage process, for an increasing load.

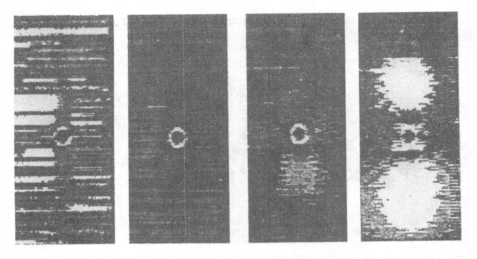

Figure 4. Spate scans of a $(0°_3,90°_4,0°)_s$ laminate loaded with 0.2 ,0.3 ,0.4 and 0.5 σ_f

The specimen was dynamically loaded in a servo hydraulic MTS testing machine with a sinusoidal loading at a frequency of 10 Hz. Results are shown for a maximum stress of 0.2, 0.3, 0.4 and 0.5 the failure stress respectively. There seems to be a certain correlation with the delamination detected by the ultrasonic c-scan, which are given in figure 5. One would expect a

Spate image conform to the contour of the delamination, as this is a place where the stress distribution is mostly affected. However, when we observe the image, we can't differentiate the lobes of the delaminations any more. Observe however, that the region above and below the hole produces a very different level of Spate output, which is physically impossible. Spate values for the region below the hole even reaches negative values. This is the reason why, at first, this effect was referred to as the "phase-shift".One can see this phenomenon better in figure 6 where the rightmost Spate scan of figure 4 is presented in a mountain plot.

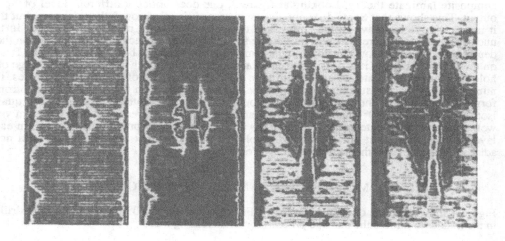

Figure 5. C-scans of a $(0°_3,90°_4,0°)_s$ laminate loaded with 0.2 ,0.3 ,0.4 and 0.5 σ_f

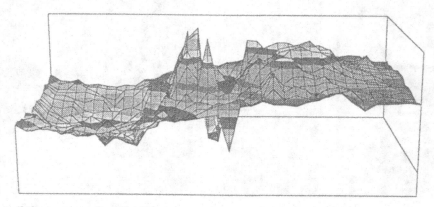

Figure 6. Spate scan of a $(0°_3,90°_4,0°)_s$ laminate dynamically stressed with $\sigma_{max} = 0.5$ σ_f

THE MOTION EFFECT

General principle

To explain this phasing problem, consider the effect of local displacements in combination with a static temperature field. When a specimen, with the top-end fixed, is being pulled down, the local displacements will start from zero and will increase towards the bottom. Here, it is assumed that this relation is linear, as schematically drawn in (1) of figure 7. Doing a whole series of tests on different laminates, it is noticed that at higher loadings, when the damage process initiates and propagates, a DC temperature field is developed beneath and on the surface of the specimen. A typical temperature field, along a vertical line going through the middle of the hole, is drawn in (2). As Spate is focused on an absolute point in space and since the specimen has local displacements, it is so that Spate will also be sensitive to the temperature gradients on the surface, this is depicted in (3), multiplied with the local displacements, which results in (4). Now, this component will contribute with the sum of the principal stresses (5) itself and gives the final Spate result in (6). When the lower part of (4) is big enough, it is easily seen that this effect can result into final negative values, in other words, the observed "180° phase-shift" between the region above and below the hole. Thus, it appears that when the temperature gradients and the local displacements are big enough, this effect can be seen. Notice that this is the reason why this effect has not yet been encountered on isotropic materials, as they establish a totally different damage behaviour in comparison with laminated anisotropic composites.

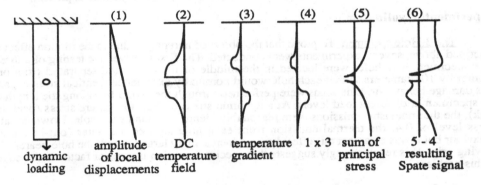

Figure 7. Schematic drawing of the motion effect

Theory

Consider a rectangular specimen of uniform cross-section. The x-axis is defined as a material axes in the longitudinal direction. The y-axis is defined as a fixed spatial axis. Taking into account the thermoelastic temperature response, the mean stress effect and plasticity, viscoelastic effects and inter-laminar fretting, it can be shown [15] that the temperature response of a specimen (in y-axis where Spate equipment is positioned) due to a sinusoidal cyclic load at frequency ω can be written as

$$\theta(y) = \left[T_0 - \frac{ky}{2}\frac{\partial T_1}{\partial x} + \frac{k^2 y^2}{2}\frac{\partial^2 T_0}{\partial x^2} \right] + \left[T_1 - ky\frac{\partial T_0}{\partial x} \right] \sin(\omega t)$$

$$+ A\sin(2\omega t) + \left[B + \frac{ky}{2}\frac{\partial T_1}{\partial x} - \frac{k^2 y^2}{2}\frac{\partial^2 T_0}{\partial x^2} \right] \cos(2\omega t)$$

(4)

where T_0 is a DC temperature component, (k y) is the local amplitude of displacement (motion of the specimen), T_1 is the basic thermoelastic response and A and B are combined coefficients based on the mean stress effect and the other non-linear effects. Consider now only the fundamental response, as this is the component which is normally measured by SPATE, and suppose that there indeed exists a steady state temperature distribution as a result of internal heat generation so that

$$\theta(y) = [\ T_1 - k\ y\ \frac{\partial T_0}{\partial x}\ ]\ \sin(\omega t)$$

(5)

It may be seen that for sufficient motion, together with an existing steady-state temperature gradient, the term $(ky)(\partial T_0/\partial x)$ can in fact become significant with respect to the actual response of interest T_1. To illustrate this problem, consider a 0° unidirectional XAS-914 composite in which the physical properties may be taken as in [5], viz., $\alpha_f = 2.8 \times 10^{-6}$, $\rho = 1630$ kg/m^3, C = 935 J/kg.°C, and that $T_0 = 300$°K. It may be shown from Eqs. 1 and 5 that a 0.1 mm motion at a location where a temperature gradient of 1°C over 10 mm exists, would be sufficient to produce a signal equivalent to approximately 180 MPa, which is about 25% of the failure stress of the specimen. This is indeed severe, and implies that the effects of motion cannot be ignored even when a relatively moderate static temperature gradient is present.

Experimental validation

The 3-Hole specimen: To prove that the observed anomaly is due to the motion effect as described above, several experiments were conducted. The first involved the testing of a three-hole specimen. The holes were located in the middle of the width and separated from one another by 10 diameters so that each hole would nominally experience identical stresses, and thus damage behaviour. Line scans were performed through the centre line along the length of the specimen at different load levels. At a dynamic stress of 0.25 the failure stress (peak-to-peak), the thermoelastic emissions were reasonably identical around each hole. However, at a stress level of 0.4, the thermal emission from each hole showed a similar "out-of-phase" behaviour but it was clear that this effect became increasingly large towards the hole nearest the moving grip. This result strongly suggests that specimen motion is a major factor in the cause of this problem.

Surface temperature measurements: Specimen motion has an effect on the fundamental response when a static temperature gradient is present. This could explain the fact that "out-of-phase" behaviour begins to occur only after a sufficient dynamic load is applied and which invariably coincides with damage growth. To investigate this aspect, surface temperatures of a single hole specimen were measured using two copper-constantan thermocouples located 20 and 70 mm above the hole. The difference between the near and far field temperatures were recorded for various load amplitudes and it was seen that a temperature difference begins to develop at a dynamic stress of 0.05 and reaches as much as 4°C at a stress of 0.3 the failure stress.

The heating wire test: So far, we have demonstrated that a cyclically loaded specimen containing a hole can generate a non-uniform static temperature distribution and that the results of the 3-hole specimen suggests that motion is responsible for corrupting the SPATE scans. To remove any possibility of abnormal stress fields (e.g., due to bending) around the delamination sites being responsible for the observed anomaly, a moving but unloaded, specimen was used as a final definitive test to show that motion and temperature gradient are sufficient to cause this "out-of-phase" phenomenon. Specimens of uniform cross-section were mounted on the moving grip and allowed to be cyclically displaced with one end free. The displacement amplitude was set at 0.35 mm which was comparable to the other, loaded, tests. A heating wire was mounted on one side of the specimens whilst a line SPATE scan was taken along the other side where the maximum surface temperature was monitored by a thermocouple. SPATE line scans on 0° and 90° unidirectional composite specimens showed the "out-of-phase" effects. It was concluded that these rather large SPATE "error" signals can certainly be responsible for corrupting any real data due to the stressing. Comparing the two specimen types, it is interesting to note that the motion effect was more confined to the heat source in the 90° specimen as the lower thermal conductivity in the longitudinal direction restricts the flow of heat to the ends. However, this also means that the temperature gradients (for a given maximum surface temperature) would also be higher and, according to the theory presented, would produce a higher maximum SPATE output than that of the 0° specimen. This was indeed found to be the case and confirmed the theory.

CONCLUSION

Experiments on isotropic materials have shown very good applicability of Spate. Agreement with analytical predictions are in general very good. However, complexity is very much increased when applying Spate on anisotropic materials. Experiments have shown primary dependency of Spate output to the orientation of the surface layer, but also to the loading frequency, the stacking sequence and the existence of delaminations. A comparison of the experiments and the expanded thermoelastic theory for anisotropic materials shows good agreements for unidirectional laminates. On the other hand, for more complex laminates, in the undamaged state, it is advisable to take into account contributions of underlying layers due to non-adiabatic conditions. In using the technique of thermoelastic stress analysis on composite laminates, it was discovered that at sufficient load levels, spurious results can occur. The anomalies usually came in the form of a gross loss of symmetry in the thermoelastic data despite symmetrical conditions being maintained. The problem was identified to be associated with specimen motion due to the cyclic loading. The basic theory of the motion effect was presented and it was shown that even small local oscillations, coupled with the presence of just moderate temperature gradient (as a result of internal heat generation), can produce a significant component which can totally corrupt the stress induced signal. A series of tests, using firstly a 3-hole specimen and later an unloaded artificially heated specimen, as well as monitoring the surface temperature gradients of a single hole specimen, were carried out and showed convincingly that motion and temperature gradient were indeed the cause of the problem.

ACKNOWLEDGEMENTS

The authors wish to express special thanks to Dr. A.K. Wong of the Aircraft Structures Division, Aeronautical Research Laboratory, Australia, for the fruitful cooperation. The authors also wish to thank A. Vrijdag for producing the illustrations and the Belgian National Science Research Fund (NFWO-FKFO) and the Research Council (OZR) of the Free University of Brussels (VUB) for the support. The authors are also indebted to Drs. R. Jones and T.G. Ryall for the many fruitful discussions.

REFERENCES

[1]   Lekhnitskii, S.G., 'Anisotropic plates', Gordon and Breach, 1968

[2]   Boresi, A.P., Sidebottom, O.M., 'Advanced mechanics of materials', Wiley, 1985

[3]   Wong A.K., Dunn S.A. and Sparrow J.G., 'Residual stress measurement by means of the thermoelastic effect', Nature, V332, 613-615 (1988)

[4]   Owens R.H., 'Application of the thermoelastic effect to typical aerospace composite materials', Proc. SPIE Conf., Stress Analysis by Thermoelastic Techniques, London, V731, 74-85 (1987)

[5]   Wong A.K., 'A non-adiabatic thermoelastic theory for composite laminates', J. Phys. Chem. Solids -- in press

[6]   Van Hemelrijck, D., Schillemans, L., et al., 'Comparative study between ultrasonic inspection and vibrothermography for the assessment of damage in Carbon Fibre Reinforced Plastics', Science and engineering of composite materials, Vol. 1. No. 4 1989, pp.109-116

[7]   Thomson W. (Lord Kelvin), 'On the dynamical theory of heat', Trans. Roy. Soc. Edinburgh, V20, 261-283 (1853)

[8]   Wong A.K., Jones R., Sparrow J.G., 'Thermoelastic constant or thermoelastic parameter ?', J. Phys. Chem. Solids, vol. 48, 1987, pp. 149 - 153

[9]   Wong A.K., Dunn S.A., Sparrow J.G., 'On the revised theory of the thermoelastic effect', J. Phys. Chem. Solids, vol. 49, 1988, pp. 395 - 400

[10]   Jones R., Tay T.E., Williams J.F., 'Thermomechanical behaviour of composites', Workshop on composite material response: constitutive relations and damage mechanisms', G.C Sih et al. (Eds), Elsevier Applied Science, 1988

[11]   Dunn S.A., Lombardo D. and Sparrow J.G., 'The mean stress effect in metallic alloys and composites', Proc. SPIE Conf., Stress and Vibration: Recent Developments in Industrial Measurement and Analysis, London, 129-142 (1989)

[12]   Henneke, E.G., Jones, T.S., 'Detection of damage in composite materials by vibrothermography. NDE and flaw criticality for composite materials', ASTM, STP 696

[13]   D. Van Hemelrijck, L. Schillemans et al., 'The use of thermoelastic emission techniques (SPATE) for damage analysis of graphite epoxy composites', International Colloquium on Durability of Polymer Based Composite Systems for Structural Applications, 27-31 august '90, Brussels, Belgium

[14]   Potter, R.T., 'Stress analysis in laminated fibre composites by thermoelastic emission', Stress analysis by thermoelastic techniques, Second international conference, London, 1987

[15]   D. Van Hemelrijck, L. Schillemans, A.H. Cardon, A.K. Wong, 'The effects of motion on thermoelastic stress analysis, to be published in Composite Structures

ACOUSTIC EMISSION SOURCE LOCATION
IN ANISOTROPIC COMPOSITE PLATES

Bernard Castagnède

Laboratoire de Mécanique Physique,
URA CNRS 867, Université de Bordeaux I,
351, Cours de la Libération, 33405 TALENCE Cedex, FRANCE

ABSTRACT

This work describes a method by which a pointlike source of acoustic emission can be located in a plate of composite material. The method is applicable to a source in an anisotropic material of arbitrary symmetry as long as the principal acoustic axes as well as the all set of elastic constants of the material are known *a priori*. A rigorous analytical solution for the general 3-D source location problem which use solely basic principles is described. A double-iterative algorithm, implemented to solve numerically such a problem, is working both with simulated and with experimental data. Numerical results demonstrating the effectiveness and robustness of the numerical method are shown on 2-D as well as 3-D versions of the algorithm. Some experimental results obtained with a unidirectional fiberglass/isophthalic polyester resin are described. The results unambiguously show that the source locations are properly recovered.

INTRODUCTION

The precise location of acoustic emission (AE) sources is a problem of importance in geophysics [1] as well as in nondestructive evaluation of materials. In terms of the characterization of engineering materials, acoustic emission source location is an essential procedure when studying the dynamics of failure processes. In any case, the AE event corresponds to a dynamic force field change and an irreversible release of energy at the source point. In turns, this process may be the result of various causes (motion of dislocations, propagation of a crack, phase transformation), or the action of simulated sources such as the thermoelastic and ablative effects accompanying the interaction of a laser on a material [2].

All source location techniques rely on the measurements of the time-of-flight of a particular mode between the source and a sensor. The measurements can be done by numerous approaches, from direct chronometry at a given threshold to indirect analysis of the digitized waveforms via Fourier and correlation techniques. There are, however, some limitations to these procedures which are related to multiple AE events, background noise, wave attenuation, geometric and material dispersion [3]. Despite these restrictions, several source location techniques dealing with homogeneous isotropic materials have been proposed in the past [4,5] including, triangulation methods [6,7], least-squares numerical techniques [8,9], and correlation methods [10,11]. Also, some numerical work was done to describe staggering in the position of the sensors [12], the presence of noise [13] and the dispersive nature of signals in guided geometries [14].

Determination of the location of a source of emission in an anisotropic material is complicated by the fact that the wavespeed of each acoustical mode is dependent on the propagation direction in the material. While it has been recognized that a solution to this problem is a prerequisite in successfully using AE measurements in anisotropic materials, no rigorous numerical scheme was available until recently. For instance in the field of rock mechanics approximate inversion techniques [15] were used to locate AE sources [16], even if it is known that such rough estimates lead to errors [17]. Also, in the area of composite materials characterization, similar approximate techniques were recently introduced to determine AE source locations [18,19], and to characterize defects [20,21]. Simultaneously, an alternative rigorous numerical scheme based upon the Christoffel wave equation was introduced [22] and tested with success on a composite material [23]. A comprehensive review of the method, including extensive numerical as well as experimental results, was recently presented [24]. We would like here to describe the approach, discuss its significance and possible extensions to other problems and geometries.

GENERAL SOLUTION OF THE SOURCE LOCATION PROBLEM

The solution of the most general three-dimensional source location problem in an anisotropic solid is a formidable task. However, restriction to the following assumptions, i.e. propagation in a plate with uniform thickness, made of a perfectly elastic and homogeneous anisotropic solid, having orthorhombic symmetry or higher, where the wave dispersion is negligible, with pointlike sources and receivers, constitutes a much more tractable problem. These assumptions are adopted here; they are discussed further in Ref. [24].

On the other hand, we restrict ourselves in this Section to the general case where the sensors are distributed randomly. Another concern is about placing sensors over one single or both surfaces of the plate. We consider here, that we have provisions to mount pointlike piezoelectric sensors onto both surfaces of a flat plate of thickness $2h$. Suppose that a pointlike AE source is located in the plate at coordinates X_1^S, X_2^S and X_3^S as shown in Fig. 1. By selecting a given pair of sensors (k,m) where k refers to the sensor positioned on the upper surface and m to a sensor mounted on the bottom surface, one can derive a simple time-difference equation between the arrivals at the two sensors:

$$\Delta t_p(k,m) = [C_p(n_k)]^{-1}\sqrt{(h+X_3^s)^2+l_k^2} - [C_p(n_m)]^{-1}\sqrt{(h-X_3^s)^2+l_m^2} , \qquad (1)$$

where $k,m = 1,...,N$, and where N is the number of sensors mounted on each surface. The wavespeeds of the quasilongitudinal wave along the two source/receiver acoustical paths are specified by $C_p(n_k)$ and $C_p(n_m)$ respectively, where n_k and n_m are the corresponding direction cosines of the propagational paths. The quantities l_k and l_m represent the distances between source and sensor k or sensor m projected onto the top or bottom surfaces of the plate respectively, and are given by the following expressions:

$$l_k^2 = R^2 + d_k^2 - 2X_1^k X_1^S - 2X_2^k X_2^S , \qquad (2)$$

where $R^2 = (X_1^S)^2 + (X_2^S)^2$, and $d_k^2 = (X_1^k)^2 + (X_2^k)^2$, and with an analogous expression for l_m. In this equation, X_1^k and X_2^k are the in-plane coordinates of sensor k.

In Eq. (1), the wavespeeds $C_P(n_k)$ and $C_P(n_m)$ are functions of the direction cosines n_k and n_m respectively whose components are provided by the relationships:

$$n^k_{1,2} = (X^k_{1,2} - X^S_{1,2})/D \quad ; \quad n^m_{1,2} = (X^m_{1,2} - X^S_{1,2})/D'$$

$$n^k_3 = (h + X^S_3)/D \quad ; \quad n^m_3 = (h - X^S_3)/D' \tag{3}$$

where
$$D = [(h + X^S_3)^2 + (X^k_1 - X^S_1)^2 + (X^k_2 - X^S_2)^2]^{(1/2)}$$
$$D' = [(h - X^S_3)^2 + (X^m_1 - X^S_1)^2 + (X^m_2 - X^S_2)^2]^{(1/2)} \ .$$

Generally, the propagation is done in non-principal planes for which there are no analytical expressions for the wavespeeds [25]. However, the wavespeeds are related to the eigenvalues of the Green-Christoffel tensor (or propagation tensor Γ_{il}) as expressed with the characteristic equation:

$$\det\left|\Gamma_{il} - \rho c^2 \delta_{il}\right| = 0, \tag{4}$$

where $\Gamma_{il} = C_{ijkl} \, n_j \, n_k$ with (i,j,k,l = 1,2,3) and where C_{ijkl} is the stiffness tensor of the material. The term ρ is the density of the material, δ_{il} is the Kronecker delta, and c are the wavespeeds of the three different bulk modes propagated in various directions of the material. The directions cosines appearing in this equation are identical to those provided in Eq. (3). From knowledge of the elastic constants for a given symmetry, the eigenvalues of the propagation tensor are then computed, by using for instance a cyclic Jacobi method [26,27]. From the values of the wavespeeds in any directions computed by the above procedure, the source coordinates can be recovered, in principle, by inverting Eq. (1). For the purpose of the inversion, Eq. (1) is rearranged and successively squared twice, and after some algebraic manipulations, the following set of nonlinear equations is obtained:

$$\Psi(X^S_1, X^S_2, X^S_3) = A_1(X^S_3)^4 + A_2(X^S_3)^3 + A_3 X^S_1(X^S_3)^2 + A_4 X^S_2(X^S_3)^2 + A_5(X^S_3)^2 + A_6(X^S_1)^2$$
$$+ A_7(X^S_2)^2 + A_8 X^S_1 X^S_2 + A_9 X^S_1 X^S_3 + A_{10} X^S_2 X^S_3 + A_{11} X^S_1 + A_{12} X^S_2 + A_{13} X^S_3 + A_{14} \approx 0 \tag{5}$$

where the coefficients A_i with $i = 1, 2,..., 14$ correspond to each (k,m) pair of sensors, and are function of Δt_p (k,m), $C_P(n_k)$, $C_P(n_m)$, h, R, d_k and d_m . The full set of the A_i coefficients was provided in Ref. [24] for the case of a circular array of sensors. Since there are only three unknown source coordinates X^S_1, X^S_2 and X^S_3 for a total of N^2 measured Δt_p , Eq. (1) or Eq. (5) are overdetermined sets of nonlinear equations, and optimization numerical techniques are used to solve optimally the problem. For that purpose, the Euclidean functional associated to Eq. (5) is introduced and then minimized by using a Newton-Raphson algorithm [26, 27]. However, significant differences exist with the standard algorithm and are discussed further in Ref. [24]. The most important modification is that the present algorithm is double-iterative. This is due to the fact that the all set of coefficients A_i in Eq. (5) are functions of the source coordinates. Thus, the problem seems to be ill-posed as the nonlinear system of Equations of the unknown source coordinates have their coefficients which are functions of the same unknowns. This paradoxal feature is inherently linked to the transcendental nature of the original set of Eq. (1), where the wavespeed are functions of the direction of propagation. In

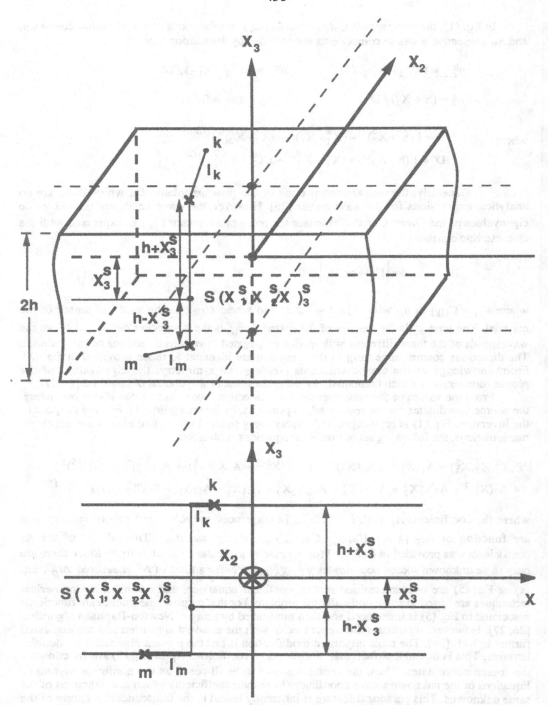

Figure 1: Geometry for the source location problem in the general case.

other words, this paradox is related to the intuitive difficulty of locating a source of emission in a material where the speed of propagation of the acoustical waves are precisely function of the unknowns of the problem (the source coordinates) themselves. The ill-posed nature of the problem is only apparent as the convergence properties are nevertheless excellent. In terms of the practical implementation of the algorithm, the all set of coefficients A_i in Eq. (5) has to be recalculated after each proper convergence, before a new outer loop of the Newton algorithm to be resumed.

NUMERICAL SIMULATIONS OF THE 2-D AND 3-D ALGORITHMS

The numerical scheme which has been described here can be specialized in various particular cases. For instance, when dealing with the circular array geometry of the sensors, and when the thickness of the plate is small enough, one can consider the further approximation of the third coordinate (X_3^s) to be fixed to either $-h$ or $+h$. In such a 2-D case, Eq. (5) reduces to a set of quadratic transcendental equations as was discussed in Ref. [24]. As a matter of fact, this limiting case is used in the Section dealing with experimental results.

Here, we would like to discuss some general properties of the 2-D and 3-D versions of the source location recovery algorithm in terms of convergence. Some common features are the following i) the algorithm is surprisingly robust to noise, ii) proper convergence is almost always independent on the initialization source coordinates vector, and iii) the convergence is obtained after a small number of iterations (typically between 5 and 30 depending on the position of the source, the number of iterations being increased in remote areas).

The analysis of the influence of anisotropy on the convergence properties of the algorithm is not very easy to achieve. A method has been proposed by using a statistical method in which the propagation of random errors in the location process is investigated as a function of the location area. This approach considers that the measurements of the time-delay differences are perturbated by some random errors, with let say a Gaussian density function at a given mean-square for the distribution. The simulation of these random errors on the computer, allows to study statistically how these errors propagate along the inverse numerical scheme of source location, by running the all process several times. An example is presented for the 2-D algorithm in Table I, for which the numerical parameters are shown in the caption. Beyond the angular dependence of $\Delta R/R$ and $\Delta\alpha$ at a given R, which may change for other sets of the parameters, it is seen that there is a dramatic decrease in the accuracy of the location determination when the source is located exterior to the array of sensors.

There are some differences between the 2-D and 3-D versions of the algorithm, however. Exhaustive numerical results obtained with the 3-D algorithm were reported in Ref. [24]. The convergence is generally more difficult to achieve when dealing with three coordinates instead of two, especially for the Z or X_3 coordinate. An example is presented in Table II.

EXPERIMENTAL RESULTS FOR A UNIDIRECTIONAL COMPOSITE

Description of the experimental procedures

The samples are fiberglass-reinforced plates of unidirectional composite from the Extren 500 serie [32], made with an isophthalic polyester resin. The plates were 305 mm square with a 6.35 mm average thickness. The glass fibers lay in plane (1,2) and were mainly oriented in direction 1. The ratio of the Young moduli along the two in-plane principal directions was 1.5.

A *point-source/point-receiver* AE system [33,34] was used to perform the experiments. The sensors used here were wide bandwidth piezoelectric transducers with a 1.3 mm aperture. They were positioned over a 50.8 mm circle on the top surface of the specimen. The detected

α (deg)	R = 5.00 cm		R = 10.00 cm		R = 15.00 cm	
	ΔR/R (%)	Δα (deg)	ΔR/R (%)	Δα (deg)	ΔR/R (%)	Δα (deg)
0	0.146	0.183	0.324	0.232	1.523	0.529
10	0.170	0.100	0.306	0.140	2.414	0.485
20	0.150	0.189	0.311	0.174	2.709	0.713
30	0.138	0.163	0.349	0.138	2.692	0.550
40	0.122	0.096	0.338	0.149	1.499	0.293
50	0.114	0.059	0.467	0.113	2.329	0.372
60	0.236	0.072	0.309	0.101	2.400	0.256
70	0.350	0.046	0.225	0.047	0.502	0.051
80	0.204	0.120	0.159	0.026	0.687	0.087
90	0.242	0.244	0.091	0.049	2.094	0.049

TABLE I. 2-D Algorithm convergence for various source locations.

Simulation parameters: d = 10.00 cm; h = 0.01 cm; N = 8; $d(\Delta t_p)$ = 1 %; ρ = 2.108 g/cm^3; C_{11} = 26.30 GPa; C_{22} = 65.50 GPa; C_{12} = 9.58 GPa; C_{66} = 10.50 GPa. Elastic constants are provided for a unidirectional composite material [28,29].

		Average	mean-square	coef. varia. (%)
N = 8	R	15.001	0.020	0.13
	a	30.026	0.149	0.50
	Z	1.028	0.127	12.36
N = 16	R	15.001	0.015	0.10
	a	29.962	0.043	0.14
	Z	1.111	0.041	3.66
N = 32	R	14.999	0.005	0.03
	a	29.984	0.033	0.11
	Z	1.004	0.019	1.90

TABLE II. 3-D Algorithm convergence for different number of sensors.

Simulation parameters: d = 10.00 cm; h = 2.00 cm; α = 30 deg; N = 8; $d(\Delta t_p)$ = 1 %; ρ = 1.28 g/cm^3; C_{11} = 7.45 GPa; C_{22} = 8.30 GPa; C_{33} = 29.30 GPa; C_{44} = 2.88 GPa; C_{55} = 2.71 GPa; C_{66} = 1.65 GPa; C_{12} = 4.36 GPa; C_{13} = 3.97 GPa; C_{23} = 4.97 GPa. Elastic constants are provided for a tropical wood [30], and were measured by using an ultrasonic spectro-interferometer [31].

AE signals were amplified 60 dB with broadband preamplifiers and recorded on transient recorders operating with a 3 MHz sampling rate and 10-bit vertical resolution. Four and eight channels were available to make the measurements. AE sources were simulated by fracturing tiny glass capillaries, producing a step force excitation with a typical 50 ns rise time.

A unidirectional composite is generally described by the hexagonal symmetry with 5 independent elastic constants [35] . Due to the small thickness of the plates used in this work, when compared to the typical source-sensors distances, the propagation can be assumed to take place approximately in plane (1,2). As a result, the 2-D version of the algorithm is sufficient, and only the in-plane elastic constants, i.e. C_{11}, C_{22}, C_{66} and C_{12} are needed. These moduli were obtained in a calibration experiment were elastic waves were launched along particular directions, i.e. along the principal axes and at 45 degrees.

Results of the isotropic and anisotropic optimization

After completion of the calibration procedure, the AE source location can proceed. The simulated AE source was located in various positions of a 2-D grid over the upper right quadrant of the circular array of sensors mounted on plane (1,2). From the recorded transients waveforms, the start of the detected signals at the different receiver stations is referenced. From the measurements of the time-arrival differences $\Delta t_p(k,m)$, the use of the 2-D version of the algorithm provided the recovered source coordinates. To ascertain if the results would differ significantly by considering the material isotropic or not, the algorithm was run twice, using for the isotropic case the additional relationships $C_{22} = C_{11}$ and $C_{66} = (C_{11} - C_{12})/2$.

Real position		Anisotropic		Isotropic	
R(mm)	α(deg.)	R(mm)	α(deg.)	R(mm)	α(deg.)
107.77	45.00	116.13	43.42	Divergence	Divergence
91.58	33.69	90.44	29.09	85.28	28.59
91.58	56.31	91.04	53.84	Divergence	Divergence
80.32	18.44	87.70	14.09	67.76	16.90
80.32	71.56	81.31	67.38	Divergence	Divergence
76.20	0.00	80.08	0.12	59.78	0.15
76.20	90.00	86.21	88.16	186.41	88.17
71.84	45.00	66.98	44.45	Divergence	Divergence
50.80	0.00	53.67	0.07	43.95	0.09
50.80	90.00	51.34	88.94	69.10	89.47
25.40	0.00	25.59	0.15	22.91	0.21
25.40	90.00	25.68	89.20	30.84	89.47

TABLE III. Experimental results

The experimental results are collected in Table III. These results clearly exhibit that the localization is much better in the anisotropic case. When using the isotropic supplementary hypothesis, there are even data points where there is no more convergence of the algorithm. This last feature constitutes a clear demonstration about the efficiency of the numerical method to locate AE sources, and to distinguish small deviations from elastic isotropy. Moreover, one can observe that the results are systematically better for locations which are close to the center of the array of receivers. This trend was well known from the work done in numerical simulation [22,24]. Several reasons, e.g. pronounced geometric dispersion of the signals, attenuation and dissipation of the bulk modes, can account for the poor accuracy of the localization process in remote areas. The amplitude of the quasilongitudinal bulk mode is negligible in comparison to the Lamb waves [36], especially for sensors in the farfield of the source where the signal strength of the bulk modes are of small amplitude or even vanish [37,38]. Other limitations exist as well. For instance, the assumption of the propagation being done in a principal plane is only an approximation, meaning that the full 3-D algorithm, which involves the complete set of elastic constants, should be considered for a more rigorous description of such a problem.

CONCLUDING REMARKS

In the present work, a general method to optimally solve the source location problem, which only utilizes basic physical principles of the propagation of bulk waves in anisotropic plates, has been described. There are several drawbacks and limitations to this approach, the most serious one being to consider only pointlike sources. Generalisations of the method to extented sources might be possible, however. Another concern is related to the use of the quasi-longitudinal mode of the bulk wave instead of other bulk modes or other type of waves, e.g. plate or surface waves. When dealing with experimental waveforms, the quasilongitudinal mode is the first arrival and for that reason is certainly the most unambiguously identifiable feature that can be easily recognized. Finally, extensions of the present method to other geometries of the array of sensors and possibly of the shape of the specimen is just a matter of additional algebra.

REFERENCES

[1] Aki, K. and Richards, P.G., *Quantitative Seismology*, Freeman, San Francisco, vol. 1, 1980, pp. 383-477.

[2] Scruby, C.B., Dewhurst, R.J., Hutchins, D.A., and Palmer, S.B., Laser generation of ultrasound in metals in *Research techniques in nondestructive testing*, Academic Press, New York, vol. 5, 1982, pp. 281-327.

[3] Baron, J.A. and Ying, S.P., Acoustic emission source location in *Nondestructive testing Handbook*:, ASNT, Colombus (Ohio), vol. 5, 1987, pp. 136-154.

[4] Sachse, W. and Sancar, S., Acoustic emission source location on platelike structures using a small array of transducers, *US Patent*, N° 4,592,034, 1986.

[5] Ying, S.P., Hamlin, D.R. and Tanneberger, D., A multichannel acoustic emission monitoring system with simultaneous multiple event data analyses. *J. Acous. Soc. Am.*, 1974, **55**, 350-356.

[6] Tobias, A., Acoustic-emission source location by an array of three sensors. *Nondest. Test. Int.*, 1976, **10**, 9-12.

[7] Asty, M., Acoustic emission source location on a spherical or plane surface. *Nondest. Test. Int.*, 1978, **12**, 223-226.

[8] Pao, Y.H., Theory of acoustic emission in *Elastic waves and nondestructive testing of materials*, ASME, New York, vol. 29, 1978, pp. 107-127.

[9] Hsieh, P.N., *Quantitative acoustic emission source characterization in an aluminum specimen.*, Ph.D. dissertation, Cornell University (New York), 1987.

[10] Grabec, I., Application of correlation techniques for localization of acoustic emission sources. *Ultrasonics*, 1978, **16**, 111-115.

[11] Bracewell, R.N., *The Fourier transform and its applications.*, McGraw Hill, New York, 1986, pp. 98-123.

[12] Carter, G.C., Passive ranging errors due to receiving hydrophone position uncertainty. *J. Acous. Soc. Am.*, 1979, **65**, 528-531.

[13] Clay, C.S., Hinich, M.J. and Shaman, P., Error analysis of velocity and direction measurements of plane waves using thick large-aperture arrays. *J. Acous. Soc. Am.*, 1973, **53**, 1161-1166.

[14] Clay, C.S. and Hinich, M.J., Use of a two-dimensional array to receive an unknown signal in a dispersive waveguide. *J. Acous. Soc. Am.*, 1970, **47**, 435-440.

[15] Rothman, R.L., Greefield, R.J. and Hardy, H.R., Errors in hypocentral location due to velocity anisotropy. *Bull. Seismol. Soc. Am.*, 1974, **14**, 1993-1996.

[16] Getting, L.C., Roecken, C. and Spetzler, H., Improved acoustic emission locations. *J. Nondest. Eval.*, 1986, **5**, 133-143.

[17] Wiggins, R.A., The general linear inverse problem: Implications of surface waves and free oscillations for earth structure. *Rev. Geophys. Space Phys.*, 1971, **10**, 251-285.

[18] Buttle, D.J. and Scruby, C.B., Acoustic emission source location in fibre reinforced plastic composites. *Harwell Report* N° AERE R 13039, 1988.

[19] Ibitolu, E.O. and Summerscales, J., Acoustic emission source location in bidirectionally reinforced composites in *Proc. 4th European Conf. on NDT*, London, 1987.

[20] Scruby, C.B., Acoustic emission measurements using point-contact transducers. *J. Acoust. Emis.*, 1985, **4**, 9-18.

[21] Scruby, C.B., Stacey, K.A. and Baldwin, G.R., Defect characterization in three dimensions by acoustic emission. *J. Phys. D*, 1986, **19**, 1597-1612.

[22] Castagnède, B., Sachse, W. and Kim, K.Y., Optimisation de la localisation d'une source ponctuelle d'émission acoustique dans un matériau anisotrope homogène. *C. R. Acad. Sci. Paris*, 1988, **II 307**, 1473-1478.

[23] Castagnède, B., Sachse, W. and Kim, K.Y., Localisation de sources d'émission acoustique dans un matériau composite. *C. R. Acad. Sci. Paris*, 1988, **II 307**, 1595-1600.

[24] Castagnède, B., Sachse, W. and Kim, K.Y., Location of pointlike acoustic emission sources in anisotropic plates. *J. Acous. Soc. Am.*, 1989, **86**, 1161-1171.

[25] Auld, B.A., *Acoustic fields and waves in solids*, Wiley, New York, vol. 1, 1973.

[26] Ciarlet, P.G., *Introduction à l'analyse numérique matricielle et à l'optimisation.*, Masson, Paris, 1982, pp. 167-207.

[27] Press, W.H., Flannery, B.P., Teukolsky, S.A. and Vetterling, W.T., *Numerical recipes: The art of scientific computing*, Cambridge U.P., New York, 1986, pp. 240-269.

[28] Hosten, B. and Castagnède, B., Optimisation du calcul des constantes élastiques à partir des mesures de vitesse d'une onde ultrasonore. *C. R. Acad. Sci. Paris*, 1983, **II 296**, 297-300.

[29] Castagnède, B., *Mesures des constantes élastiques de solides anisotropes par une méthode ultrasonore*, Doctorat d'Acoustique, Université de Bordeaux I, 1984.

[30] Hosten, B. and Castagnède, B., Mesure des constantes élastiques du bois à l'aide d'un interféromètre ultrasonore numérique et leur optimisation. *C. R. Acad. Sci. Paris*, 1983, **II 296**, 1761-1764.

[31] Roux, J., Hosten, B., Castagnède, B. and Deschamps, M., Caractérisation mécanique des solides par spectro-interférométrie ultrasonore". *Rev. Phys. Appl.*, 1985, **20**, 351-358.

[32] *Technical booklet on fiber glass reinforced materials*, Extren 500 series. Chicago: Ryerson (Plastic division), 1987.

[33] Sachse, W. and Kim, K.Y., Point source / point receiver materials testing in *Review of quantitative nondestructive evaluation*, Plenum, New York, vol. 6A, 1986, pp. 311-320.

[34] Sachse, W. and Kim, K.Y., Quantitative acoustic emission and failure mechanics of composite materials". *Ultrasonics*, 1987, **25**, 195-203.

[35] Dieulesaint, E. and Royer, D., *Ondes élastiques dans les solides.*, Masson, Paris, 1974, pp. 113-166.

[36] Pao, Y.H. and Kaul, R.K., Waves and vibrations in isotropic and anisotropic plates in *R.D. Mindlin and Applied Mechanics*, Pergamon, New York, 1974.

[37] Ceranoglu, A.N. and Pao, Y.H., Propagation of elastic pulses and acoustic emission in a plate. *ASME J. Appl. Mech.*, 1981, **48**, 125-147.

[38] Weaver, R.L. and Pao, Y.H., Axisymmetric elastic waves excited by a point source in a plate. *ASME J. Appl. Mech.*, 1981, **49**, 821-836.

EUROMECH 269 3-6 DEC 1990

PARTICIPANTS

Name		Institution	Department	Address	Postal code	City	Country
Albertini	C.	CEC Joint Research Center	Department of Engineering	ISPRA Establishment	21020	ISPRA	I
Beaumont	P.	Cambridge University	Department of Engineering	Trumpington Street	CB2 1PZ	CAMBRIDGE	UK
Bogolyubov	V.	Res. Inst. for Aviation Tech		Petrovka street, 24	103051	MOSCOU	USSR
Bouipaep	F.	Vrije Universiteit Brussel		Pleinlaan 2	1050	BRUXELLES	B
Brandt	A.M.	Inst. of Fund. Technological Research		Swietokrzyska 21	00-049	WARSZAWA	PL
Calmelly	H.	DRET	Technological Research		00460	PARIS-ARMEES	F
Cambier	F.	SOLVAY	SDCE	310, rue de Ransbeek	1120	BRUXELLES	B
Cerdon	A.	Vrije Universiteit Brussel	Faculty of Applied Sciences	Pleinlaan 2	1050	BRUXELLES	B
Caron	J.-F.	E.N.P.C.	C.E.R.A.M.	Central IV, 1 avenue Montaigne	93167	NOISY-LE-GRAND	F
Castagede	B.	Université de Bordeaux 1	Lab. de Mécanique Physique	351,cours de la Libération	33405	TALENCE	F
Chapuis	M.	IUT A	Dep. Génie Méca. et Productique	Domaine Universitaire	33405	TALENCE	F
Cluzel	C.	ENS Cachan	Lab. de Méc. et Technologie	61, av. du Président Wilson	94230	CACHAN	F
Courard	L.	Université de Liège	Lab. Matériaux de Construction	6, quai Banning	4000	LIEGE	B
Courade	R.-M.	ENSE		58 rue Jean Parot	42023	ST-ETIENNE	F
Deerden	I.	Vrije Universiteit Brussel	Faculty of Applied Sciences	Pleinlaan 2	1050	BRUXELLES	B
De Visscher	J.	Vrije Universiteit Brussel	Faculty of Applied Sciences	Pleinlaan 2	B-1050	BRUXELLES	B
De Wilde	W.P.	Vrije Universiteit Brussel	Faculty of Applied Sciences	Pleinlaan 2	B-1050	BRUXELLES	B
Degrieck	J.	Rijksuniversiteit Gent	Sem. voor Mech. van de Mach.	Grotesteenweg Noord 2	B-9052	ZWIJNAARDE	B
Dual	J.	Swiss Federal Institute of Technology	Institut für Mechanik	ETH Zentrum	8092	ZURICH	CH
El Guerjouma	R.	INSA	Lab. Trait. Signal et Ultrasons	20, Av. A. Einstein - Bât. 502	69621	VILLEURBANNE	F
Faignet	S.	Vrije Universiteit Brussel	Faculty of Applied Sciences	Pleinlaan 2	B-1050	BRUXELLES	B
Fällström	K.-E.	Luleå University of Technology	Div. of Exp. Mech.		S-95187	LULEÅ	S
Férage	I.-M.	Dassault Aviation	DGT/DRM	78, quai Marcel Dassault	92214	ST-CLOUD	F
Férent	B.	Ecole des Mines de St-Etienne	Dpt. Mécanique et Matériaux	158, Cours Fauriel	42023	ST ETIENNE	F
Foret	G.	E.N.P.C.	C.E.R.A.M.	Central IV, 1 avenue Montaigne	93167	NOISY-LE-GRAND	F
Frederiksen	P.S.	The Technical University of Denmark	Department of Solid Mechanics	Building 404	2800	LYNGBY	DK
Grédiac	M.	ENISE		58 rue Jean Parot	42023	ST-ETIENNE	F
Guénaud	A.	INSA-UCB Lyon	CERMAC	20, Av. A. Einstein - Bât. 304	69621	VILLEURBANNE	F
Guz	A.	Ac. Sc. Ukr. SSR	Inst. of Mechanics	Nesterov Str. 3	252057	KIEV	USSR
Hamelin	P.	INSA-UCB Lyon	CERMAC	20, Av. A. Einstein - Bât. 304	69621	VILLEURBANNE	F
Harding	J.	University of Oxford	Dept. of Engineering Science	Parks Road	OX1-3PJ	OXFORD	UK
Huchon	R.	Université de Bordeaux I	Lab. de Mécanique Physique	351, cours de la Libération	33405	TALENCE	F
Kawata	K.	Science University of Tokyo		Noda, Chiba	278	YAMAZAKI	J

Name		Institution	Department	Address	Postal	City	Country
Ladeveze	P.	ENS Cachan	Lab. de Méc. et Technologie	61, av. du Président Wilson	94230	CACHAN	F
Lallement	G.	Université de Besançon	Lab. de Mécanique Appliquée	UPRST, route de Gray	25030	BESANCON	F
Maier	G.	Dipartimento di Ingegneria Strut.	Politecnico di Milano	Piazza Leonardo da Vinci 32	20133	MILANO	I
Mathia	T.	Ecole Centrale de Lyon	Labo. de Techn. des Surfaces	BP 163	69131	ECULLY	F
Milosavljevic	D.	University "Svetozar Markovic"	Faculty of Mechanical Eng.	Sestre Janjic 6	34000	KRAGUJEVAC	YU
Minster	J.	Inst. of Theo. and Appl. Mech.	Czechoslovak Acad.of Sciences	Vyschradska 49	128 49	PRAHA	CS
Moor	E.	Swiss Federal Inst. of Technology	Institue for Mechanics	ETH-Zentrum	8092	ZURICH	CH
Morais	I.J.L.	Univer. de Tras os Mon. e Al. Dou.			5000	VILA REAL	P
Nesa	D.	Renault	Direction de la Recherche	9-11, avenue du 18 juin 1940	92500	RUEIL MALMAISON	F
Oomens	C.	Eindhoven University of Technology		P.O. BOX 513	5600 MB	EINDHOVEN	NL
Pagliitis	A.	Riga Polytechnic Institute		1, Kalku, str.	226355	RIGA	USSR
Petit	C.	IUT	Dpt of Architek. & Civil Eng. Lab. de Génie Civil - ERIN		19300	EGLETONS	F
Pinon	L.			58, rue Fénelon BP 555	92542	MONTROUGE	F
Post	D.	Virginia Polytechnic Institute	Eng.Science and Mech. Dpt.		24061 VA	BLACKSBURG	USA
Pouyet	J.	Université de Bordeaux I	Lab. de Mécanique Physique	351, cours de la Libération	33405	TALENCE	F
Povolny	J.	Resrch Inst. for Sound and Picture		Provaznicka 8	110 00	PRAHA	CS
Ramade	C.	SNECMA (YKOM 2)	Etablissement de Corbeil	BP 81	91003	EVRY	F
Rouger	F.	CTBA		10, avenue de Saint-Mandé	75012	PARIS	F
Sancaktar	E.	Clarkson University	Dept. of Mech. and Aer. Engi.	Potsdam 13699-5725		NEW-YORK	USA
Sayers	C.	Billiton Research B.V.		Westervoortsedijk 67d - PO Box 40	6800 AA	ARNHEM	NL
Schillemans	L.	Vrije Universiteit Brussel	Faculty of Applied Sciences	Pleinlaan 2	1050	BRUXELLES	B
Solomos	G.	CEC Joint Research Center	Inst. for Systems Eng. & Info.	ISPRA Establishment	21020	ISPRA	I
Surrel	Y.	Ecole des Mines de St-Etienne	Dept. Mécanique et Matériaux	158, cours Fauriel	42023	ST ETIENNE	F
ten Busschen	A.	Technical University Delft	Lab. for Engineering Mechanics	Mekelweg 2	2628 CD	DELFT	NL
Van Busschen D.		Vrije Universiteit Brussel	Faculty of Applied Sciences	Pleinlaan 2	1050	BRUXELLES	B
Van Vinckenroy	G.	Vrije Universiteit Brussel	Faculty of Applied Sciences	Pleinlaan 2	B-1050	BRUXELLES	B
Vautrin	A.	Ecole des Mines de St-Etienne	Dpt. Mécanique et Matériaux	158, Cours Fauriel	42023	ST ETIENNE	F
Veidt	M.	Swiss Federal Institue of Technology	Institute of Mechanics	ETH-Zentrum	8092	ZURICH	CH
Verchery	G.	Ecole des Mines de St-Etienne	Dpt. Mécanique et Matériaux	158, cours Fauriel	42023	ST ETIENNE	F
Virgolini	M.P.	Centro Ricerche MonteDipe		Via della Chimica, 5	30175	PORTO MARGHERA	I
Xia	Y.	University of Oxford	Dept. of Engineering Science	Parks Road	OX1 3PJ	OXFORD	UK

INDEX OF CONTRIBUTORS